Student's Solutions Manual

Nancy S. Boudreau
Bowling Green State University

AF506105

A First Course in Statistics
Eleventh Edition

James T. McClave

Info Tech, Inc.
University of Florida

Terry Sincich

University of South Florida

PEARSON

Boston Columbus Indianapolis New York San Francisco Upper Saddle River
Amsterdam Cape Town Dubai London Madrid Milan Munich Paris Montreal Toronto
Delhi Mexico City Sao Paulo Sydney Hong Kong Seoul Singapore Taipei Tokyo

www.pearsonhighered.com

Table of Contents

Chapter 1: Statistics, Data, and Statistical Thinking   1

Chapter 2: Methods for Describing Sets of Data   5

Chapter 3: Probability   52

Chapter 4: Discrete Random Variables   79

Chapter 5: Inferences Based on a Single Sample *Estimation with Confidence Intervals*   130

Chapter 6: Inferences Based on a Single Sample *Tests of Hypothesis*   162

Chapter 7: Comparing Population Means   195

Chapter 8: Comparing Population Proportions   239

Chapter 9: Simple Linear Regression   291

Preface

This solutions manual is designed to accompany the text, *A First Course in Statistics*, Eleventh Edition, by James T. McClave and Terry Sincich. It provides solutions to the odd-numbered exercises for each chapter in the text.

Other methods of solution may also be appropriate; however, the author has presented one that she believes to be the most instructive to the beginning statistics student. The student should first attempt to solve the assigned exercises without help from this manual. Then, if unsuccessful, the solution in the manual will clarify points necessary to the solution. The student who successfully solves an exercise should still refer to the manual's solution. Many points are clarified and expanded upon to provide maximum insight into and benefit from each exercise.

Instructors will also benefit from the use of this manual. It will save time in preparing presentations of the solutions and possibly provide another point of view regarding their meaning.

Some of the exercises are subjective in nature and thus omitted from the Answer Key at the end of *A First Course in Statistics*. The subjective decisions regarding these exercises have been made and are explained by the author. Solutions based on these decisions are presented; the solution to this type of exercise is often most instructive. When an alternative interpretation of an exercise may occur, the author has often addressed it and given justification for the approach taken.

Nancy S. Boudreau
Bowling Green State University
Bowling Green, Ohio

Chapter

1

Statistics, Data, and Statistical Thinking

1.1   Statistics is a science that deals with the collection, classification, analysis, and interpretation of information or data. It is a meaningful, useful science with a broad, almost limitless scope of applications to business, government, and the physical and social sciences.

1.3   The first element of inferential statistics is the population of interest. The population is a set of existing units. The second element is one or more variables that are to be investigated. A variable is a characteristic or property of an individual population unit. The third element is the sample. A sample is a subset of the units of a population. The fourth element is the inference about the population based on information contained in the sample. A statistical inference is an estimate, prediction, or generalization about a population based on information contained in a sample. The fifth and final element of inferential statistics is the measure of reliability for the inference. The reliability of an inference is how confident one is that the inference is correct.

1.5   Quantitative data are measurements that are recorded on a meaningful numerical scale. Qualitative data are measurements that are not numerical in nature; they can only be classified into one of a group of categories.

1.7   A population is a set of existing units such as people, objects, transactions, or events. A sample is a subset of the units of a population.

1.9   An inference without a measure of reliability is nothing more than a guess. A measure of reliability separates statistical inference from fortune telling or guessing. Reliability gives a measure of how confident one is that the inference is correct.

1.11   The data consisting of the classifications A, B, C, and D are qualitative. These data are nominal and thus are qualitative. After the data are input as 1, 2, 3, and 4, they are still nominal and thus qualitative. The only differences between the two data sets are the names of the categories. The numbers associated with the four groups are meaningless.

1.13   a.   The experimental units for this study are the 15 earthquake sites.

        b.   From 1940 to 1995, there were many more than 15 earthquakes. Thus, these 15 earthquakes represent a sample.

        c.   There are 3 variables in this problem. The variable 'type of ground motion' has 3 levels (short, long, or forward directive). Thus, this variable is qualitative. The variable 'earthquake magnitude' is measured on a Richter scale which results in a meaningful number. Thus this variable is quantitative. The variable peak ground acceleration is measured in feet per second and is quantitative.

1.15   a.   For question 1, the type of data collected is qualitative. The answer to the question will be either 'yes' or 'no'. For question 2, the type of data collected is quantitative. The answer to the question will be a number. For question 3, the type of data collected is qualitative. The answer to the question will be either 'yes' or 'no'.

b.   The data collected from the 1,010 adults represents a sample. The population of interest is all adults in the U.S. Only 1,010 adults in the U.S. were actually questioned.

1.17.   This study represents a descriptive statistical study. Data were collected and used to portray or describe the condition of the U.S. Treasury in 1861. No inferences were made.

1.19   a.   The town where the sample was collected is a name and thus, is a qualitative variable.

b.   The type of water supply is a category. This is a qualitative variable.

c.   The acidic level is a meaningful number. Thus, this is a quantitative variable.

d.   The turbidity level is measured on a numerical scale and is a quantitative variable.

e.   The temperature is measured on a numerical scale and is a quantitative variable.

f.   The number of fecal coliforms per 100 milliliters is measured on a numerical scale and is thus, a quantitative variable.

g.   The free chlorine-residual (milligrams per liter) is measured on a numerical scale and is a quantitative variable.

h.   The presence of hydrogen sulphide is measured on a categorical scale (yes or no). This is a qualitative variable.

1.21   a.   Length of maximum span can take on values such as 15 feet, 50 feet, 75 feet, etc. Therefore, it is quantitative.

b.   The number of vehicle lanes can take on values such as 2, 4, etc. Therefore, it is quantitative.

c.   The answer to this item is "yes" or "no," which are not numeric. Therefore, it is qualitative.

d.   Average daily traffic could take on values such as 150 vehicles, 3,579 vehicles, 53,295 vehicles, etc. Therefore, it is quantitative.

e.   Condition can take on values "good," "fair," or "poor," which are not numeric. Therefore, it is qualitative.

f.   The length of the bypass or detour could take on values such as 1 mile, 4 miles, etc. Therefore, it is quantitative.

g.   Route type can take on values "interstate," U.S.," "state," "county," or "city," which are not numeric. Therefore, it is qualitative.

1.23   a.   The data collection method used by the cancer researchers is a designed experiment. Those participating in the study were randomly assigned to one of the two screening methods.

b. The experimental units in this study are the 50,000 smokers.

c. The variable measured in this study is the age at which the scanning method first detects a tumor. Since age is represented by a meaningful number, it is quantitative.

d. The population of interest is all smokers in the U.S. The sample of interest is the set of 50,000 smokers participating in the study.

e. The inference of interest is the difference in mean age at which each of the scanning methods first detects a tumor.

1.25   a. The data collection method used by the researchers was a designed experiment. Half of the boxers received the massage and half did not.

b. The experimental units are the amateur boxers.

c. There are two variables measured on the boxers – heart rate and blood lactate level. Both of these variables are measured on a numeric scale and thus, are quantitative.

d. The inferences drawn from the analysis are: there is no difference in the mean heart rates between the two groups of boxers (those receiving massage and those not receiving massage) and there is no difference in the mean blood lactate levels between the two groups of boxers. Thus, massage did not affect the recovery rate of the boxers.

e. No. Only amateur boxers were used in the experiment. Thus, all inferences relate only to boxers.

1.27.   a. Since only students were sampled, the population of interest would be all students. The sample would be the 155 students who participated in the experiment.

b. The data were collected using a designed experiment. Each of the participants was randomly assigned to a treatment (emotional state).

c. The researchers inferred that a higher proportion of students in the guilty-state group chose not to repair the car than those in the neutral-state and anger-state groups.

d. One factor that could affect the reliability of the inference from this study is how representative the sample is. The sample was 155 volunteer students. Students in psychology classes typically have to volunteer for a certain number of experiments. These students may not take the study seriously or may not be representative of all students.

1.29.   a. Data for age would be quantitative. Age is measured on a numerical scale. Data for gender would be qualitative. Gender is either 'male' or 'female'. Data for dating experience is quantitative. The number of dates is measured on a numerical scale. Data for the extent to which a student was willing to tell parents about a dating issue would be qualitative. The responses could be 'never tell', 'rarely tell', 'sometimes tell', 'almost always tell', and 'always tell'.

b. If students were recruited from a primarily European American middle-class school district, then the responses would probably not be typical of all high school students. Upper-class students and lower-class students' behaviors might be quite different from middle-class students' behaviors. In addition, the behaviors of minority students might be quite different from that of European American students. Thus, the inferences from this study may not be reflective of the entire population of high school students.

1.31. a. The data collection method for this study is observational. Specifically, a survey was conducted to collect the data.

b. The data collected for this study are qualitative. The responses were either 'Obama' or 'Palin'.

c. The results of this study are very suspect. The sample used was self-selected. Only those readers who elected to respond to the question were included. People who respond to these types of surveys typically have very strong views and may not be representative of the entire population. In addition, readers who might want to respond but who do not have access to an online computer would not be able to participate in the survey. Thus, the results would very likely not be representative of the population in general.

1.33 a. Although eating oat bran can reduce cholesterol, oat bran must be the only thing eaten. Reporting that eating oat bran is an easy and cheap way to lower your cholesterol implies that if you add oat bran to your diet, you can reduce your cholesterol, which may not be the case at all.

b. One would need to collect data on a sample of women who gave birth to babies with birth defects. Then, each woman in the study would also be asked whether she was a victim of domestic violence while pregnant or not. The data collection method for this study would be an observational study. Specifically, one would use a survey to collect the data.

c. In this study, only the results of the most positive response were reported. Not that many high school girls would 'Always' be happy with the way they were. To be more representative of those who are happy with the way they were, one should combine the results of those who responded 'Always true', 'Sort of true', and 'Sometimes true'.

d. In this study, the parents were asked questions about whether they ever cut the size of meals, whether they ever ate less than they felt they should, and whether they ever relied on a limited number of foods to feed their children because they were running out of money. From the responses to these questions, the researchers concluded that one in four American children under age 12 is hungry or at risk of hunger. The children were never asked if they were hungry or not. Just because one eats less than one feels he/she should eat does not necessarily mean the children are hungry. To collect the appropriate data, the researchers should survey children under age 12 and ask them if they ever start their day hungry and if they ever go to bed hungry.

Methods for Describing Set of Data

2.1   The class frequency is the number of observations in the data set falling in a particular class. The class relative frequency is the class frequency divided by the total number of observations in the data set. The class percentage is the class relative frequency multiplied by 100.

2.3   In a bar graph, a bar or rectangle is drawn above each class of the qualitative variable corresponding to the class frequency or class relative frequency. In a Pareto diagram, the bars of the bar graph are arranged in order from the largest to the smallest from left to right.

2.5   a.   To find the frequency for each class, count the number of times each letter occurs. The frequencies for the three classes are:

Class	Frequency
X	8
Y	9
Z	3
Total	20

b.   The relative frequency for each class is found by dividing the frequency by the total sample size. The relative frequency for the class X is $8/20 = .40$. The relative frequency for the class Y is $9/20 = .45$. The relative frequency for the class Z is $3/20 = .15$.

Class	Frequency	Relative Frequency
X	8	.40
Y	9	.45
Z	3	.15
Total	20	1.00

c.   The frequency bar chart is:

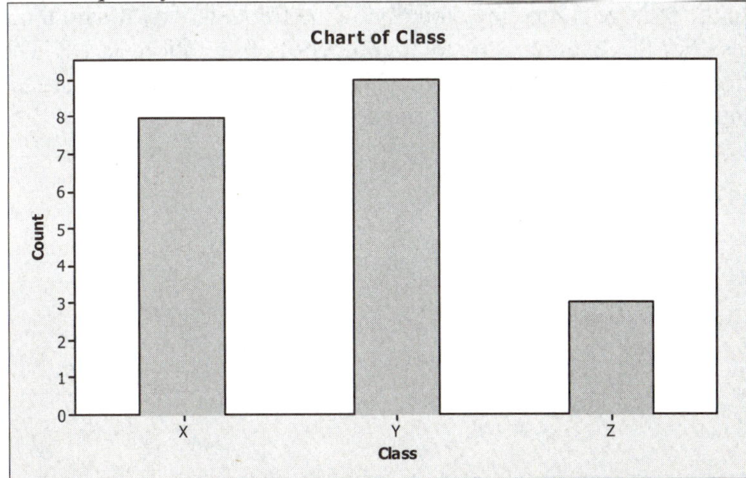

d.   The pie chart for the relative frequency distribution is:

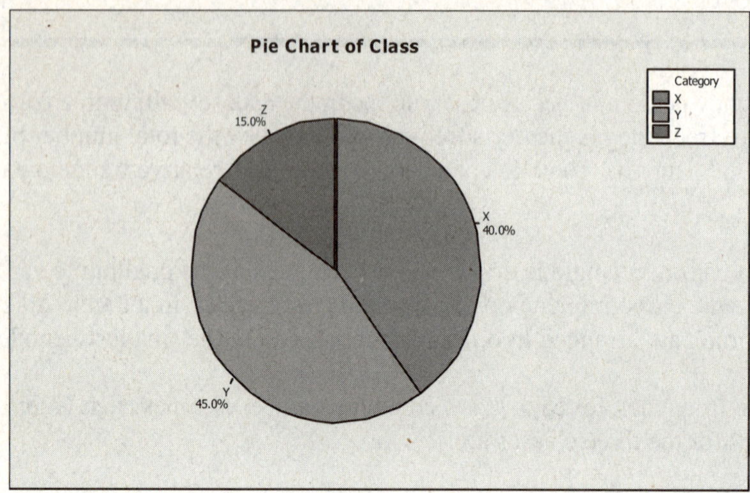

2.7   a.   The variable of interest in this study is degree of tooth wear. The variable is qualitative. Possible responses include 'Unknown', 'Heavy', Moderate-heavy', etc.

b.   The number of cheek teeth in each wear category is:

Degree of Wear	Frequency	Relative Frequency
Unknown	5	5/18 = .278
Unworn	2	2/18 = .111
Slight	4	4/18 = .222
Light-moderate	2	2/18 = .111
Moderate	3	3/18 = .167
Moderate-heavy	1	1/18 = .056
Heavy	1	1/18 = .056
Total	**18**	**18/18 = 1**

c.   To compute the relative frequency, you divide the frequency by the total number of observations. For this example, the total number of observations is 18. The relative frequency for 'Unknown' is 5/18 = .278. The rest of the calculations are done in a similar manner and appear in the table in part b.

d.   Using MINITAB, a relative frequency bar graph is:

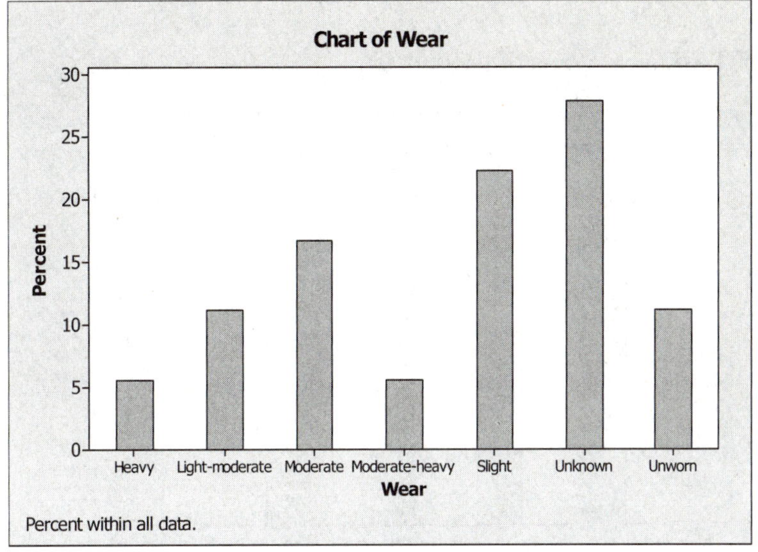

e.   A Pareto diagram is a bar graph with the categories arranged in order from the largest to the smallest from left to right. The Pareto diagram for this data is:

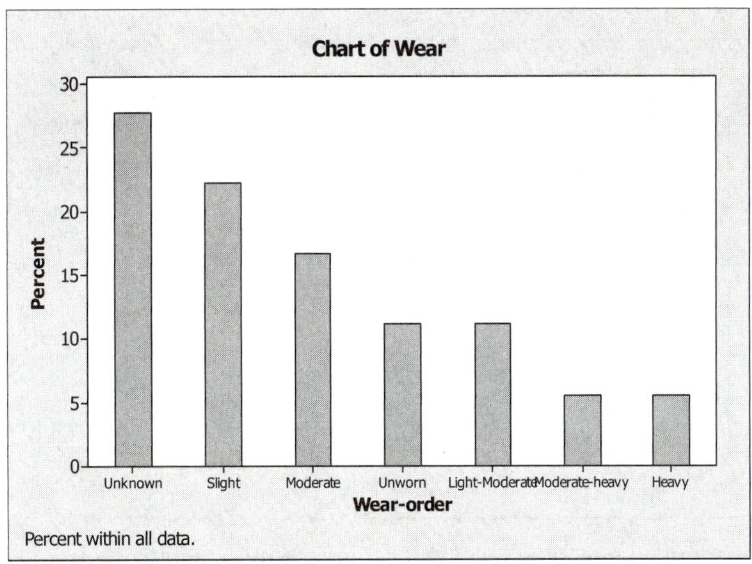

f.   The degree of wear category that occurred most often is 'Unknown'.

2.9   a.   The proportion of books at reading level 1 is found by dividing the number of books at level 1 by the total number of books, or $39/266 = .147$.

b.   For reading level 2, the proportion of books is $76/266 = .286$. For reading level 3, the proportion of books is $50/266 = .188$. For reading level 4, the proportion of books is $87/266 = .327$. For reading level 5, the proportion of books is $11/266 = .041$. For reading level 6, the proportion of books is $3/266 = .011$.

c.   The sum of the proportions is $.147 + .286 + .188 + .327 + .041 + .011 = 1.000$.

d.   Using MINITAB, a bar graph of the data is:

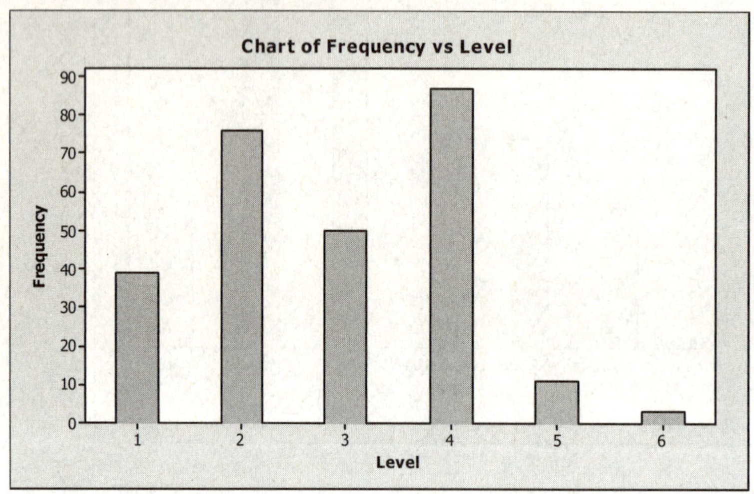

e.   Using MINITAB, the Pareto diagram is:

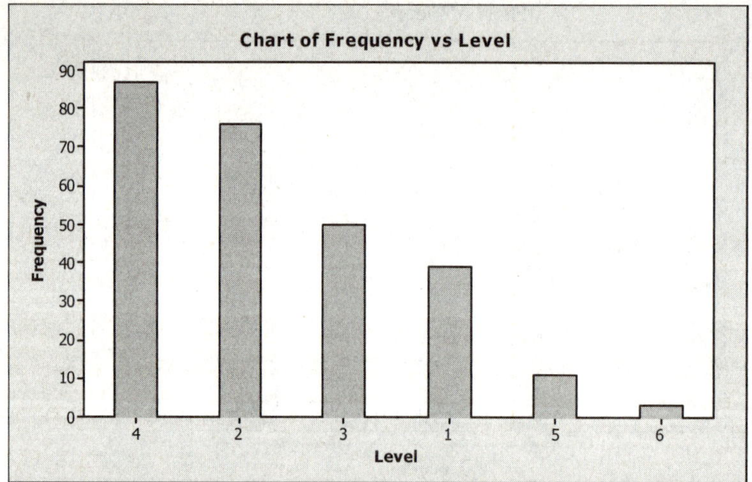

From this diagram, the reading level that occurs the most frequently is level 4.

2.11   a.   To construct a relative frequency table for data, we must find the relative frequency for each species of rhinos. To find the relative frequency, divide the frequency by the total population size, 17,800. The relative frequency for African Black rhinos is $3{,}610/17{,}800 = .203$. The rest of the relative frequencies are found in a similar manner and are reported in the table.

Rhino species	Population Estimate	Relative Frequency
African Black	3,610	3610 / 17800 = .203
African White	11,330	11330 / 17800 = .637
(Asian) Sumatran	300	300 / 17800 = .017
(Asian) Javan	60	60 / 17800 = .003
(Asian) Indian	2,500	2500 / 17800 = .140
Total	**17,800**	**1.000**

b.   The relative frequency bar chart is:

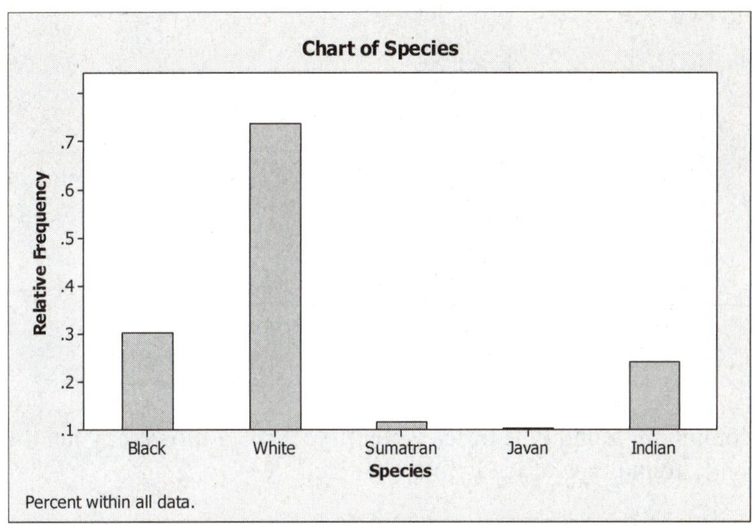

c.   The proportion of rhinos that are African is $(3{,}610 + 11{,}330)/17{,}800 = .839$.

The proportion of rhinos that are Asian is $(300 + 60 + 2{,}500)/17{,}800 = .161$.

2.13   a.   From the summary table, the proportion of melt ponds that had landfast ice is $38.89 / 100 = .3889$.

b.   Yes. From the summary table 17.46% of the melt ponds had first-year ice.

c.   Using MINITAB, the Pareto diagram is:

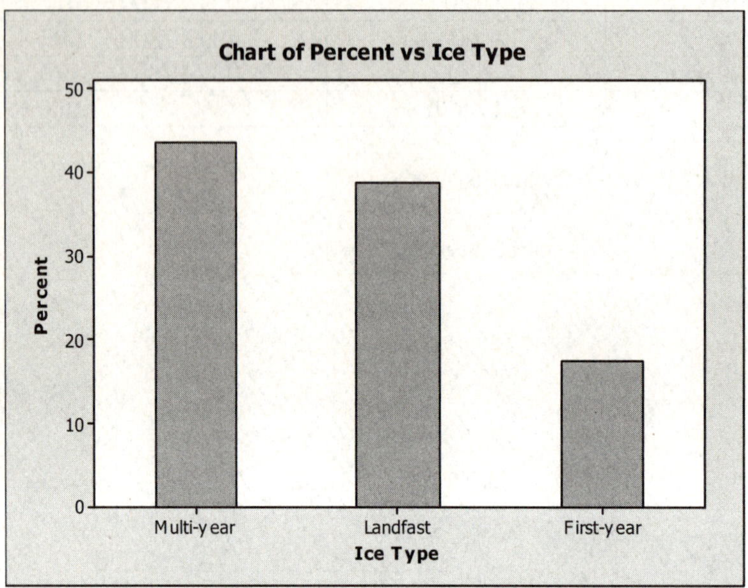

The most commonly found type of ice is multi-year ice, followed by landfast ice. The least frequently found type of ice is first-year ice.

2.15   Using MINITAB, pie charts for the two sets of data are:

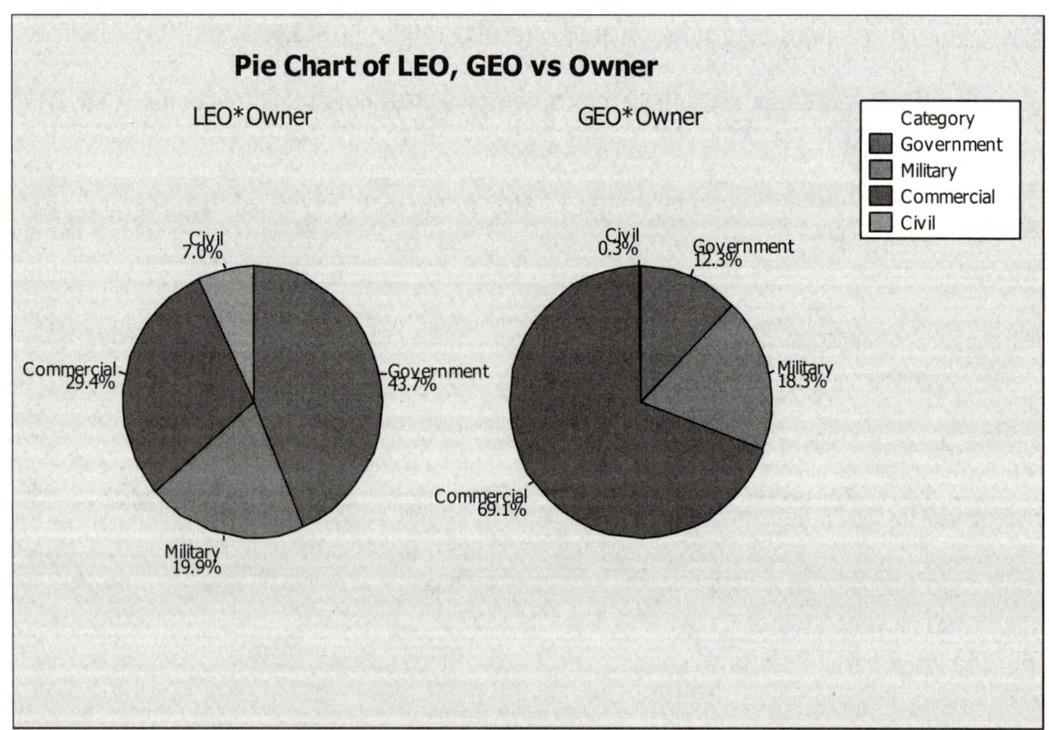

In the LEO, most of the satellites are owned by the government (43.7%), while in GEO, most of the satellites are commercially owned (69.1%). About the same percentage of both GEO and LEO satellites are owned by the military (18.3% and 19.9% respectively).

2.17   First, construct a frequency table. There are 5 categories of highest degree obtained. The relative frequencies are found by dividing the frequencies by the total number of CEO's in the sample, which is 40. The table is:

Highest Degree Obtained	Frequency	Relative Frequency
None	6	6 / 40 = .150
Bachelors	14	14 / 40 = .350
MBA	11	11 / 40 = .275
Masters	3	3 / 40 = .075
PhD	3	3 / 40 = .075
Law	3	3 / 40 = .075
Total	**40**	**1.00**

Using MINITAB, a pie chart of the data is:

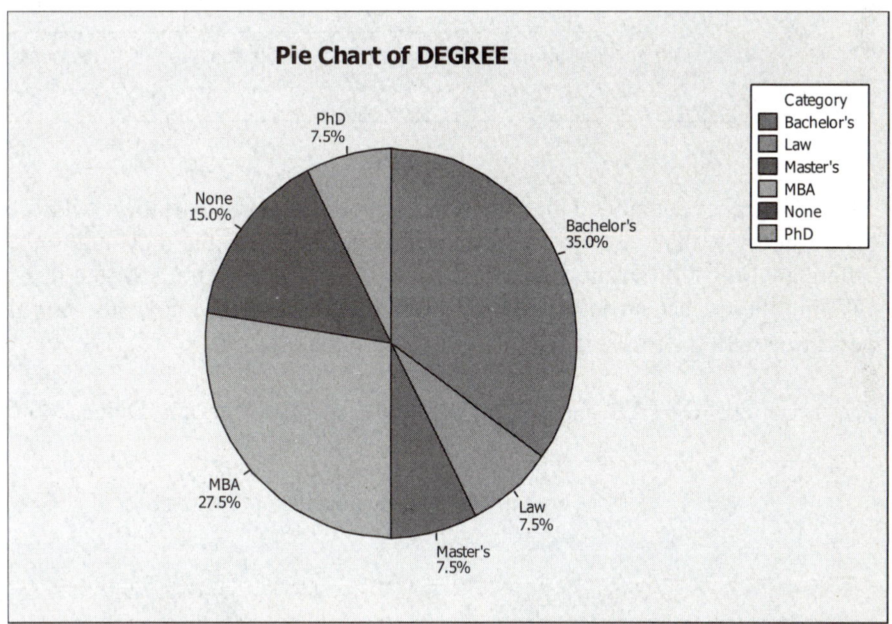

Half of the CEO's have some type of advanced degree (.275 + .075 + .075 + .075 = .500), while half do not (.350 + .150 = .500).

2.19   Using MINITAB, a bar graph of the data is:

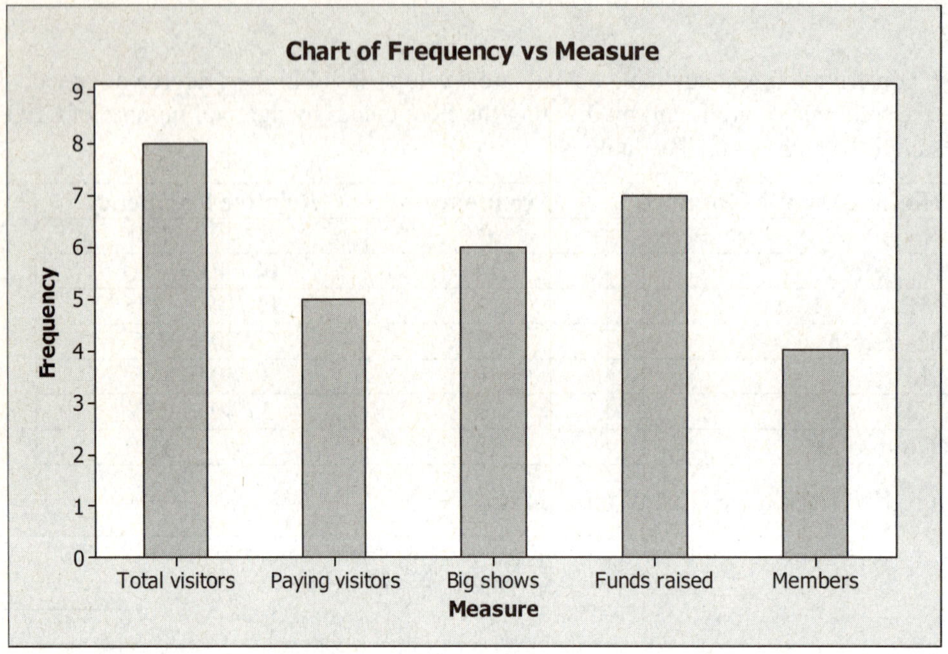

The researcher concluded that "there is a large amount of variation within the museum community with regards to . . . performance measurement and evaluation". From the data, there are only 5 different performance measures. I would not say that this is a large amount. Within these 5 categories, the number of times each is used does not vary that much. I would disagree with the researcher. There is not much variation.

2.21   a.   Using MINITAB, pie charts for Well Class, Aquifer, and Detect MTBE are:

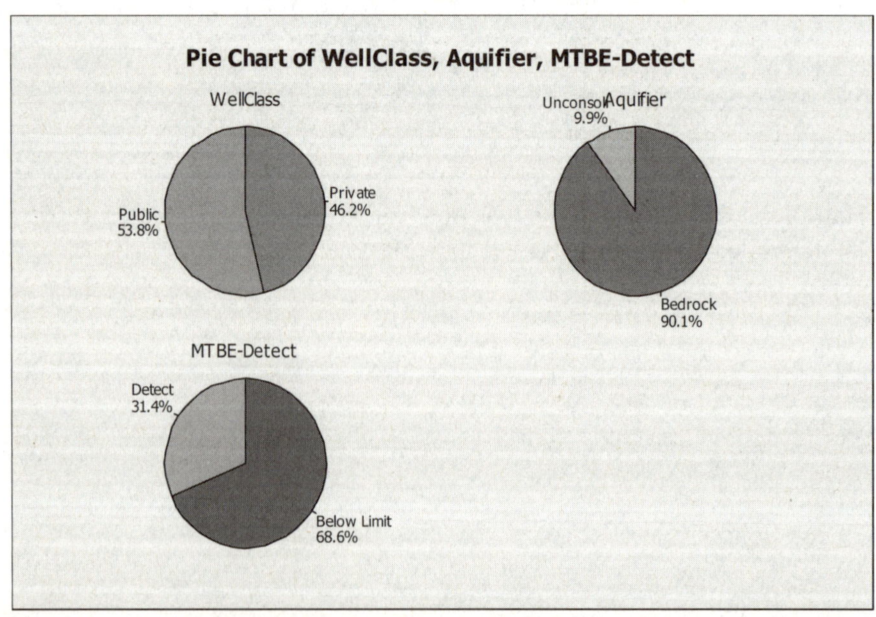

b.   Using MINITAB, the side-by-side bar charts to compare the proportions of contaminated wells for private and public wells classes are:

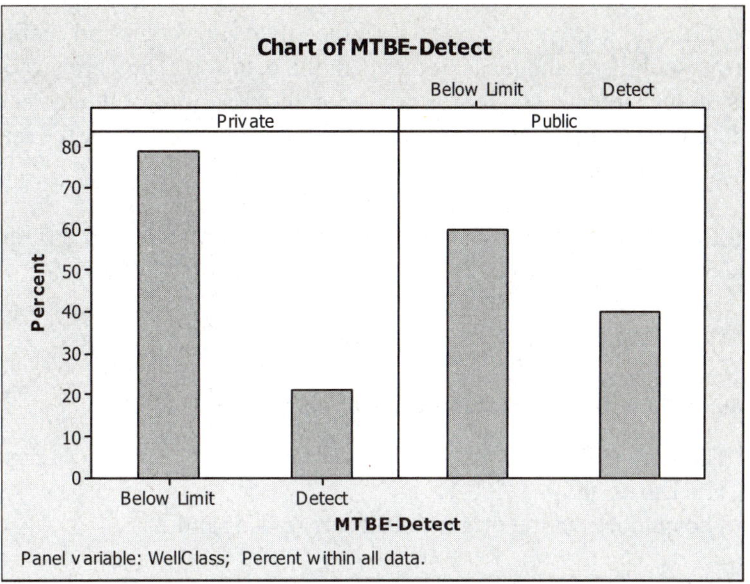

c.   Using MINITAB, the side-by-side bar charts to compare the proportions of contaminated wells for bedrock and unconsolidated aquifers are:

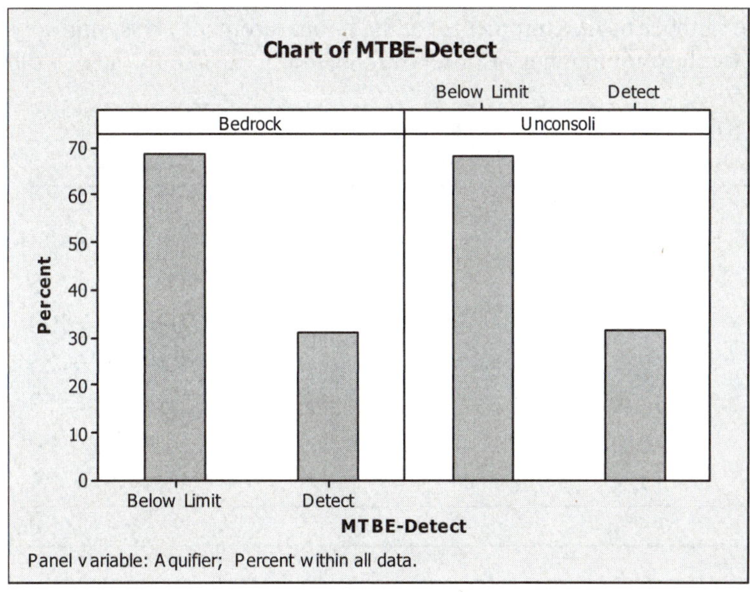

d.   For the pie charts in part **a**, a little more than half of the wells are public (53.8%) and a little less than half are private (46.2%). Most of the aquifers are bedrock (90.1%) and very few are unconsolidated (9.9%). About two-thirds of the wells are not contaminated (below limit – 68.6%) and about one-third are contaminated (detect – 31.4%).

For part **b**, a larger proportion of public wells are contaminated than private wells.

For part **c**, about the same proportion of bedrock and unconsolidated aquifers are contaminated.

2.23   In a dot plot, the values are placed on the horizontal axis. The numerical value of each measurement is located on the horizontal axis by a dot above the respective value. When values repeat, the dots are placed one above the other. In a stem-and-leaf display, the stems are the left-most digits of the measurements in the data set. They are arranged vertically from the smallest to the largest. The leaves are the right-most digit of the measurements in the data set. Each leaf is placed beside the appropriate stem. For each stem, the leaves are arranged in order from the smallest to the largest..

2.25   In a histogram, a class interval is a range of numbers above which the frequency of the measurements or relative frequency of the measurements is plotted.

2.27   a.   The original data set has $1 + 3 + 5 + 7 + 4 + 3 = 23$ observations.

        b.   For the bottom row of the stem-and-leaf display:

                The stem is 0.
                The leaves are 0, 1, 2.
                The numbers in the original data set are 0, 1, and 2.

        c.   The dot plot corresponding to all the data points is:

2.29   To find the number of measurements for each measurement class, multiply the relative frequency by the total number of observations, $n = 500$. The frequency table is:

Measurement Class	Relative Frequency	Frequency
.5 − 2.5	.10	500(.10) = 50
2.5 − 4.5	.15	500(.15) = 75
4.5 − 6.5	.25	500(.25) = 125
6.5 − 8.5	.20	500(.20) = 100
8.5 − 10.5	.05	500(.05) = 25
10.5 − 12.5	.10	500(.10) = 50
12.5 − 14.5	.10	500(.10) = 50
14.5 − 16.5	.05	500(.05) = 25
		500

The frequency histogram is:

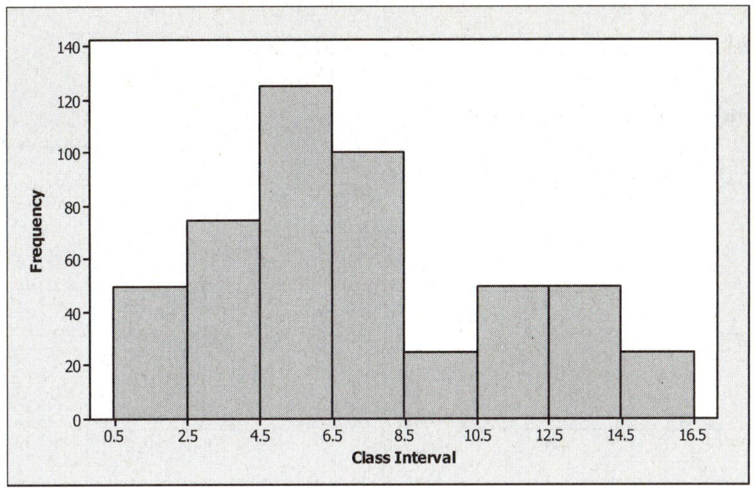

2.31   a.   This graph is a frequency histogram.

b.   There are $16 + 22 = 38$ armadillos with body lengths between 87 and 91 centimeters.

c.   The proportion of 38 armadillos with body lengths between 87 and 91 centimeters is the number observed divided by the total number of captured turtles or $38/80 = .475$.

d.   The number of illegal armadillos is $2 + 4 + 3 + 1 + 1 = 11$. The proportion is $11/80 = .1375$.

2.33   a.   Using MINITAB, the stem-and-leaf display is as follows.

Character Stem-and-Leaf Display

```
Stem-and-leaf of No. Book   N  = 14
Leaf Unit = 1.0

      1       1 6
      5       2 0124
      6       2 8
     (3)      3 044
      5       3 9
      4       4 002
      1       4
      1       5 3
```

b.   The leaves that correspond to students who earned an "A" grade are underlined in the graph above. Those students who earned A's tended to read the most books.

2.35   a.   Using MINITAB, the stem-and-leaf display of the data is:

Stem-and-Leaf Display: Score

```
Stem-and-leaf of Score   N  = 186
Leaf Unit = 1.0

     1     6    9
     1     7
     2     7    3
     3     7    4
     3     7
     4     7    8
     4     8
     5     8    3
     7     8    44
    11     8    6667
    17     8    888999
    24     9    0001111
    41     9    22222222222333333
    62     9    4444445555555555555555
   (37)    9    6666666666666667777777777777777777777
    87     9    888888888888888888888888888888999999999999999999999999999999
    27    10    000000000000000000000000000
```

The stems are 60, 70, 80, … 100. The leaves are units from 0 to 9.

b.   From the stem-and-leaf display, we see that there are only 7 observations with sanitation scores less than the acceptable score of 86. The proportion of ships that have an accepted sanitation standard would be (186 – 7) / 186 = .962.

c.   The sanitation score of 69 is highlighted in the stem-and-leaf display in part **a**.

2.37   Using MINITAB, a histogram of the data is:

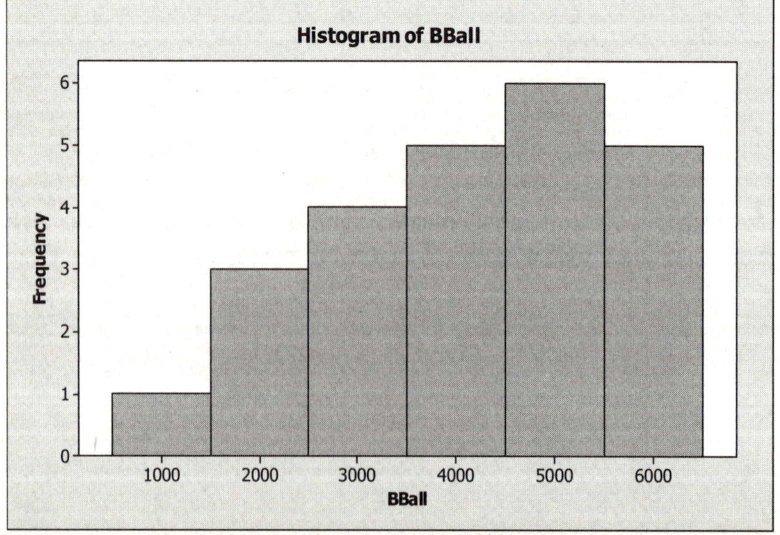

2.39   The dot plot for these data is:

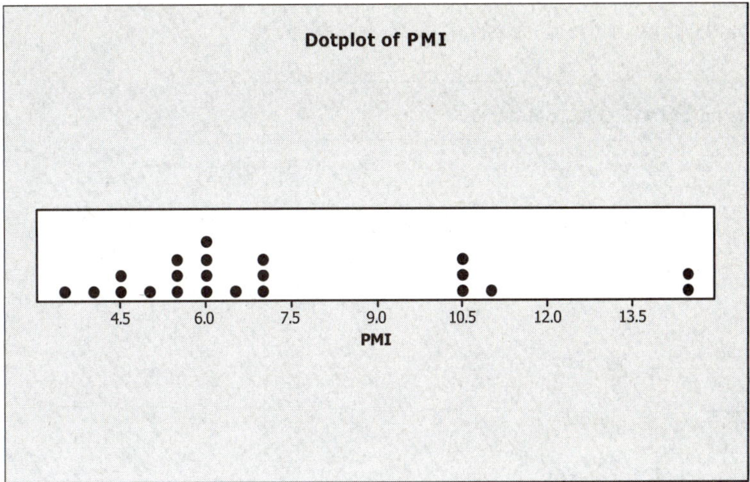

From the dot plot, most of the PMI's range from 3 to 7.5 (16 of the 22). The rest of the PMI's range from 10 to 15.

2.41   Using MINITAB, the stem-and-leaf display for the duration is:

Character Stem-and-Leaf Display

Stem-and-leaf of Duration N = 18
Leaf Unit = 0.10

```
    7    1 0000000
    8    2 0
   (2)   3 00
    8    4 000
    5    5
    5    6
    5    7
    5    8
    5    9 0
    4   10 0
    3   11 0
    2   12 00
```

The leaves that correspond to sit-ins where at least one arrest was made are highlighted in the graph above. The pattern revealed does not support the theory that sit-ins of longer duration are more likely to lead to arrests.

2.43   Using MINITAB, a histogram of the data is:

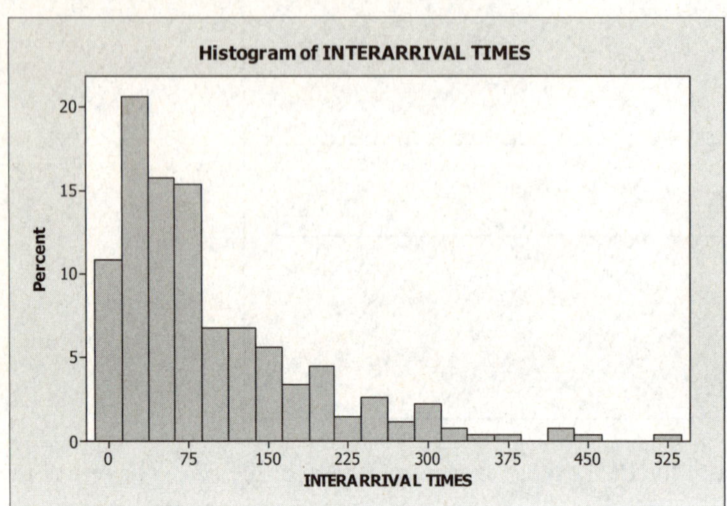

This histogram looks very similar to the one shown in the problem. Thus, there appears that there was minimal or no collaboration or collusion from within the company. We could conclude that the phishing attack against the organization was probably not an "inside job".

2.45   a. $\sum x = 3 + 8 + 4 + 5 + 3 + 4 + 6 = 33$

b. $\sum x^2 = 3^2 + 8^2 + 4^2 + 5^2 + 3^2 + 4^2 + 6^2 = 175$

c. $\sum (x-5)^2 = (3-5)^2 + (8-5)^2 + (4-5)^2 + (5-5)^2 + (3-5)^2 + (4-5)^2 + (6-5)^2 = 20$

d. $\sum (x-2)^2 = (3-2)^2 + (8-2)^2 + (4-2)^2 + (5-2)^2 + (3-2)^2 + (4-2)^2 + (6-2)^2 = 71$

e. $\left(\sum x\right)^2 = (3 + 8 + 4 + 5 + 3 + 4 + 6)^2 = 33^2 = 1089$

2.47   a. $\sum x = 6 + 0 + (-2) + (-1) + 3 = 6$

b. $\sum x^2 = 6^2 + 0^2 + (-2)^2 + (-1)^2 + 3^2 = 50$

c. $\sum x^2 - \dfrac{\left(\sum x\right)^2}{5} = 50 - \dfrac{6^2}{5} = 50 - 7.2 = 42.8$

2.49   Three measures of central tendency are the mean, the median, and the mode.

2.51   The two factors that impact the accuracy of the sample mean as an estimate of the population mean are sample size and variability or spread of the data. For the sample size, the larger the sample size, the more accurate the estimate. For variability, all other factors remaining constant, the more variable the data the less accurate the estimate of the mean.

2.53   a.   For a distribution that is skewed to the left, the mean is less than the median.

     b.   For a distribution that is skewed to the right, the mean is greater than the median.

     c.   For a symmetric distribution, the mean and median are equal.

2.55   Assume the data are a sample. The mode is the observation that occurs most frequently. For this sample, the mode is 15, which occurs 3 times.

The sample mean is:

$$\bar{x} = \frac{\sum x}{n} = \frac{18+10+15+13+17+15+12+15+18+16+11}{11} = \frac{160}{11} = 14.545$$

The median is the middle number when the data are arranged in order. The data arranged in order are: 10, 11, 12, 13, 15, 15, 15, 16, 17, 18, 18. The middle number is the 6th number, which is 15.

2.57   a. $\bar{x} = \dfrac{\sum x}{n} = \dfrac{85}{10} = 8.5$

     b. $\bar{x} = \dfrac{400}{16} = 25$

     c. $\bar{x} = \dfrac{35}{45} = .78$

     d. $\bar{x} = \dfrac{242}{18} = 13.44$

2.59   The sample mean is:

$$\bar{x} = \frac{\sum_{i=1}^{13} x_i}{n} = \frac{10.94+13.71+11.38+...+6.77}{13} = \frac{126.32}{13} = 9.72$$

The average rebound length of these 13 observations is 9.72.

The sample median is found by finding the middle observation once the data are arranged in order. The middle number is 10.94. Thus, the sample median is 10.94. Half of all rebound lengths are longer than 10.94 and half are shorter.

2.61   a.   The mean is $\bar{x} = \dfrac{\sum\limits_{i=1}^{n} x_i}{n} = \dfrac{53 + 42 + 40 + \ldots + 16}{14} = \dfrac{443}{14} = 31.64$. This is the average

number of books read per student.

To find the median, the data must be arranged in order. In this problem, the data are already arranged in order. There are a total of 14 observations, which is an even number. The median is the average of the middle 2 numbers which are 30 and 34. The

median is $\dfrac{34 + 30}{2} = \dfrac{64}{2} = 32$. Half of the students read more than 32 books and half read

fewer.

The mode is the observation appearing the most. In this data set, the mode is 34 and 40 because each appears 2 times in the data set. The most frequent number of books read is either 34 or 40.

   b.   Since the mean and the median are almost the same, the distribution of the data set is approximately symmetric. This can be verified by the stem-and-leaf display of Exercise 2.33.

2.63   a.   The sample mean radioactivity level is:

$$\bar{x} = \frac{\sum x}{n} = \frac{-43.75}{9} = -4.861$$

The median is the middle observation once they have been ordered. The 5th observation is -4.85. Thus the median is -4.85.

The mode is -5.00.

   b.   The average radioactivity level is -4.861. Half of the radioactivity levels are less than -4.85 and half are greater. The modal observation occurred 2 times.

2.65   a.   The mean response was 5.87. The average rating (across the 15 respondents) of the statement "The Training Game is a great way for students to understand the animal's perspective during training" was 5.87. Since the highest rating the statement could receive was 7, this indicated that on the average, the students rather strongly agreed with this statement.

The mode response was 6. More students rated the statement with a 6 than any other rating. A rating of 6 indicated a rather strong agreement to the statement.

   b.   Since the mode is only slightly larger than the mean, there probably is no skewness present. If any skewness exists, the data would probably be skewed to the left since the mode is larger than the mean and because the mode is very close to the largest value possible.

2.67   a.   The sample mean is $\bar{x} = \dfrac{\sum_{i=1}^{n} x_i}{n} = \dfrac{18.12 + 19.48 + 19.36 + \ldots + 16.20}{18} = \dfrac{296.99}{18} = 16.499$.

The average depth of all the teeth collected is 16.499 mm. If the largest depth measurement were doubled, the mean would increase.

b.   There are 18 measurements in the data set so the median is the average of the middle 2 numbers once the data have been arranged in order. The 9th and 10th observations when arranged in order are 16.12 and 16.20. The average is $\dfrac{16.12 + 16.20}{2} = \dfrac{32.32}{2} = 16.16$.

Thus, the median is 16.16. Half of the depths are larger than 16.16 and half are smaller. If the largest depth measurement were doubled, the median would not change. It would stay the same.

c.   Since no observation occurs more than once, then we say there is either no mode or we say that all of the observations are modes.

2.69   a.   The mean cylinder power measurements for student #11 is:

$$\bar{x} = \frac{\sum x}{n} = \frac{-.08 + (-.06) + \ldots + (-.16)}{25} = \frac{-3.86}{25} = -.1544$$

The median is the middle number once the data have been arranged in order:
$-1.07, -.21, -.20, -.17, -.17, -.17, -.16, -.16, -.16, -.15, -.12, -.12, -.11, -.10,$
$-.09, -.09, -.09, -.08, -.08, -.07, -.07, -.06, -.06, -.06, -.04$

The median is $-.11$.

The mode is the value with the highest frequency. Since $-.17, -.16, -.09, -.06$ each occur 3 times, all are modes.

From the printout, the mean is -0.1544 and the median is -0.11. (All the numbers in the table are negative numbers.) Since these two values are very close together, both are good representatives of the middle of the data set.

b.   The outlier is -1.07.

c.   After deleting -1.07, the mean is:

$$\bar{x} = \frac{\sum x}{n} = \frac{(-.08) + (-.06) + (-.15) + \cdots + (-.16)}{24} = \frac{-2.79}{24} = -.11625$$

The median is the average of the middle two numbers once the data are arranged in order.

The middle two numbers are 10 and 11. The median is $\dfrac{(-.10) + (-.11)}{2} = -.105$

The modes remain the same.

The mean changes from $-.1544$ to $-.11625$ while the median changes from $-.11$ to $-.105$. The mean changes much more than the median when the outlier is removed.

2.71   The range of a data set is the difference between the largest and smallest measurements.

2.73   The sample variance is the sum of the squared deviations from the sample mean divided by the sample size minus 1. The population variance is the sum of the squared deviations from the population mean divided by the population size.

2.75   If the standard deviation increases, this implies that the data are more variable.

2.77   a.   Range $= 42 - 37 = 5$

$$s^2 = \frac{\sum x^2 - \dfrac{\left(\sum x\right)^2}{n}}{n-1} = \frac{7935 - \dfrac{199^2}{5}}{5-1} = 3.7 \qquad s = \sqrt{3.7} = 1.92$$

b.   Range $= 100 - 1 = 99$

$$s^2 = \frac{\sum x^2 - \dfrac{\left(\sum x\right)^2}{n}}{n-1} = \frac{25,795 - \dfrac{303^2}{9}}{9-1} = 1,949.25 \qquad s = \sqrt{1,949.25} = 44.15$$

c.   Range $= 100 - 2 = 98$

$$s^2 = \frac{\sum x^2 - \dfrac{\left(\sum x\right)^2}{n}}{n-1} = \frac{20,033 - \dfrac{295^2}{8}}{8-1} = 1,307.84 \qquad s = \sqrt{1,307.84} = 36.16$$

2.79   This is one possibility for the two data sets.

Data Set 1: 1, 1, 2, 2, 3, 3, 4, 4, 5, 5
Data Set 2: 1, 1, 1, 1, 1, 5, 5, 5, 5, 5

$$\bar{x}_1 = \frac{\sum x}{n} = \frac{1+1+2+2+3+3+4+4+5+5+}{10} = \frac{30}{10} = 3$$

$$\bar{x}_2 = \frac{\sum x}{n} = \frac{1+1+1+1+1+5+5+5+5+5}{10} = \frac{30}{10} = 3$$

Therefore, the two data sets have the same mean. The variances for the two data sets are:

$$s_1^2 = \frac{\sum x^2 - \dfrac{\left(\sum x\right)^2}{n}}{n-1} = \frac{110 - \dfrac{30^2}{10}}{9} = \frac{20}{9} = 2.2222$$

$$s_2^2 = \frac{\sum x^2 - \dfrac{\left(\sum x\right)^2}{n}}{n-1} = \frac{130 - \dfrac{30^2}{10}}{9} = \frac{40}{9} = 4.4444$$

The dot diagram for the two data sets are shown below.

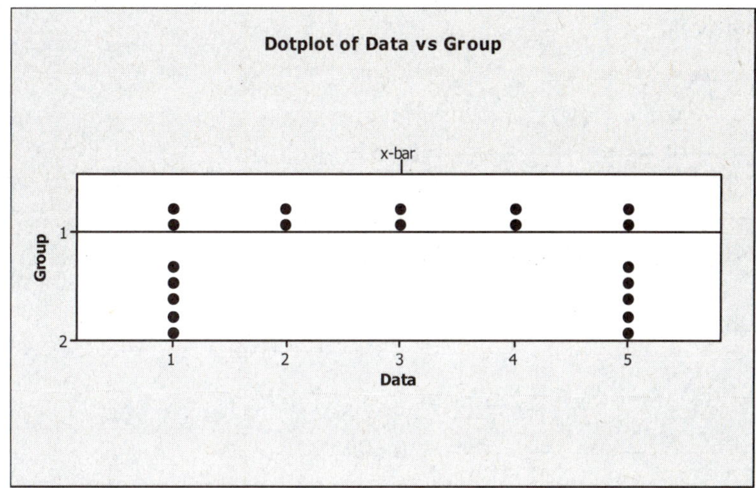

2.81 a. Range $= 3 - 0 = 3$

$$s^2 = \frac{\sum x^2 - \dfrac{\left(\sum x\right)^2}{n}}{n-1} = \frac{15 - \dfrac{7^2}{5}}{5-1} = 1.3 \qquad s = \sqrt{1.3} = 1.1402$$

  b. After adding 3 to each of the data points,

    Range $= 6 - 3 = 3$

$$s^2 = \frac{\sum x^2 - \dfrac{\left(\sum x\right)^2}{n}}{n-1} = \frac{102 - \dfrac{22^2}{5}}{5-1} = 1.3 \qquad s = \sqrt{1.3} = 1.1402$$

  c. After subtracting 4 from each of the data points,

    Range $= -1 - (-4) = 3$

$$s^2 = \frac{\sum x^2 - \dfrac{\left(\sum x\right)^2}{n}}{n-1} = \frac{39 - \dfrac{(-13)^2}{5}}{5-1} = 1.3 \qquad s = \sqrt{1.3} = 1.1402$$

  d. The range, variance, and standard deviation remain the same when any number is added to or subtracted from each measurement in the data set.

2.83 a. From the printout, the range is 51.26.

  b. From the printout, the variance is 128.57.

c.   From the printout, the standard deviation is 11.34.

d.   If we are interested in just these 76 captured turtles, then the data would represent a population. The variance would be represented by σ^2 and the standard deviation would be represented by σ.

2.85   a.   The sample variance is:

$$s^2 = \frac{\sum x^2 - \dfrac{\left(\sum x\right)^2}{n}}{n-1} = \frac{4295 - \dfrac{375^2}{35}}{35-1} = 8.151$$

The standard deviation is $s = \sqrt{8.151} = 2.855$

b.   The sample variance is:

$$s^2 = \frac{\sum x^2 - \dfrac{\left(\sum x\right)^2}{n}}{n-1} = \frac{2631 - \dfrac{275^2}{33}}{33-1} = 10.604$$

The standard deviation is $s = \sqrt{10.604} = 3.256$

c.   The sample variance is:

$$s^2 = \frac{\sum x^2 - \dfrac{\left(\sum x\right)^2}{n}}{n-1} = \frac{1881 - \dfrac{241^2}{37}}{37-1} = 8.646$$

The standard deviation is $s = \sqrt{8.646} = 2.940$

d.   The DM dosage group appears to have the most variability since the variance of the DM dosage group is larger than the variances of the other 2 groups. The honey dosage group appears to have the least variability since the variance of the honey dosage group is smaller than the variances of the other 2 groups. However, the variance of the honey dosage group and the control group are very similar.

2.87   a.   The range is the difference between the largest and smallest observations and is 19.70 – 13.25 = 6.45 mm. If the largest depth measurement were doubled, the range would increase. The new range would be 39.40 – 13.25 = 26.15.

b.　The variance is:

$$s^2 = \frac{\sum x^2 - \dfrac{\left(\sum x\right)^2}{n}}{n-1} = \frac{4966.173 - \dfrac{296.99^2}{18}}{18-1} = 3.883$$

If the largest depth measurement were doubled, the variance would increase. The new variance would be:

$$s^2 = \frac{\sum x^2 - \dfrac{\left(\sum x\right)^2}{n}}{n-1} = \frac{6130.443 - \dfrac{316.69^2}{18}}{18-1} = 32.861$$

c.　The standard deviation is $s = \sqrt{3.883} = 1.970$. If the largest depth measurement were doubled, the standard deviation would increase. The new standard deviation would be $s = \sqrt{32.861} = 5.732$

2.89　a.　Range $= 11 - 1 = 10$

$$s^2 = \frac{\sum x^2 - \dfrac{\left(\sum x\right)^2}{n}}{n-1} = \frac{450 - \dfrac{78^2}{20}}{20-1} = 7.674$$

$$s = \sqrt{s^2} = \sqrt{7.674} = 2.770$$

b.　Dropping the largest measurement:

Range $= 9 - 1 = 8$

$$s^2 = \frac{\sum x^2 - \dfrac{\left(\sum x\right)^2}{n}}{n-1} = \frac{329 - \dfrac{67^2}{19}}{19-1} = 5.152$$

$$s = \sqrt{s^2} = \sqrt{5.152} = 2.270$$

By dropping the largest observation from the data set, the range decreased from 10 to 8, the variance decreased from 7.674 to 5.1520 and the standard deviation decreased from 2.770 to 2.270.

c.　Dropping the largest and smallest measurements:

Range $= 9 - 1 = 8$

$$s^2 = \frac{\sum x^2 - \dfrac{\left(\sum x\right)^2}{n}}{n-1} = \frac{328 - \dfrac{66^2}{18}}{18-1} = 5.059$$

$$s = \sqrt{s^2} = \sqrt{5.059} = 2.249$$

By dropping the largest and smallest observations from the data set, the range decreased from 10 to 8, the variance decreased from 7.674 to 5.059 and the standard deviation decreased from 2.770 to 2.249.

2.91   a.   The unit of measurement of the variable of interest is dollars (the same as the mean and standard deviation). Based on this, the data are quantitative.

b.   Since no information is given about the shape of the data set, we can only use Chebyshev's rule.

$900 is 2 standard deviations below the mean, and $2100 is 2 standard deviations above the mean. Using Chebyshev's rule, at least 3/4 of the measurements (or $3/4 \times 200 = 150$ measurements) will fall between $900 and $2100.

$600 is 3 standard deviations below the mean and $2400 is 3 standard deviations above the mean. Using Chebyshev's rule, at least 8/9 of the measurements (or $8/9 \times 200 = 178$ measurements) will fall between $600 and $2400.

$1200 is 1 standard deviation below the mean and $1800 is 1 standard deviation above the mean. Using Chebyshev's rule, nothing can be said about the number of measurements that will fall between $1200 and $1800.

$1500 is equal to the mean and $2100 is 2 standard deviations above the mean. Using Chebyshev's rule, at least 3/4 of the measurements (or $3/4 \times 200 = 150$ measurements) will fall between $900 and $2100. It is possible that all of the 150 measurements will be between $900 and $1500. Thus, nothing can be said about the number of measurements between $1500 and $2100.

2.93   According to the Empirical Rule:

a.   Approximately 68% of the measurements will be contained in the interval $\bar{x} - s$ to $\bar{x} + s$.

b.   Approximately 95% of the measurements will be contained in the interval $\bar{x} - 2s$ to $\bar{x} + 2s$.

c.   Essentially all the measurements will be contained in the interval $\bar{x} - 3s$ to $\bar{x} + 3s$.

2.95   Using Chebyshev's rule, at least 8/9 of the measurements will fall within 3 standard deviations of the mean. Thus, the range of the data would be around 6 standard deviations. Using the Empirical Rule, approximately 95% of the observations are within 2 standard deviations of the mean. Thus, the range of the data would be around 4 standard deviations. We would expect the standard deviation to be somewhere between Range/6 and Range/4.

For our data, the range $= 760 - 135 = 625$.

The Range/6 $= 625/6 = 104.17$ and Range/4 $= 625/4 = 156.25$.

Therefore, I would estimate that the standard deviation of the data set is between 104.17 and 156.25.

It would not be feasible to have a standard deviation of 25. If the standard deviation were 25, the data would span $625/25 = 25$ standard deviations. This would be extremely unlikely.

2.97 a. $\bar{x} = \dfrac{\sum x}{n} = \dfrac{17800}{186} = 95.699$

$$s^2 = \dfrac{\sum x^2 - \dfrac{\left(\sum x\right)^2}{n}}{n-1} = \dfrac{1707998 - \dfrac{17800^2}{186}}{186-1} = 24.633$$

$$s = \sqrt{s^2} = \sqrt{24.633} = 4.963$$

b. $\bar{x} \pm s \Rightarrow 95.699 \pm 4.963 \Rightarrow (90.736, 100.662)$

$\bar{x} \pm 2s \Rightarrow 95.699 \pm 2(4.963) \Rightarrow 95.699 \pm 9.926 \Rightarrow (85.773, 105.625)$

$\bar{x} \pm 3s \Rightarrow 95.699 \pm 3(4.963) \Rightarrow 95.699 \pm 14.889 \Rightarrow (80.810, 110.588)$

c.   There are 166 out of 186 observations in the first interval. This is $(166/186)*100\% = 89.2\%$. There are 179 out of 186 observations in the second interval. This is $(179/186)*100\% = 96.2\%$. There are 182 out of 186 observations in the second interval. This is $(182/186)*100\% = 97.8\%$.

The percentages for the first 2 intervals are somewhat larger than what we would expect using the Empirical Rule. The Empirical Rule indicates that approximately 68% of the observations will fall within 1 standard deviation of the mean. It also indicates that approximately 95% of the observations will fall within 2 standard deviations of the mean. Chebyshev's Theorem says that at least ¾ or 75% of the observations will fall within 2 standard deviations of the mean and at least 8/9 or 88.8% of the observations will fall within 3 standard deviations of the mean. It appears that our observed percentages agree with Chebyshev's Theorem better that the Empirical Rule.

2.99 a.   The mean is 79 and the standard deviation is 23. Thus, a measurement of 102 is 1 standard deviation above the mean. If nothing is known about the shape of the distribution, we must use Chebyshev's Rule to describe the distribution. Using this, we know nothing about the number of observations within 1 standard deviation of the mean, and thus, know nothing about the percentage of observations less than 102.

b.   If we assume that the distribution is mound-shaped, we can use the Empirical Rule to describe the distribution. If the distribution is mound-shaped, then it is symmetric. Thus, we know that half of the observations will be below 79. We know that about 68%

of the observations are within 1 standard deviation of the mean. Therefore, about 34% of the observations will be between the mean and 1 standard deviation above the mean, or between 79 and 102. Thus, the percentage of observations less than 102 is about $50\% + 34\% = 84\%$.

2.101   For parts a and b, we must use Chebyshev's Rule to describe the data sets. For both the handrubbers and the handwashers, the standard deviation is greater than the mean. Since no observations can be less than 0, we know that the smallest observation in the data sets is less than 1 standard deviation from the mean. Thus, the distributions cannot be symmetric, but rather skewed to the right. We cannot use the Empirical Rule.

    a.   Using Chebyshev's Rule, we know that at least $1 - \dfrac{1}{2^2} = 1 - \dfrac{1}{4} = \dfrac{3}{4} = .75$ of the observations fall within 2 standard deviations of the mean.
Thus, the interval $\bar{x} \pm 2s$ will contain at least .75 of the observations.

        For handrubbers: $\bar{x} \pm 2s \Rightarrow 35 \pm 2(59) \Rightarrow 35 \pm 118 \Rightarrow (0,\ 153)$

    b.   For handwashers: $\bar{x} \pm 2s \Rightarrow 69 \pm 2(106) \Rightarrow 69 \pm 212 \Rightarrow (0,\ 281)$

    c.   Since the interval for handwashers is much larger that that for handrubbers and the sample mean for handrubbers is less than that for handwashers, we can conclude that handrubbing is more effective than handwashing for reducing bacterial counts.

2.103   a.   Since no information is given about the distribution of the velocities of the Winchester bullets, we can only use Chebyshev's rule to describe the data. We know that at least 3/4 of the velocities will fall within the interval:

        $\bar{x} \pm 2s \Rightarrow 936 \pm 2(10) \Rightarrow 936 \pm 20 \Rightarrow (916,\ 956)$

        Also, at least 8/9 of the velocities will fall within the interval:

        $\bar{x} \pm 3s \Rightarrow 936 \pm 3(10) \Rightarrow 936 \pm 30 \Rightarrow (906,\ 966)$

    b.   Since a velocity of 1,000 is much larger than the largest value in the second interval in part **a**, it is very unlikely that the bullet was manufactured by Winchester.

2.105   a.   Regardless of the shape of the distribution, most of the observations will fall within 3 standard deviations of the mean. Thus, for the SAT-Math scores, an interval likely to contain a student's change in score is:

        $\bar{x} \pm 3s \Rightarrow 19 \pm 3(65) \Rightarrow 19 \pm 195 \Rightarrow (-176,\ 214)$

    b.   Regardless of the shape of the distribution, most of the observations will fall within 3 standard deviations of the mean. Thus, for the SAT-Verbal scores, an interval likely to contain a student's change in score is:

        $\bar{x} \pm 3s \Rightarrow 7 \pm 3(49) \Rightarrow 7 \pm 147 \Rightarrow (-140,\ 154)$

c.   For the SAT-Verbal, the maximum increase in scores is about 154 points. For the SAT-Math, the maximum increase in scores is approximately 214. Thus, a student is more likely to get a 140-point increase on the SAT-Math test.

2.107   Since we do not know if the distribution of the heights of the trees is mound-shaped, we need to apply Chebyshev's rule. We know $\mu = 30$ and $\sigma = 3$. Therefore,

$$\mu \pm 3\sigma \Rightarrow 30 \pm 3(3) \Rightarrow 30 \pm 9 \Rightarrow (21,\ 39)$$

According to Chebyshev's rule, at least 8/9 or .89 of the tree heights on this piece of land fall within this interval and at most $\dfrac{1}{9}$ or .11 of the tree heights will fall above the interval.

However, the buyer will only purchase the land if at least $\dfrac{1000}{5000}$ or .20 of the tree heights are at least 40 feet tall. Therefore, the buyer should not buy the piece of land.

2.109   Using the definition of a percentile:

	Percentile	Percentage Above	Percentage Below
a.	75th	25%	75%
b.	50th	50%	50%
c.	20th	80%	20%
d.	84th	16%	84%

2.111   a. $z = \dfrac{x - \bar{x}}{s} = \dfrac{40 - 30}{5} = 2$ (sample)   2 standard deviations above the mean.

b. $z = \dfrac{x - \mu}{\sigma} = \dfrac{90 - 89}{2} = .5$ (population)   .5 standard deviations above the mean.

c. $z = \dfrac{x - \mu}{\sigma} = \dfrac{50 - 50}{5} = 0$ (population)   0 standard deviations above the mean.

d. $z = \dfrac{x - \bar{x}}{s} = \dfrac{20 - 30}{4} = -2.5$ (sample)   2.5 standard deviations below the mean.

2.113   Since the element 40 has a z-score of -2 and 90 has a z-score of 3,

$$-2 = \frac{40 - \mu}{\sigma} \text{ and } 3 = \frac{90 - \mu}{\sigma}$$

$$\Rightarrow -2\sigma = 40 - \mu \qquad \Rightarrow 3\sigma = 90 - \mu$$

$$\Rightarrow \mu - 2\sigma = 40 \qquad \Rightarrow \mu + 3\sigma = 90$$

$$\Rightarrow \mu = 40 + 2\sigma$$

By substitution,
$$40 + 2\sigma + 3\sigma = 90$$
$$\Rightarrow 5\sigma = 50$$
$$\Rightarrow \sigma = 10$$

By substitution, $\mu = 40 + 2(10) = 60$

Therefore, the population mean is 60 and the standard deviation is 10.

2.115   The percentile ranking for the age of 25 years in the distribution of all ages of licensed drivers stopped by police is $100 - 74 = 26$ percentile.

2.117   a.   The z-score associated with a score of 30 is $z = \dfrac{x - \bar{x}}{s} = \dfrac{30 - 39}{6} = -1.50$. This means that a score of 30 is 1.5 standard deviations below the mean.

b.   The z-score associated with a score of 39 is $z = \dfrac{x - \bar{x}}{s} = \dfrac{39 - 39}{6} = 0$. Half or .5 of the observations are below a score of 39.

2.119   a.   Using SAS, the output is:

```
                        EGG LENGTH

                   The UNIVARIATE Procedure
                   Variable:   length

                          Moments

N                            130      Sum Weights              130
Mean                  60.6538462      Sum Observations        7885
Std Deviation         43.9861168      Variance          1934.77847
Skewness               2.1091362      Kurtosis           4.70026314
Uncorrected SS            727842      Corrected SS       249586.423
Coeff Variation       72.5199136      Std Error Mean     3.85783765

                 Basic Statistical Measures

        Location                        Variability

     Mean     60.65385      Std Deviation            43.98612
     Median   49.50000      Variance                     1935
     Mode     35.00000      Range                220.00000
                            Interquartile Range   32.00000
```

```
                    Tests for Location: Mu0=0

        Test              -Statistic-      -----p Value------

        Student's t     t  15.72224       Pr > |t|    <.0001
        Sign            M        65       Pr >= |M|   <.0001
        Signed Rank     S    4257.5       Pr >= |S|   <.0001


                    Quantiles (Definition 5)

                    Quantile        Estimate

                    100% Max          236.0
                    99%               218.0
                    95%               160.0
                    90%               122.0
                    75% Q3             67.0
                    50% Median         49.5
                    25% Q1             35.0
                    10%                23.0
                    5%                 19.0
                    1%                 16.0
                    0% Min             16.0


                         EGG LENGTH

                    The UNIVARIATE Procedure
                      Variable:  length

                    Extreme Observations

          ----Lowest----            ----Highest---

          Value        Obs          Value        Obs

           16.0        102           195         128
           16.0         90           205         123
           17.0        101           216         122
           18.0        103           218         129
           18.5        100           236         130


                    Missing Values

                              -----Percent Of-----
        Missing                            Missing
        Value        Count     All Obs        Obs

           .            2        1.52       100.00
```

The 10th percentile egg length is 23. This means that 10% of all egg lengths are less than 23 and 90% are greater than 23.

b.   From the printout, $\bar{x} = 60.65$ and $s = 43.99$. The z-score corresponding to the moas (*P. australis*) bird species' egg length is:

$$z = \frac{x - \bar{x}}{s} = \frac{205 - 60.65}{43.99} = 3.28$$

The z-score for moas is 3.28 standard deviations above the mean. This is a fairly large value for a z-score. This indicates that the egg length for moas could be very unusual.

2.121   a.   The z-score for Harvard is 5.08. This means that the productivity score for Harvard was 5.08 standard deviations above the mean. This is an unusually high z-score.

b.   The z-score for Howard is -0.81. This means that the productivity score for Howard was 0.81 standard deviations below the mean.

c.   Yes. If only 44 out of 129 schools had positive z-scores, this indicates that the data are skewed to the right. Other indications that the data are skewed to the right are the z-scores associated with the smallest and largest scores. The largest z-score is 5.08 and the smallest is -0.81. The largest observation is much further from the mean than the smallest observation.

Using MINITAB, a histogram of the data is:

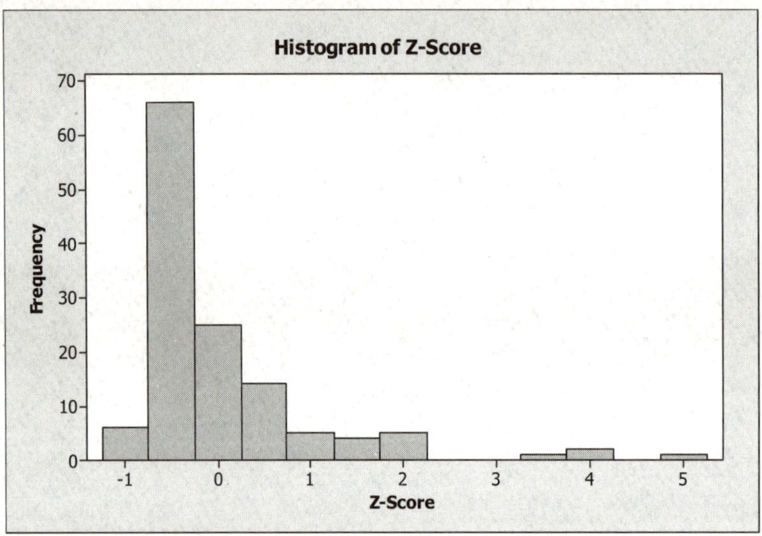

The data are, in fact, skewed to the right.

2.123   a.   From the problem, $\mu = 2.7$ and $\sigma = .5$

$$z = \frac{x - \mu}{\sigma} \Rightarrow z\sigma = x - \mu \Rightarrow x = \mu + z\sigma$$

For $z = 2.0$, $x = 2.7 + 2.0(.5) = 3.7$

For $z = -1.0$, $x = 2.7 - 1.0(.5) = 2.2$

For $z = .5$, $x = 2.7 + .5(.5) = 2.95$

For $z = -2.5$, $x = 2.7 - 2.5(.5) = 1.45$

b.   For $z = -1.6$, $x = 2.7 - 1.6(.5) = 1.9$

c.   If we assume the distribution of GPAs is approximately mound-shaped, we can use the Empirical Rule.

From the Empirical Rule, we know that $\approx .025$ or $\approx 2.5\%$ of the students will have GPAs above 3.7 (with $z = 2$). Thus, the GPA corresponding to summa cum laude (top 2.5%) will be greater than 3.7 ($z > 2$).

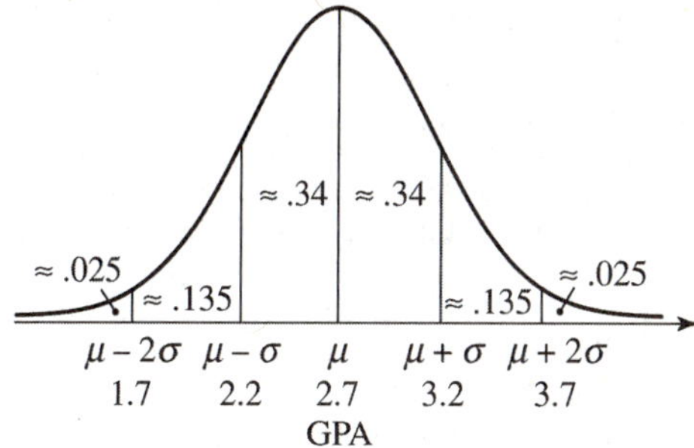

We know that $\approx .16$ or $\approx 16\%$ of the students will have GPAs above 3.2 ($z = 1$).
Thus, the limit on GPAs for cum laude (top 16%) will be greater than 3.2 ($z > 1$).

We must assume the distribution is mound-shaped.

2.125   An observation that is unusually large or small relative to the data values we want to describe is an outlier.

2.127   The hinges of a box plot are the upper quartile and the lower quartile (top and bottom of the rectangle).

2.129   To determine if the measurements are outliers, compute the z-score.

a. $z = \dfrac{x - \bar{x}}{s} = \dfrac{65 - 57}{11} = .727$ Since this z-score is less than 3 in magnitude, 65 is not an outlier.

b. $z = \dfrac{x - \bar{x}}{s} = \dfrac{21 - 57}{11} = -3.273$ Since this z-score is more than 3 in magnitude, 21 is an outlier.

c. $z = \dfrac{x - \bar{x}}{s} = \dfrac{72 - 57}{11} = 1.364$ Since this z-score is less than 3 in magnitude, 72 is not an outlier.

d. $z = \dfrac{x - \bar{x}}{s} = \dfrac{98 - 57}{11} = 3.727$ Since this z-score is more than 3 in magnitude, 98 is an outlier.

2.131   a.   The median is approximately 4.

b. Q_L (Lower quartile) is approximately 3 and Q_U (Upper quartile) is approximately 6.

c.   The interquartile range is IQR $= Q_U - Q_L \approx 6 - 3 = 3$.

d.   The data set is skewed to the right since the right whisker is longer than the left and there are outlying observations to the right.

e.   50% of the observations lie to the right of the median. 75% of the observations lie to the left of the upper quartile.

f.   The outliers in the data set are 12, 13, and 16.

2.133   a.   The approximate 25th percentile PASI score before treatment is 10. The approximate median before treatment is 15. The approximate 75th percentile PASI score before treatment is 27.5.

   b.   The approximate 25th percentile PASI score after treatment is 3.5. The approximate median after treatment is 5. The approximate 75th percentile PASI score after treatment is 7.5.

   c.   Since the 75th percentile after treatment is lower than the 25th percentile before treatment, it appears that the ichthyotherapy is effective in treating psoriasis.

2.135   a.   The z-score is $z = \dfrac{x - \bar{x}}{s} = \dfrac{3.3 - 7.3}{3.18} = -1.26$.

   b.   A PMI score of 3.3 would not be considered an outlier. The z-score is -1.26. A z-score this small is not considered unusual.

2.137   a.   Using MINITAB, the boxplot is:

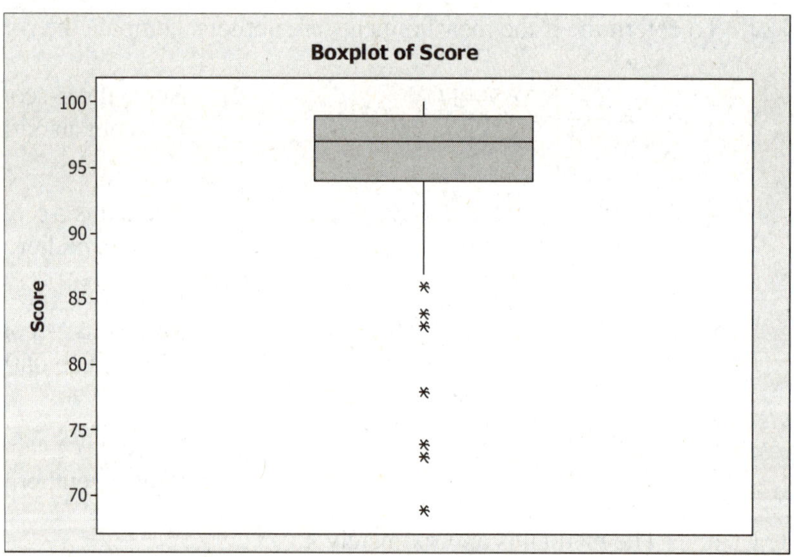

   From the boxplot, there appears to be 7 outliers: 69, 73, 74, 78, 83, 84, and 86. However, there are 2 observations with a value of 84 and 3 observations with a value of 86. Thus, there appear to be 10 outliers.

   b.   From Exercise 2.97, $\bar{x} = 95.699$ and $s = 4.963$. Since the data are skewed to the left, we will consider observations more than 3 standard deviations from the mean to be outliers. An observation with a z-score of 3 would have the value:

$$z = \frac{x - \bar{x}}{s} \Rightarrow 3 = \frac{x - 95.699}{4.963} \Rightarrow 3(4.963) = x - 95.699 \Rightarrow 14.889 = x - 95.699 \Rightarrow x = 110.588$$

An observation with a z-score of -3 would have the value:

$$z = \frac{x - \bar{x}}{s} \Rightarrow -3 = \frac{x - 95.699}{4.963} \Rightarrow -3(4.963) = x - 95.699 \Rightarrow -14.889 = x - 95.699 \Rightarrow x = 80.810$$

Observations greater than 10.588 or less than 80.81 would be considered outliers. Using this criterion, the following observations would be outliers: 69, 73, 74, and 78.

c.   No, these methods do not agree. Using the boxplot, 10 observations were identified as outliers. Using the z-score method, only 4 observations were identified as outliers. Since the data are very highly skewed to the left, the z-score method may not be appropriate.

2.139   Using MINITAB, the boxplots of the 3 groups are:

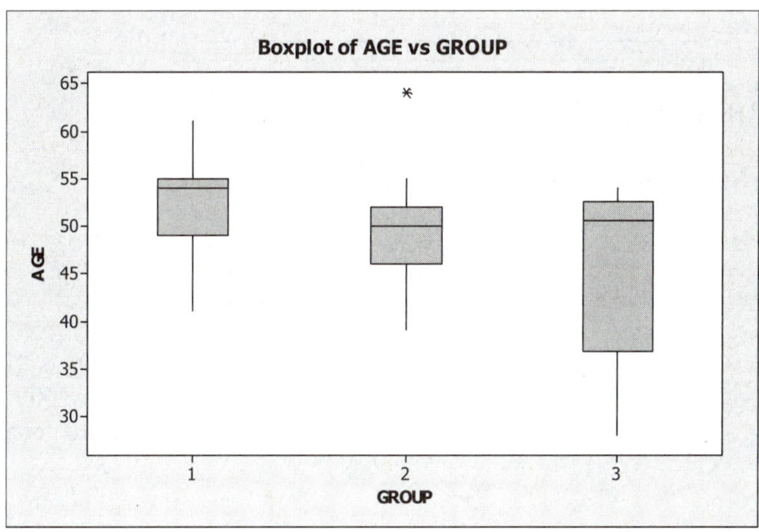

From the boxplots, there appears to be only 1 outlier. That outlier is in the second group.

2.141   a.   Using MINITAB, the side-by-side box plots are:

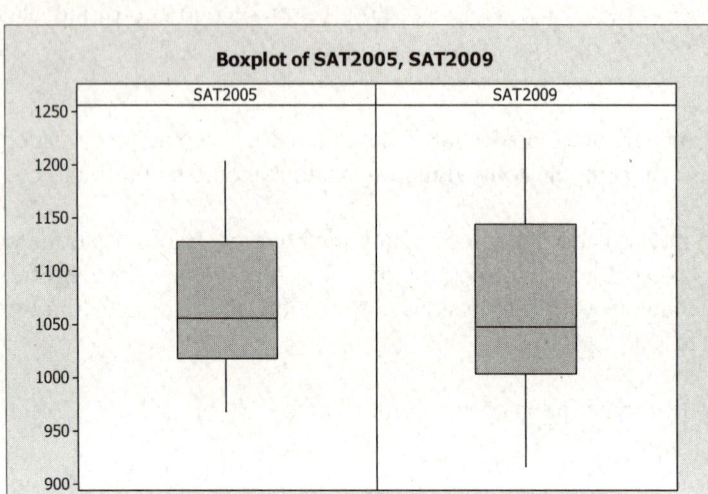

b.   Using MINITAB, the descriptive statistics are:

Descriptive Statistics: SAT2005, SAT2009

```
Variable    N    Mean  StDev  Minimum      Q1  Median      Q3  Maximum
SAT2005    51  1077.0   67.9    968.0  1018.0  1056.0  1127.0   1204.0
SAT2009    51  1073.5   81.4    917.0  1003.0  1048.0  1144.0   1225.0
```

The standard deviation for 2005 is 67.9 and the standard deviation for 2009 is 81.4. Also, the IQR for 2005 is $Q_U - Q_L = 1127 - 1018 = 109$ while the IQR for 2009 is $Q_U - Q_L = 1144 - 1003 = 141$. Thus, the variability for 2009 is slightly greater than that for 2005.

c.   Since there are no observations outside the inner fences for either year, there are no outliers.

2.143   A bivariate relationship is a relationship between 2 quantitative variables.

2.145   A positive association between two variables means that as one variable increases, the other variable tends to also increase. A negative association between two variables means that as one variable increases, the other variable tends to decrease.

2.147   Using MINITAB, the scatterplot is:

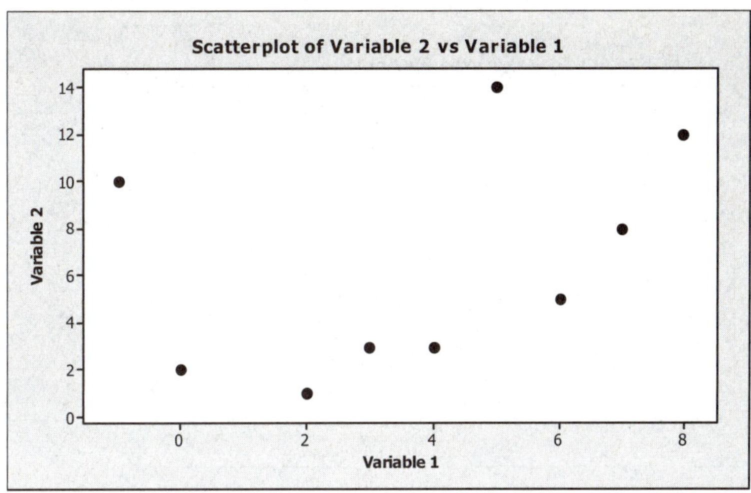

From the scatterplot, there does not appear to be much of a trend between variable 1 and variable 2. There is a slight positive linear trend - as variable 1 increases, variable 2 tends to increase. However, this relationship appears to be very weak.

2.149   If we constructed a scatterplot of the 2009 SAT scores versus the 2005 SAT scores, we would expect there to be a positive linear trend to the data. If a state had a high mean SAT score in 2005, we would expect that the state would also have a high mean SAT score in 2009. Using MINITAB, the scatterplot of the data is:

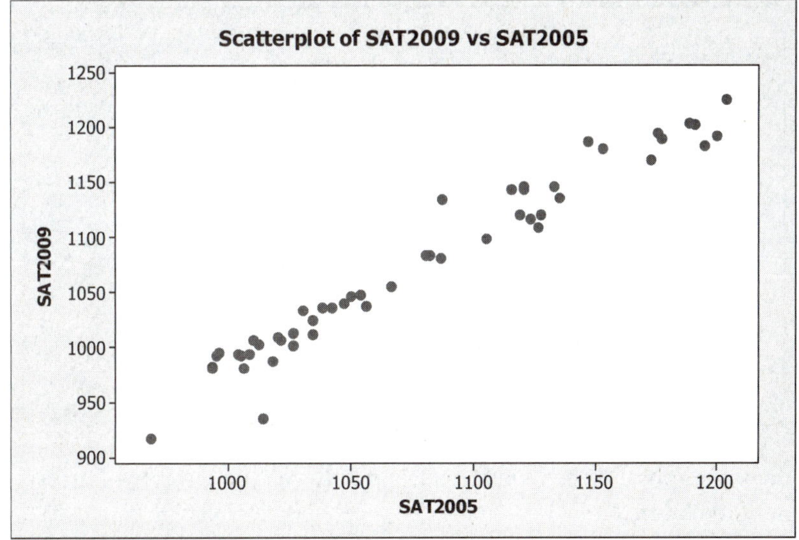

2.151   Using MINITAB, the scatterplot of the data is:

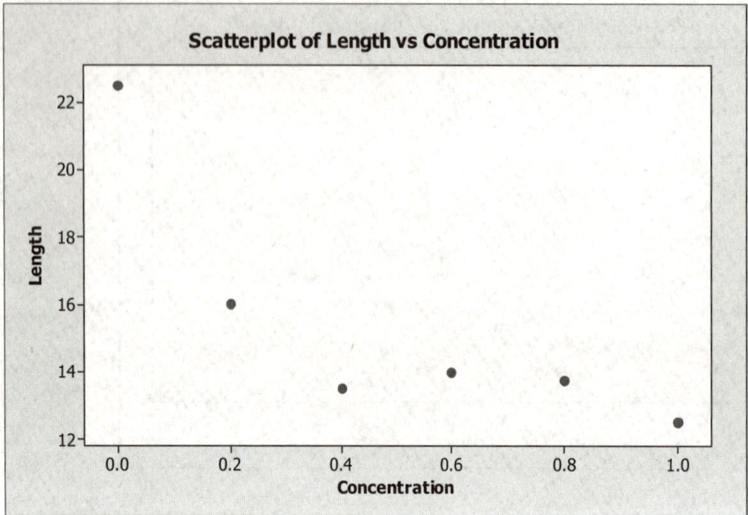

There appears to be a negative trend to the data. As the concentration increases, the wicking length tends to decrease. It appears that the relationship may not be linear, but rather curvilinear.

2.153   a.   Using MINITAB, a scatterplot of the data is:

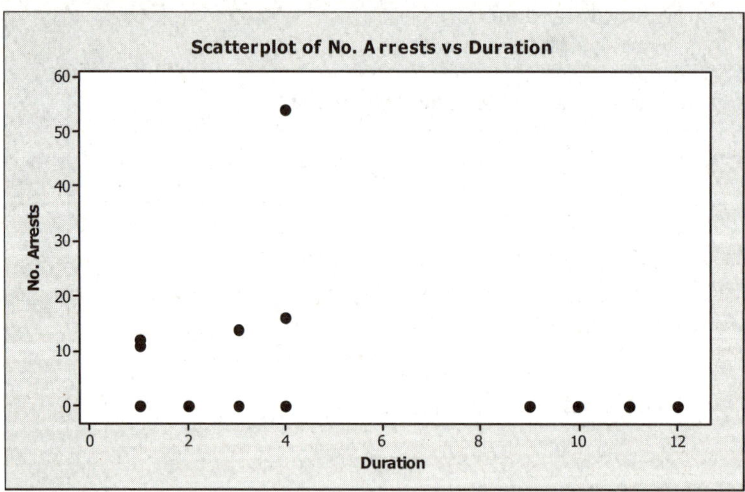

There does not appear to be much of a trend between the number of arrests and the duration of the sit-in.

b.   Using MINITAB, a scatterplot of the data using only those observations where there was at least one arrest is:

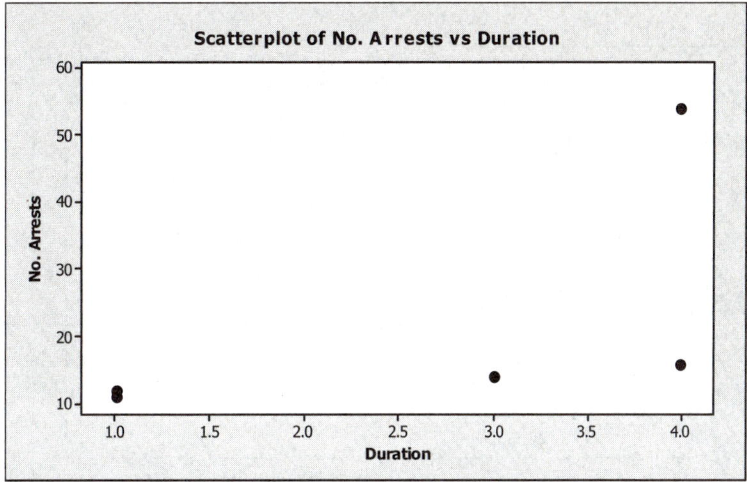

There appears to be a positive relationship between the number of arrests and the duration of the sit-in. As the duration increases, the number of arrests tends to increase.

c.   Since there are only 5 observations used in the scatterplot in part **b**, the reliability is suspect.

2.155   a.   Using MINITAB, a scatterplot of the data is:

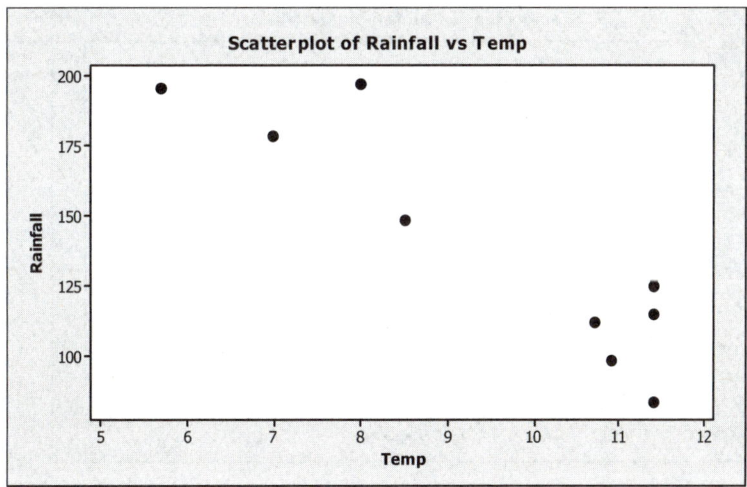

There appears to be a negative relationship between the annual rainfall and the maximum daily temperature. As the maximum daily temperature increases, the amount of annual rainfall tends to decrease.

b.   Using MINITAB, a scatterplot of the relationship between annual rainfall and total plant cover is:

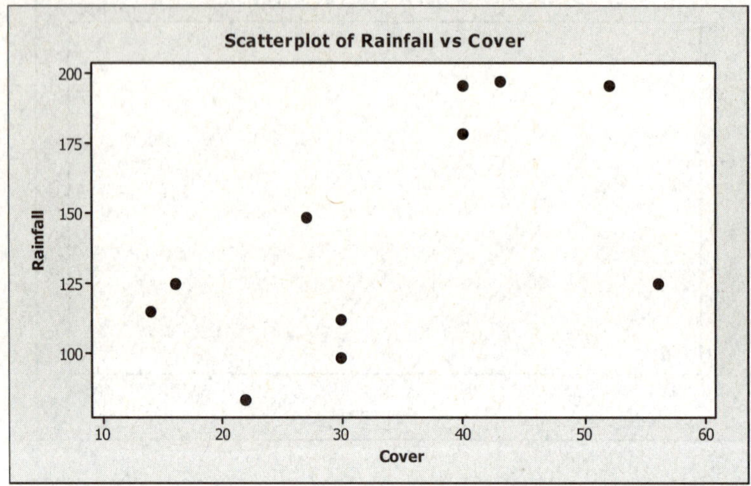

There appears to be a positive relationship between annual rainfall and total plant cover. As the total plant cover increases, the annual rainfall tends to increase.

Using MINITAB, a scatterplot of the relationship between annual rainfall and number of ant species is:

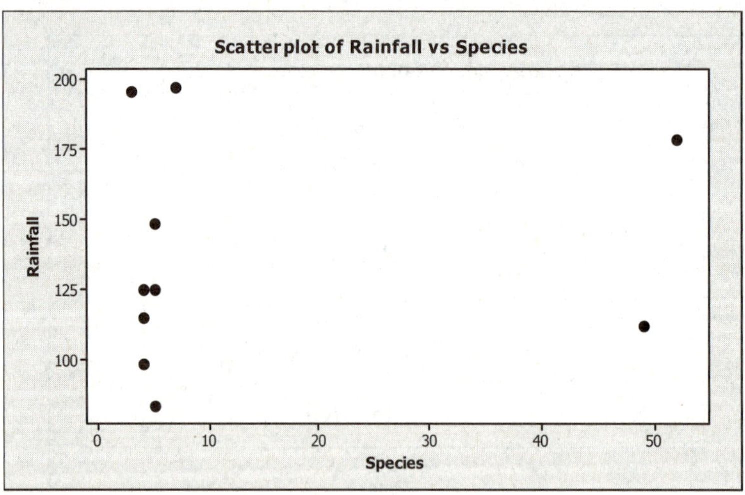

There does not appear to be a relationship between annual rainfall and number of ant species.

Using MINITAB, a scatterplot of the relationship between annual rainfall and species diversity index is:

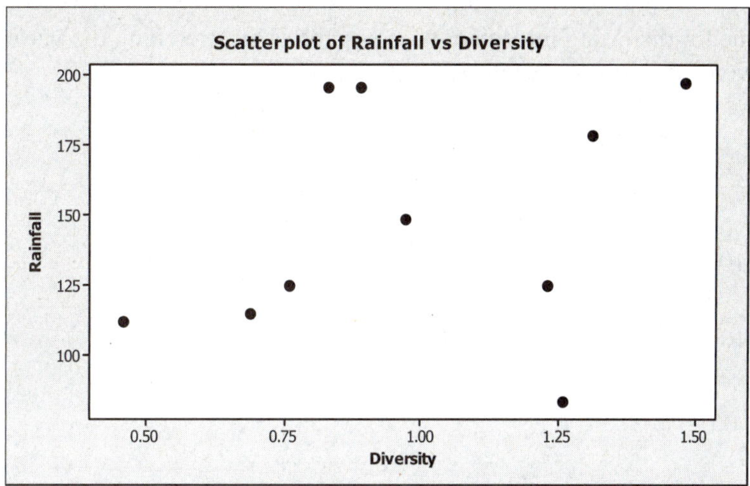

There appears to be a positive relationship between annual rainfall and species diversity index. As the species diversity index increases, the annual rainfall tends to increase. However, this relationship appears to be very weak.

2.157   Using MINITAB, a graph of the accuracy scores versus the driving distance is as follows:

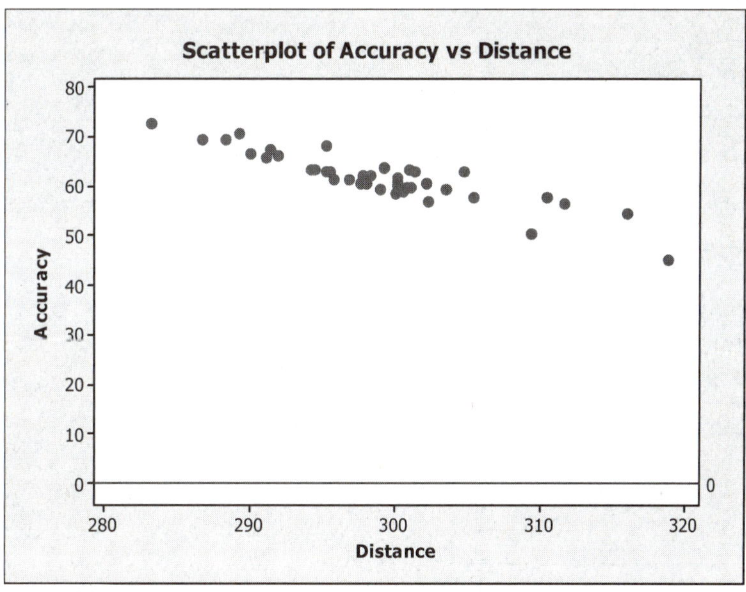

Yes, the golfer's concern is valid. From the graph, as the driving distance increases, the accuracy decreases.

2.159   The median is preferred to the mean as a measure of central tendency when the data set is skewed either right or left. If the data set has some extreme observations, the median better measure of central tendency than the mean.

2.161   A stem-and-leaf display is generally preferred over a histogram when the data set is relatively small.

2.163   One technique for distorting information on a graph is by stretching the vertical axis by starting the vertical axis somewhere above 0.

2.165   a. $z = \dfrac{x-\mu}{\sigma} = \dfrac{50-60}{10} = -1$

$z = \dfrac{70-60}{10} = 1$

$z = \dfrac{80-60}{10} = 2$

b. $z = \dfrac{x-\mu}{\sigma} = \dfrac{50-60}{5} = -2$

$z = \dfrac{70-60}{5} = 2$

$z = \dfrac{80-60}{5} = 4$

c. $z = \dfrac{x-\mu}{\sigma} = \dfrac{50-40}{10} = 1$

$z = \dfrac{70-40}{10} = 3$

$z = \dfrac{80-40}{10} = 4$

d. $z = \dfrac{x-\mu}{\sigma} = \dfrac{50-40}{100} = .1$

$z = \dfrac{70-40}{100} = .3$

$z = \dfrac{80-40}{100} = .4$

2.167   a. $s^2 = \dfrac{\sum x^2 - \dfrac{\left(\sum x\right)^2}{n}}{n-1} = \dfrac{246 - \dfrac{63^2}{22}}{22-1} = 3.1234$

b. $s^2 = \dfrac{\sum x^2 - \dfrac{\left(\sum x\right)^2}{n}}{n-1} = \dfrac{666 - \dfrac{106^2}{25}}{25-1} = 9.0233$

c. $\quad s^2 = \dfrac{\sum x^2 - \dfrac{\left(\sum x\right)^2}{n}}{n-1} = \dfrac{76 - \dfrac{11^2}{7}}{7-1} = 9.7857$

2.169   a. $\quad \sum x = 4 + 6 + 6 + 5 + 6 + 7 = 34$

$\sum x^2 = 4^2 + 6^2 + 6^2 + 5^2 + 6^2 + 7^2 = 198$

$\bar{x} = \dfrac{\sum x}{n} = \dfrac{34}{6} = 5.67$

$s^2 = \dfrac{\sum x^2 - \dfrac{\left(\sum x\right)^2}{n}}{n-1} = \dfrac{198 - \dfrac{34^2}{6}}{6-1} = \dfrac{5.3333}{5} = 1.0667$

$s = \sqrt{1.0667} = 1.03$

b. $\quad \sum x = -1 + 4 + (-3) + 0 + (-3) + (-6) = -9$

$\sum x^2 = (-1)^2 + 4^2 + (-3)^2 + 0^2 + (-3)^2 + (-6)^2 = 71$

$\bar{x} = \dfrac{\sum x}{n} = \dfrac{-9}{6} = -\$1.5$

$s^2 = \dfrac{\sum x^2 - \dfrac{\left(\sum x\right)^2}{n}}{n-1} = \dfrac{71 - \dfrac{(-9)^2}{6}}{6-1} = \dfrac{57.5}{5} = 11.5$

$s = \sqrt{11.5} = \$3.39$

c. $\quad \sum x = \dfrac{3}{5} + \dfrac{4}{5} + \dfrac{2}{5} + \dfrac{1}{5} + \dfrac{1}{16} = 2.0625$

$\sum x^2 = \left(\dfrac{3}{5}\right)^2 + \left(\dfrac{4}{5}\right)^2 + \left(\dfrac{2}{5}\right)^2 + \left(\dfrac{1}{5}\right)^2 + \left(\dfrac{1}{16}\right)^2 = 1.2039$

$\bar{x} = \dfrac{\sum x}{n} = \dfrac{2.0625}{5} = .4125\%$

$s^2 = \dfrac{\sum x^2 - \dfrac{\left(\sum x\right)^2}{n}}{n-1} = \dfrac{1.2039 - \dfrac{2.0625^2}{5}}{5-1} = \dfrac{.3531}{4} = .0883\% \text{ squared}$

$s = \sqrt{.0883} = .297\%$

d. (a)   Range $= 7 - 4 = 3$

(b)   Range $= \$4 - (\$-6) = \$10$

(c)   Range $= \dfrac{4}{5}\% - \dfrac{1}{16}\% = \dfrac{64}{80}\% - \dfrac{5}{80}\% = \dfrac{59}{80}\% = .7375\%$

2.171   Using MINITAB, the scatterplot is:

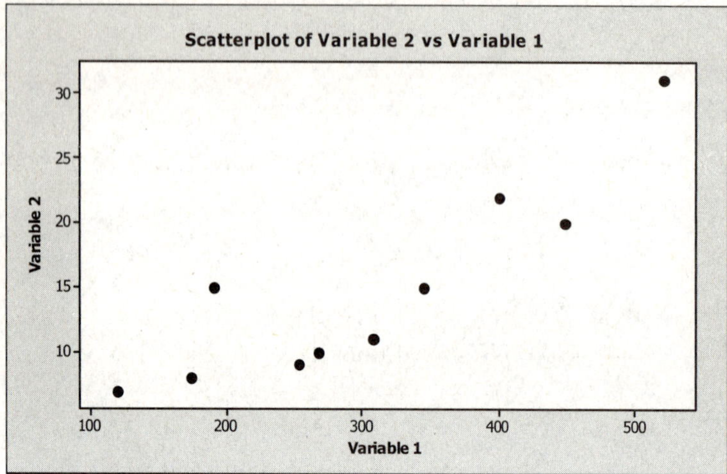

From the scatterplot, it appears that there is a positive trend. As variable 1 increases, variable 2 also tends to increase.

2.173   Using MINITAB, the dot plot of the data is:

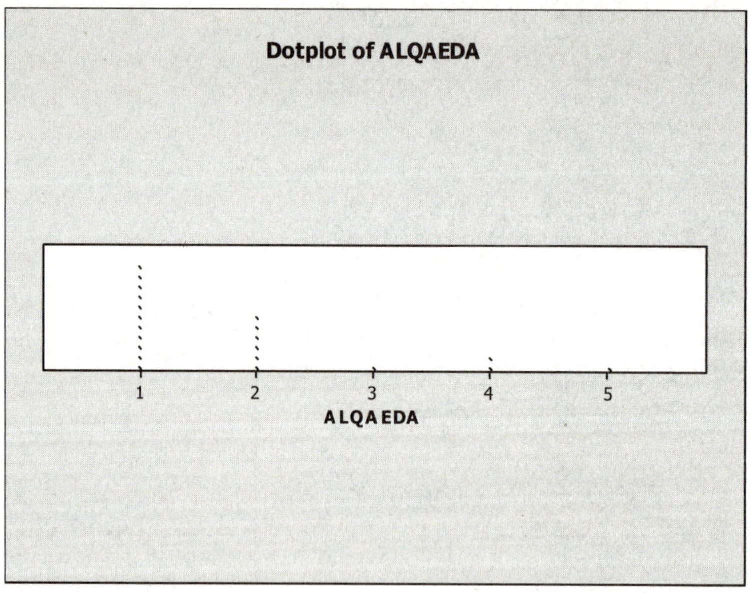

The most frequent number of attacks per incident is 1. There is only 1 incident with 3 attacks and only 1 incident with 5 attacks.

2.175   a.   The sample mean is 603.7. This is the average of the 98 observations. The standard deviation is StDev = 185.4. We would expect most of the observations to fall within 2 standard deviations of the mean. The minimum value is 216. This is the smallest observation in the data set. The Q1 value is 475. This is the 25th percentile. Twenty-five percent of all observations in the data set are less than or equal to 475. The median is 605.0. Half of the observations in the data set are above 605 and half are below. The Q3 value is 724.3. This is the 75th percentile. Seventy-five percent of all observations in the data set are less than or equal to 724.3. The maximum value is 1240.0. This is the value of the largest observation in the data set.

b. The z-score is: $z = \dfrac{x - \bar{x}}{s} = \dfrac{408 - 603.7}{185.4} = -1.06$

A head-injury rating of 408 is less than the mean head-injury rating. It is a little more than one standard deviation below the mean.

2.177 a. The data collection method was a survey.

b. Since the data were numbers (percentage of US labor and materials), the variable is quantitative. Once the data were collected, they were grouped into 4 categories.

c. Using MINITAB, a pie chart of the data is:

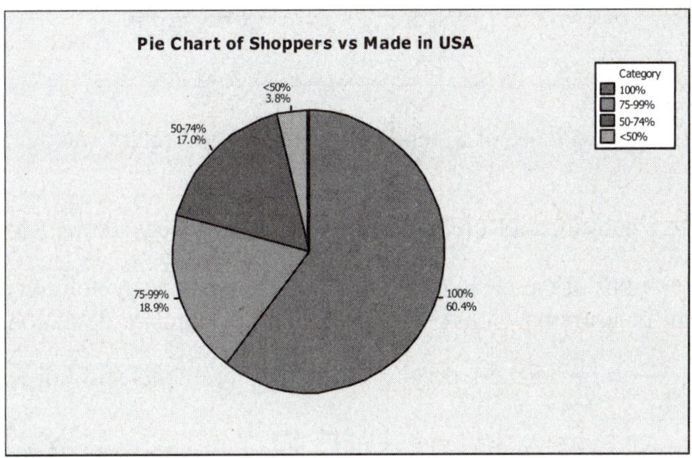

About 60% of those surveyed believe that "Made in USA" means 100% US labor and materials.

2.179 a. If the distribution of scores was symmetric, the mean and median would be equal. The fact that the mean exceeds the median is an indication that the distribution of scores is skewed to the right.

b. It means that 90% of the scores are below 660, and 10% are above 660. (This ignores the possibility of ties, i.e., other people obtaining a score of 660.)

c. If you scored at the 94th percentile, 94% of the scores are below your score, while 6% exceed your score.

2.181   a.   Using MINITAB, a scatterplot of the data is:

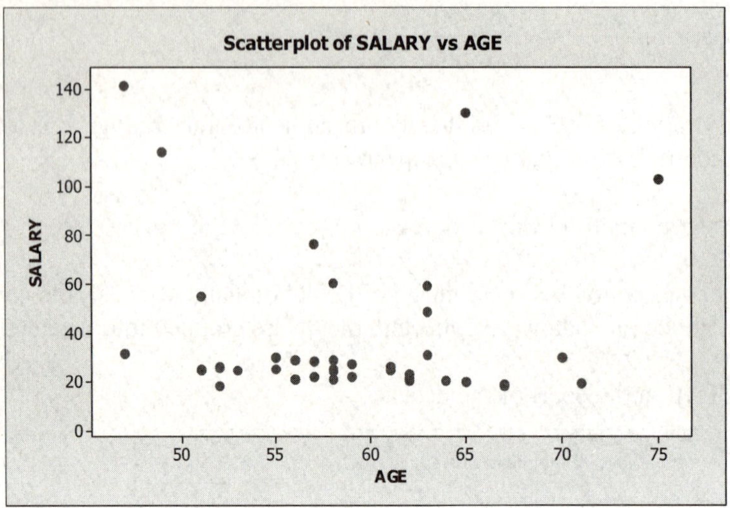

There is not much evidence of a trend between a CEO's salary and age.

b.   According to Chebyshev's Rule, at least $1 - \dfrac{1}{k^2}$ of the measurements will fall within k standard deviations of the mean (for k > 1). Thus, we can say nothing about the percentage of measurements that will fall within .75 standard deviations of the mean.

At least $1 - \dfrac{1}{2.5^2} = 1 - .16 = .84$ or 84% of the measurements will fall within 2.5 standard deviations of the mean. At least $1 - \dfrac{1}{4^2} = 1 - .0625 = .9375$ or 93.75% of the measurements will fall within 4 standard deviations of the mean.

c.   Using MINITAB, the descriptive statistics are:

Descriptive Statistics: AGE, SALARY

Variable	N	Mean	StDev	Minimum	Q1	Median	Q3	Maximum
AGE	40	58.68	6.51	47.00	53.50	58.00	63.00	75.00
SALARY	40	38.58	31.34	18.39	21.45	25.36	31.28	141.36

Next, we compute the intervals for age:

$$\bar{x} \pm .75s \Rightarrow 58.68 \pm .75(6.51) \Rightarrow 58.68 \pm 4.88 \Rightarrow (53.80, 63.56)$$

Twenty-two of the 40 or (22/40)100% = 55% of the ages fall within .75 standard deviations of the mean.

$$\bar{x} \pm 2.5s \Rightarrow 58.68 \pm 2.5(6.51) \Rightarrow 58.68 \pm 16.28 \Rightarrow (42.41, 74.96)$$

Thirty-nine of the 40 or $(39/40)100\% = 97.5\%$ of the ages fall within 2.5 standard deviations of the mean. This is at least 84% that we found in part b.

$$\bar{x} \pm 4s \Rightarrow 58.68 \pm 4(6.51) \Rightarrow 58.68 \pm 26.04 \Rightarrow (32.64, 84.72)$$

Forty of the 40 or $(40/40)100\% = 100\%$ of the ages fall within 4 standard deviations of the mean. This is at least 93.7% that we found in part b.

d. Now, we compute the intervals for salary:

$$\bar{x} \pm .75s \Rightarrow 38.58 \pm .75(31.34) \Rightarrow 38.58 \pm 23.51 \Rightarrow (15.08, 62.09)$$

Thirty-five of the 40 or $(35/40)100\% = 87.5\%$ of the ages fall within .75 standard deviations of the mean.

$$\bar{x} \pm 2.5s \Rightarrow 38.58 \pm 2.5(31.34) \Rightarrow 38.58 \pm 78.35 \Rightarrow (-39.77, 116.93)$$

Thirty-eight of the 40 or $(38/40)100\% = 95\%$ of the salaries fall within 2.5 standard deviations of the mean. This is at least 84% that we found in part b.

$$\bar{x} \pm 4s \Rightarrow 38.58 \pm 4(31.34) \Rightarrow 38.58 \pm 125.36 \Rightarrow (-86.78, 163.94)$$

Forty of the 40 or $(40/40)100\% = 100\%$ of the salaries fall within 4 standard deviations of the mean. This is at least 93.7% that we found in part b.

2.183 A relative frequency bar graph is used to depict the data:

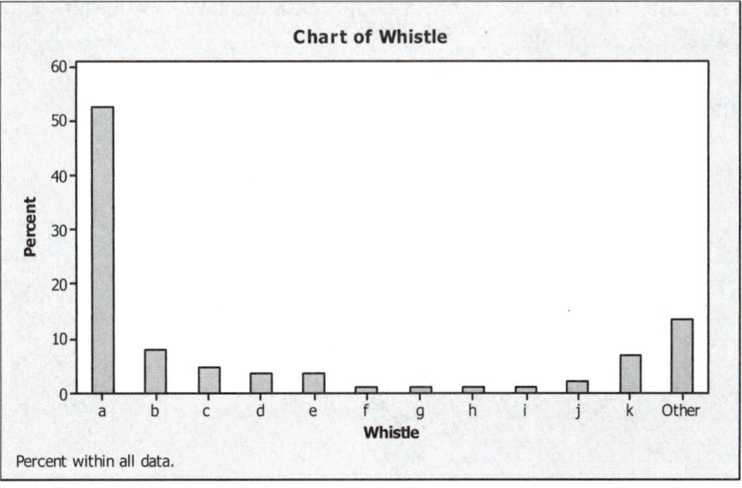

From the bar graph, over half of the whistle types were "Type a." The next most frequent category was the "Other types" with a relative frequency of about .14. Whistle types b and k were the next most frequent. None of the other whistle types had relative frequencies higher than .05.

2.185   a.   Using MINITAB, the boxplot is:

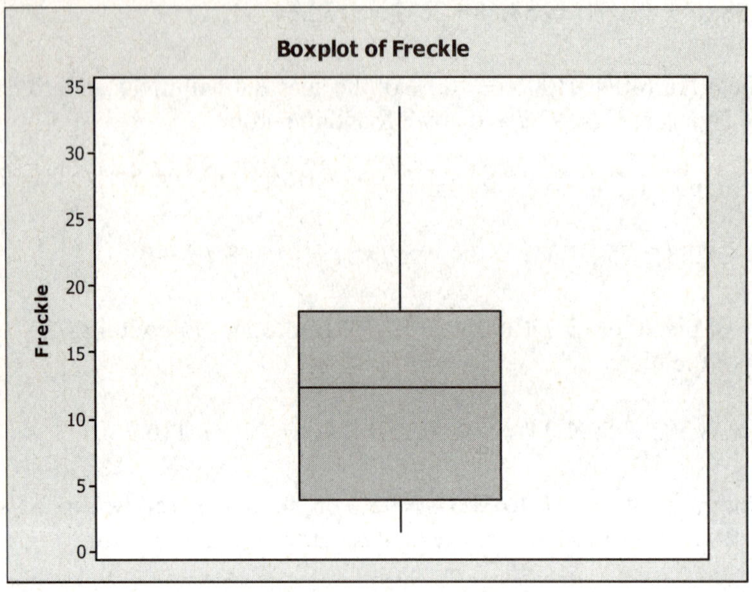

From the boxplot, there are no outliers detected.

b.   Since there are no outliers, we cannot find the z-score for the points identified as outliers.

2.187   a.   For each transect, three variables were measured. The number of seabirds found is quantitative.   The length of the transect is also quantitative. Whether or not the transect was in an oiled area is qualitative.

b.   The experimental unit is the transect.

c.   A pie chart of the oiled and unoiled areas is:

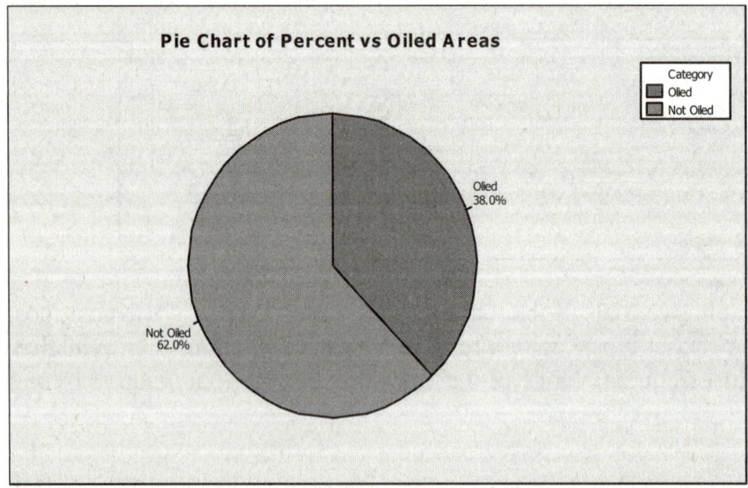

d.   Using MINITAB, a scattergram of the data is:

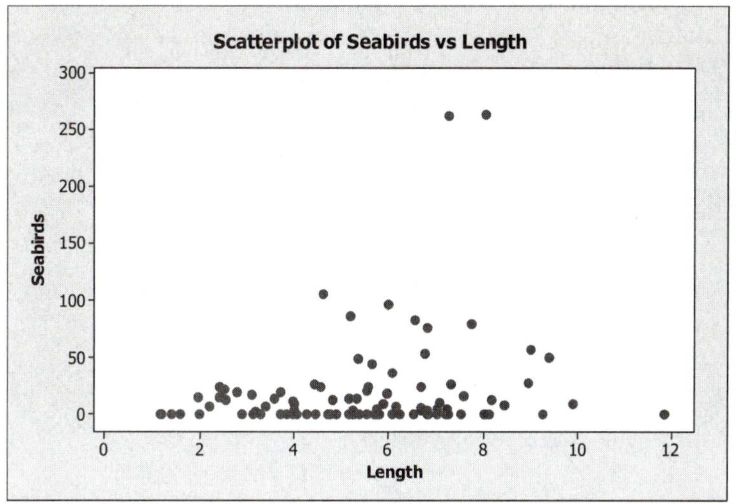

e.   The mean density for the unoiled area is 3.27, while the mean for the oiled area is 3.495. The median for the unoiled area is .89 and is .70 for the oiled area. These are both fairly similar.

f.   Using Chebyshev's Theorem, at least 75% of the observations will fall within 2 standard deviations of the mean. This interval for unoiled areas would be:

$$\bar{x} \pm 2s \Rightarrow 3.27 \pm 2(6.7) \Rightarrow 3.27 \pm 13.4 \Rightarrow (-10.13, 16.67)$$

g.   Using Chebyshev's Theorem, at least 75% of the observations will fall within 2 standard deviations of the mean. This interval for oiled areas would be:

$$\bar{x} \pm 2s \Rightarrow 3.495 \pm 2(5.968) \Rightarrow 3.495 \pm 11.939 \Rightarrow (-8.441, 15.431)$$

h.   From the above two intervals, we know that at least 75% of the observations for the unoiled area will fall between −10.31 and 16.67 and at most 25% of the observations will fall above 15.431 for the oiled areas. Thus, the unoiled areas would be more likely to have a seabird density of 16.

2.189   a.   Using Minitab, a stem-and-leaf display of the data is:

```
Stem-and-leaf of VELOCITY        N = 51
Leaf Unit = 100

   1     18  4
   3     18  79
  12     19  001112444
  18     19  566788
  20     20  12
  21     20  7
  21     21
  23     21  99
  (5)    22  11344
  23     22  56667777777889
  10     23  001222344
   1     23
   1     24
   1     24  9
```

  b.   From this stem-and-leaf display, it is fairly obvious that there are two different distributions since there are two groups of data.

  c.   Since there appears to be two distributions, we will compute two sets of numerical descriptive measures. We will call the group with the smaller velocities A1775A and the group with the larger velocities A1775B.

  For A1775A:

  $$\bar{x} = \frac{\sum x}{n} = \frac{408,707}{21} = 19,462.2$$

  $$s^2 = \frac{\sum x^2 - \frac{\left(\sum x\right)^2}{n}}{n-1} = \frac{7,960,019,531 - \frac{408,707^2}{21}}{21-1} = 283,329.3$$

  $$s = \sqrt{283,329.3} = 532.29$$

  For A1775B:

  $$\bar{x} = \frac{\sum x}{n} = \frac{685,154}{30} = 22,838.5$$

  $$s^2 = \frac{\sum x^2 - \frac{\left(\sum x\right)^2}{n}}{n-1} = \frac{15,656,992,942 - \frac{685,154^2}{30}}{30-1} = 314,694.88$$

  $$s = \sqrt{314,694.88} = 560.98$$

d.   To determine which of the two clusters this observation probably belongs to, we will compute z-scores for this observation for each of the two clusters.

For A1775A:
$$z = \frac{20,000 - 19,462.2}{532.29} = 1.01$$

Since this z-score is so small, it would not be unlikely that this observation came from this cluster.

For A1775B:

$$z = \frac{20,000 - 22,838.5}{560.98} = -5.06$$

Since this z-score is so large (in magnitude), it would be very unlikely that this observation came from this cluster.

Thus, this observation probably came from the cluster A1775A.

2.191   First, compute the z-score for 1.80: $z = \frac{x - \mu}{\sigma} = \frac{1.80 - 2.00}{.08} = -2.5$

If the data are actually mound-shaped, it would be very unusual (less than 2.5%) to observe a batch with 1.80% zinc phosphide if the true mean is 2.0%. Thus, if we did observe 1.8%, we would conclude that the mean percent of zinc phosphide in today's production is probably less than 2.0%.

2.193   a.   Due to the "elite" superstars, the salary distribution is skewed to the right. Since this implies that the median is less than the mean, the players' association would want to use the median.

b.   The owners, by the logic of part **a**, would want to use the mean.

Probability

Chapter
3

3.1   An experiment is an act or process of observation that leads to a single outcome that cannot be predicted with certainty.

3.3   A sample space of an experiment is a collection of all its sample points.

3.5   Two probability rules for sample points are:

1.   All sample point probabilities **must** lie between 0 and 1, i.e. $0 \le p_i \le 1$.

2.   The probabilities of all sample points within a sample space **must** sum to 1, i.e., $\sum p_i = 1$.

3.7   The probability of an event A is calculated by summing the probabilities of the sample points in the sample space for A.

3.9   a.   Since the probabilities must sum to 1,

$$P(E_3) = 1 - P(E_1) - P(E_2) - P(E_4) - P(E_5) = 1 - .1 - .2 - .1 - .1 = .5$$

b. $P(E_3) = 1 - P(E_1) - P(E_2) - P(E_4) - P(E_5) = 1 - P(E_3) - P(E_2) - P(E_4) - P(E_5)$
$$\Rightarrow 2P(E_3) = 1 - P(E_2) - P(E_4) - P(E_5)$$
$$\Rightarrow 2P(E_3) = 1 - .1 - .2 - .1 \Rightarrow 2P(E_3) = .6 \Rightarrow P(E_3) = .3$$

c. $P(E_3) = 1 - P(E_1) - P(E_2) - P(E_4) - P(E_5) = 1 - .1 - .1 - .1 - .1 = .6$

3.11 $P(A) = P(1) + P(2) + P(3) = .05 + .20 + .30 = .55$

$P(B) = P(1) + P(3) + P(5) = .05 + .30 + .15 = .50$

$P(C) = P(1) + P(2) + P(3) + P(5) = .05 + .20 + .30 + .15 = .70$

3.13   a. $\dbinom{5}{2} = \dfrac{5!}{2!(5-2)!} = \dfrac{5!}{2!3!} = \dfrac{5 \cdot 4 \cdot 3 \cdot 2 \cdot 1}{2 \cdot 1 \cdot 3 \cdot 2 \cdot 1} = 10$

b. $\dbinom{6}{3} = \dfrac{6!}{3!(6-3)!} = \dfrac{6!}{3!3!} = \dfrac{6 \cdot 5 \cdot 4 \cdot 3 \cdot 2 \cdot 1}{3 \cdot 2 \cdot 1 \cdot 3 \cdot 2 \cdot 1} = 20$

c. $\dbinom{20}{5} = \dfrac{20!}{5!(20-5)!} = \dfrac{20!}{5!15!} = \dfrac{20 \cdot 19 \cdot 18 \cdots 2 \cdot 1}{5 \cdot 4 \cdot 3 \cdot 2 \cdot 1 \cdot 15 \cdot 14 \cdot 13 \cdots 2 \cdot 1} = 15,504$

3.15   a.   If we denote the marbles as B_1, B_2, R_1, R_2, R_3, then the ten equally likely sample points in the sample space would be:

$$S: \begin{bmatrix} (B_1, B_2), (B_1, R_1), (B_1, R_2), (B_1, R_3), (B_2, R_1) \\ (B_2, R_2), (B_2, R_3), (R_1, R_2), (R_1, R_3), (R_2, R_3) \end{bmatrix}$$

Notice that order is ignored, as the only concern is whether or not a marble is selected.

b.   Each of these ten would be equally likely, implying that each occurs with a probability 1/10.

c. $P(A) = \dfrac{1}{10}$ $P(B) = 6\left(\dfrac{1}{10}\right) = \dfrac{6}{10} = \dfrac{3}{5}$ $P(C) = 3\left(\dfrac{1}{10}\right) = \dfrac{3}{10}$

3.17   a.   There are 6 sample points for this experiment: Blue, Orange, Green, Yellow, Brown, and Red.

b.   According to the Mars Corporation, reasonable probabilities would be:

P(Blue) = .24, P(Orange) = .20, P(Green) = .16, P(Yellow) = .14, P(Brown) = .13, and P(Red) = .13.

c. P(Brown) = .13.

d. P(Red, Green, or Yellow) = P(Red) + P(Green) + P(Yellow) = .13 + .16 + .14 = .43.

e. P(Not Blue) = 1 – P(Blue) = 1 - .24 = .76.

3.19   a.   Let A = {chicken passes inspection with fecal contamination}. $P(A) = 1 / 100 = .01$.

b.   Yes. The relative frequency of passing inspection with fecal contamination is 306/32,075 = .0095 ≈ .01.

3.21   a.   There are 5 sample points for this experiment: None, 1 or 2, 3-5, 6-9, and 10 or more.

b.   We can use the percentages given as reasonable estimates of the probabilities: P(None) = .25, P(1 or 2) = .31, P(3-5) = .25, P(6-9) = .05, and P(10 or more) = .14.

c. P(more than 2) = P(3-5) + P(6-9) + P(10 or more) = .25 + .05 + .14 = .44.

3.23 P(Interview or Grounded Theory) = (5,079 + 537) / 7,506 = 5,616 / 7,506 = .748

3.25 P(tooth shows slight or moderate amount of wear)

= P(tooth shows slight amount of wear) + P(tooth shows moderate amount of wear)

= 4/18 + 3/18 = 7/18 = .389.

3.27   a.   There are a total of $\dbinom{5}{3} = \dfrac{5!}{3!(5-3)!} = \dfrac{5!}{3!2!} = \dfrac{5 \cdot 4 \cdot 3 \cdot 2 \cdot 1}{3 \cdot 2 \cdot 1 \cdot 2 \cdot 1} = 10$ ways to get 3-grill displays.

These 10 display combinations are:

1, 2, 3	1, 3, 4	2, 3, 4	3, 4, 5
1, 2, 4	1, 3, 5	2, 3, 5	
1, 2, 5	1, 4, 5	2, 4, 5	

However, since grill #2 must be selected, there are only 6 possibilities:

1, 2, 3	2, 3, 4
1, 2, 4	2, 3, 5
1, 2, 5	2, 4, 5

   b.   We can estimate the probabilities by using the relative frequency for each sample point. The relative frequency is found by dividing the frequency by the total sample size of 124. These estimates are contained in the following table:

Grill Display Combination	Number of Students	Probability
1-2-3	35	.282
1-2-4	8	.065
1-2-5	42	.339
2-3-4	4	.032
2-3-5	1	.008
2-4-5	34	.274
TOTAL	124	1.0000

   c.   Of the 6 sample points, only 3 of them contain Grill #1.

P(Grill #1 chosen) = P(1-2-3 or 1-2-4 or 1-2-5) = .282 + .065 + .339 = .686.

3.29   a.   There are a total of $\binom{8}{2} = \dfrac{8!}{2!(8-2)!} = \dfrac{8!}{2!6!} = \dfrac{8 \cdot 7 \cdot 6 \cdot 5 \cdot 4 \cdot 3 \cdot 2 \cdot 1}{2 \cdot 1 \cdot 6 \cdot 5 \cdot 4 \cdot 3 \cdot 2 \cdot 1} = 28$ possible Quinella bets.

   b.   If all the players are of equal ability, then the probability of getting any combination is 1/28.

3.31   The total number of pairs of bullets is a combination of 1,837 bullets taken 2 at a time or:

$$\binom{1,837}{2} = \frac{1,837!}{2!\,1,835!} = \frac{1,837(1,836)}{2} = 1,686,366$$

The probability of finding a match or a false positive is 693 / 1,686,366 = .000411.

This probability is very small. The confidence in the FBI's forensic evidence should be very high.

3.33   a. $\binom{6}{2} = \dfrac{6!}{2!(6-2)!} = \dfrac{6!}{2!4!} = \dfrac{6 \cdot 5 \cdot 4 \cdot 3 \cdot 2 \cdot 1}{2 \cdot 1 \cdot 4 \cdot 3 \cdot 2 \cdot 1} = 15$

b. $\dbinom{6}{3} = \dfrac{6!}{3!(6-3)!} = \dfrac{6!}{3!3!} = \dfrac{6 \cdot 5 \cdot 4 \cdot 3 \cdot 2 \cdot 1}{3 \cdot 2 \cdot 1 \cdot 3 \cdot 2 \cdot 1} = 20$

c. $\dbinom{6}{4} = \dfrac{6!}{4!(6-4)!} = \dfrac{6!}{4!2!} = \dfrac{6 \cdot 5 \cdot 4 \cdot 3 \cdot 2 \cdot 1}{4 \cdot 3 \cdot 2 \cdot 1 \cdot 2 \cdot 1} = 15$

d. $\dbinom{6}{5} = \dfrac{6!}{5!(6-5)!} = \dfrac{6!}{5!1!} = \dfrac{6 \cdot 5 \cdot 4 \cdot 3 \cdot 2 \cdot 1}{5 \cdot 4 \cdot 3 \cdot 2 \cdot 1 \cdot 1} = 6$

e. In addition to the combinations computed in parts a – d, we need to find the number of 0-drug combinations, 1-drug combination, and 6-drug combinations.

$\dbinom{6}{0} = \dfrac{6!}{0!(6-0)!} = \dfrac{6!}{0!6!} = \dfrac{6 \cdot 5 \cdot 4 \cdot 3 \cdot 2 \cdot 1}{1 \cdot 6 \cdot 5 \cdot 4 \cdot 3 \cdot 2 \cdot 1} = 1$

$\dbinom{6}{1} = \dfrac{6!}{1!(6-1)!} = \dfrac{6!}{1!5!!} = \dfrac{6 \cdot 5 \cdot 4 \cdot 3 \cdot 2 \cdot 1}{1 \cdot 5 \cdot 4 \cdot 3 \cdot 2 \cdot 1} = 6$

$\dbinom{6}{6} = \dfrac{6!}{6!(6-6)!} = \dfrac{6!}{6!0!} = \dfrac{6 \cdot 5 \cdot 4 \cdot 3 \cdot 2 \cdot 1}{6 \cdot 5 \cdot 4 \cdot 3 \cdot 2 \cdot 1 \cdot 1} = 1$

The total number of ways the 6 drugs can be combined is $15 + 20 + 15 + 6 + 1 + 6 + 1 = 64$.

3.35 The union of 2 events A and B is the event that occurs if either A or B or both occur on a single performance of the experiment.

3.37 The complement of an event A is the event that A does not occur – that is, the event consisting of all sample points that are not in event A.

3.39 The Additive Rule of Probability is: The probability of the union of events A and B is the sum of the probability of events A and B minus the probability of the intersection of events A and B, that is $P(A \cup B) = P(A) + P(B) - P(A \cap B)$.

3.41 The Additive Rule of Probability for mutually exclusive events is: If 2 events A and B are mutually exclusive, the probability of the union of A and B equals the sum of the probabilities of A and B; that is, $P(A \cup B) = P(A) + P(B)$.

3.43   a.

$$A: \{HHH, HHT, HTH, THH, TTH, THT, HTT\}$$
$$B: \{HHH, TTH, THT, HTT\}$$
$$A \cup B: \{HHH, HHT, HTH, THH, TTH, THT, HTT\}$$
$$A^c: \{TTT\}$$
$$A \cap B: \{HHH, TTH, THT, HTT\}$$

b.   If the coin is fair, then each of the 8 possible outcomes are equally likely, with probability 1/8.

$$P(A) = \frac{7}{8} \qquad P(B) = \frac{4}{8} = \frac{1}{2} \qquad P(A \cup B) = \frac{7}{8}$$

$$P(A^c) = \frac{1}{8} \qquad P(A \cap B) = \frac{4}{8} = \frac{1}{2}$$

c. $P(A \cup B) = P(A) + P(B) - P(A \cap B) = \dfrac{7}{8} + \dfrac{1}{2} - \dfrac{1}{2} = \dfrac{7}{8}$

d.   No. $P(A \cap B) = \dfrac{1}{2}$ which is not 0.

3.45   a. $P(A) = P(E_1) + P(E_2) + P(E_3) + P(E_5) + P(E_6) = \dfrac{1}{5} + \dfrac{1}{5} + \dfrac{1}{5} + \dfrac{1}{20} + \dfrac{1}{10} = \dfrac{15}{20} = \dfrac{3}{4}$

b. $P(B) = P(E_2) + P(E_3) + P(E_4) + P(E_7) = \dfrac{1}{5} + \dfrac{1}{5} + \dfrac{1}{20} + \dfrac{1}{5} = \dfrac{13}{20}$

c. $P(A \cup B) = P(E_1) + P(E_2) + P(E_3) + P(E_4) + P(E_5) + P(E_6) + P(E_7)$

$$= \dfrac{1}{5} + \dfrac{1}{5} + \dfrac{1}{5} + \dfrac{1}{20} + \dfrac{1}{20} + \dfrac{1}{10} + \dfrac{1}{5} = 1$$

d. $P(A \cap B) = P(E_2) + P(E_3) = \dfrac{1}{5} + \dfrac{1}{5} = \dfrac{2}{5}$

e. $P(A^c) = 1 - P(A) = 1 - \dfrac{3}{4} = \dfrac{1}{4}$

f. $P(B^c) = 1 - P(B) = 1 - \dfrac{13}{20} = \dfrac{7}{20}$

g. $P(A \cup A^c) = P(E_1) + P(E_2) + P(E_3) + P(E_4) + P(E_5) + P(E_6) + P(E_7)$

$$= \dfrac{1}{5} + \dfrac{1}{5} + \dfrac{1}{5} + \dfrac{1}{20} + \dfrac{1}{20} + \dfrac{1}{10} + \dfrac{1}{5} = 1$$

h. $P(A^c \cap B) = P(E_4) + P(E_7) = \dfrac{1}{20} + \dfrac{1}{5} = \dfrac{5}{20} = \dfrac{1}{4}$

3.47   a. $P(A) = .50 + .10 + .05 = .65$

b. $P(B) = .10 + .07 + .50 + .05 = .72$

c. $P(C) = .25$

d. $P(D) = .05 + .03 = .08$

e. $P(A^c) = .25 + .07 + .03 = .35$ (Note: $P(A^c) = 1 - P(A) = 1 - .65 = .35$)

f. $P(A \cup B) = P(B) = .10 + .07 + .50 + .05 = .72$

g. $P(A \cap B) = P(A) = .50 + .10 + .05 = .655$

h.   Two events are mutually exclusive if they have no sample points in common or if the probability of their intersection is 0.

$P(A \cap B) = .50 + .10 + .05 = .65$. Since this is not 0, A and B are not mutually exclusive.

$P(A \cap C) = 0$. Since this is 0, A and C are mutually exclusive.

$P(A \cap D) = .05$. Since this is not 0, A and D are not mutually exclusive.

$P(B \cap C) = 0$. Since this is 0, B and C are mutually exclusive.

$P(B \cap D) = .05$. Since this is not 0, B and D are not mutually exclusive.

$P(C \cap D) = 0$. Since this is 0, C and D are mutually exclusive.

3.49   a.   A Venn diagram of the data is:

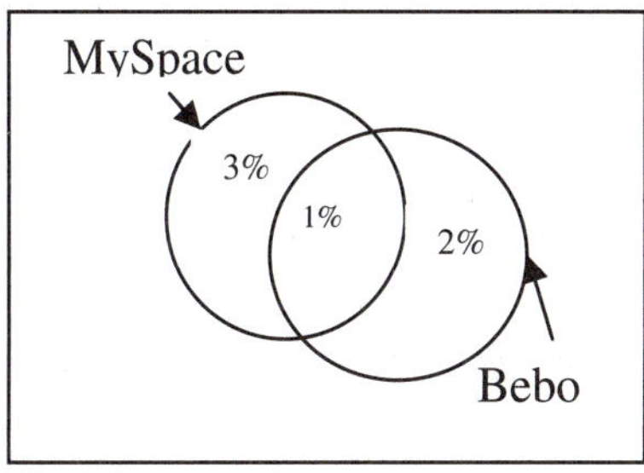

b.   Define the following events:
A: {UK citizen visits MySpace}
B: {UK citizen visits Bebo}

$P(A \cup B) = P(A) + P(B) - P(A \cap B) = .04 + .03 - .01 = .06$

c. $P\left[A^c \cap B^c\right] = 1 - P(A \cup B) = 1 - .06 = .94$

3.51   a.   The sample points of this experiment are the locations where toxic chemical incidents occurred. They are:

School laboratory, In Transit, Chemical plant, Non-chemical plant, and Other.

b.   Reasonable probabilities would be the percents of the incidents changed to proportions.

P(School laboratory) = .06, P(In Transit) = .26, P(Chemical plant) = .21, P(Non-chemical plant) = .35, P(Other) = .12

c. P(School laboratory) = .06

d. P(Chemical plant or Non-chemical plant) = .21 + .35 = .56

e. P(Not occur In Transit) = 1 − P(In Transit) = 1 − .26 = .74

3.53   a.   A Venn Diagram that illustrates the results of the gene profiling analysis is:

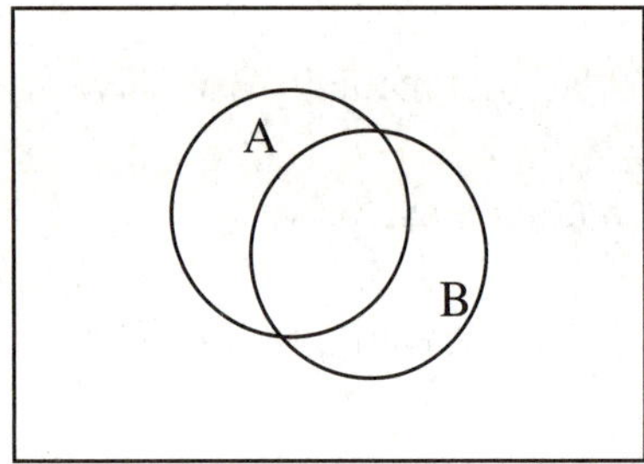

b.   From the problem, we know $P(A) = .41$, $P(B) = .42$, and $P(A \cap B) = .40$
$P(A \cup B) = P(A) + P(B) - P(A \cap B) = .41 + .42 - .40 = .43.$

c. P(Neither) $= P(A \cup B)^c = 1 - P(A \cup B) = 1 - .43 = .57$

3.55   a.   Define the following events:
A: {Student choose stated option}
B: {Student did not choose stated option}
C: {Emotion state is Guilt}
D: {Emotion state is Anger}
E: {Emotion state is Neutral}

$$P(C) = \frac{45 + 12}{171} = \frac{57}{171} = .333$$

b. $$P(A) = \frac{45 + 8 + 7}{171} = \frac{60}{171} = .351$$

c. $P(C \cap A) = \dfrac{45}{171} = .263$

d. $P(C \cup A) = P(C) + P(A) - P(C \cap A) = .333 + .351 - .263 = .421$

3.57   a.   Define the following events:

    F: {Fight}
    N: {No fight}
    W: {Initiator Wins}
    T: {No Clear Winner}
    L: {Initiator Loses}

    $P(F \cap W) = 26/167 = .156$

b. $P(N) = 103/167 = .617$

c. $P(T) = 35/167 = .210$

d. $P(F \cup L) = (26 + 23 + 15 + 11)/167 = 75/167 = .449$

e.   Yes. $P(T \cap L) = 0$.

3.59   a. $P(A) = \dfrac{1,465}{2,143} = .684$

b. $P(B) = \dfrac{265}{2,143} = .124$

c.   No. $P(A \cap B) = \dfrac{194}{2,143} = .091$. Since this is not 0, events A and B are not mutually exclusive.

d. $P(A^c) = 1 - P(A) = 1 - .684 = .316$.

e. $P(A \cup B) = P(A) + P(B) - P(A \cap B) = .684 + .124 - .091 = .717$

f. $P(A \cap B) = \dfrac{194}{2,143} = .091$

3.61   a. $P(\text{responds to all 3 source odors}) = .19$

b. $P(\text{responds to kairomone}) = .19 + .025 + .025 + .165 = .405$

c. $P(\text{responds to Mups A and Mups B but not kairomone}) = .21 + .275 + .11 = .595$

3.63   A conditional probability is the probability of an event based on another event, while an unconditional probability is not based on another event.

3.65 $P(A|B) = \dfrac{P(A \cap B)}{P(B)}$

3.67   a. $P(A|B) = \dfrac{P(A \cap B)}{P(B)} = \dfrac{.1}{.2} = .5$

     b. $P(B|A) = \dfrac{P(A \cap B)}{P(A)} = \dfrac{.1}{.4} = .25$

     c.   No, events A and B are not independent. If A and B were independent, then $P(A|B) = P(A)$. For this problem, $P(A|B) = .5$ and $P(A) = .4$.

3.69   a. $P(A \cap B) = P(A)P(B) = .4(.2) = .08$ (Since A and B are independent)

     b. $P(A|B) = \dfrac{P(A \cap B)}{P(B)} = \dfrac{.08}{.2} = .4$

     c. $P(A \cup B) = P(A) + P(B) - P(A \cap B) = .4 + .2 - .08 = .52$

3.71   a. $P(A) = P(E_1) + P(E_2) + P(E_3) = .1 + .1 + .2 = .4$

        $P(B) = P(E_2) + P(E_3) + P(E_5) = .1 + .2 + .1 = .4$

        $P(A \cap B) = P(E_2) + P(E_3) = .1 + .2 = .3$

     b. $P(E_1|A) = \dfrac{P(E_1 \cap A)}{P(A)} = \dfrac{.1}{.4} = .25$

        $P(E_2|A) = \dfrac{P(E_2 \cap A)}{P(A)} = \dfrac{.1}{.4} = .25$

        $P(E_3|A) = \dfrac{P(E_3 \cap A)}{P(A)} = \dfrac{.2}{.4} = .5$

        We are given that $P(E_1) = .1$, $P(E_2) = .1$, and $P(E_3) = .2$

        Thus, $P(E_1) = P(E_2)$ and $P(E_3) = 2P(E_1)$

        From above, $P(E_1|A) = P(E_2|A)$ and $P(E_3|A) = 2P(E_1|A)$

        $P(E_1|A) + P(E_2|A) + P(E_3|A) = .25 + .25 + .50 = 1.00$

     c.   Using the sum of the conditional probabilities,

        $P(B|A) = P(E_2|A) + P(E_3|A) = .25 + .50 = .75$

Using the formula,

$$P(B\,|\,A)=\frac{P(A\cap B)}{P(A)}=\frac{.3}{.4}=.75$$

3.73   The 36 possible outcomes obtained when tossing two dice are listed below:

(1, 1) (1, 2) (1, 3) (1, 4) (1, 5) (1, 6)
(2, 1) (2, 2) (2, 3) (2, 4) (2, 5) (2, 6)
(3, 1) (3, 2) (3, 3) (3, 4) (3, 5) (3, 6)
(4, 1) (4, 2) (4, 3) (4, 4) (4, 5) (4, 6)
(5, 1) (5, 2) (5, 3) (5, 4) (5, 5) (5, 6)
(6, 1) (6, 2) (6, 3) (6, 4) (6, 5) (6, 6)

A: {(1, 2), (1, 4), (1, 6), (2, 1), (2, 3), (2, 5), (3, 2), (3, 4), (3, 6), (4, 1), (4, 3),
(4, 5), (5, 2), (5, 4), (5, 6), (6, 1), (6, 3), (6, 5)}

B: {(3, 6), (4, 5), (5, 4), (5, 6), (6, 3), (6, 5), (6, 6)}

$A\cap B$: {(3, 6), (4, 5), (5, 4), (5, 6), (6, 3), (6, 5)}

If A and B are independent, then $P(A)P(B)=P(A\cap B)$.

$$P(A)=\frac{18}{36}=\frac{1}{2}\qquad P(B)=\frac{7}{36}\qquad P(A\cap B)=\frac{6}{36}=\frac{1}{6}$$

$P(A)P(B)=\frac{1}{2}\cdot\frac{7}{36}=\frac{7}{72}\neq\frac{1}{6}=P(A\cap B)$. Thus, A and B are not independent.

3.75   Let W_1 and W_2 represent the two white chips, R_1 and R_2 represent the two red chips, and B_1
and B_2 represent the two blue chips. The sample space is:

$$
\begin{array}{lll}
W_1W_2 & W_2R_1 & R_1B_1 \\
W_1R_1 & W_2R_2 & R_1B_2 \\
W_1R_2 & W_2B_1 & R_2B_1 \\
W_1B_1 & W_2B_2 & R_2B_2 \\
W_1B_2 & R_1R_2 & B_1B_2
\end{array}
$$

Assuming each event is equally likely, each event will have a probability of 1/15.

Then, $P(A)=P(W_1W_2)+P(R_1R_2)+P(B_1B_2)=3\left(\frac{1}{15}\right)=\frac{3}{15}=\frac{1}{5}$

$$P(B)=P(R_1R_2)=\frac{1}{15}$$

$$P(C) = P(W_1W_2) + P(W_1R_1) + P(W_1R_2) + P(W_1B_1) + P(W_1B_2) + P(W_2R_1)$$
$$+ P(W_2R_2) + P(W_2B_1) + P(W_2B_2) + P(R_1R_2) + P(R_1B_1) + P(R_1B_2)$$
$$+ P(R_2B_1) + P(R_2B_2)$$
$$= 14\left(\frac{1}{15}\right) = \frac{14}{15}$$

$$P(A \cap B) = P(R_1R_2) = \frac{1}{15}$$

$$P(A^c) = 1 - P(A) = 1 - \frac{1}{5} = \frac{4}{5}$$

$$P(A^c \cap B) = 0$$

$$P(B \cap C) = P(R_1R_2) = \frac{1}{15}$$

$$P(A \cap C) = P(W_1W_2) + P(R_1R_2) = 2\left(\frac{1}{15}\right) = \frac{2}{15}$$

$$P(A^c \cap C) = P(W_1R_1) + P(W_1R_2) + P(W_1B_1) + P(W_1B_2) + P(W_2R_1)$$
$$+ P(W_2R_2) + P(W_2B_1) + P(W_2B_2) + P(R_1B_1) + P(R_1B_2)$$
$$+ P(R_2B_1) + P(R_2B_2)$$
$$= 12\left(\frac{1}{15}\right) = \frac{12}{15} = \frac{4}{5}$$

$$P(B \mid A) = \frac{P(A \cap B)}{P(A)} = \frac{\frac{1}{15}}{\frac{1}{5}} = \frac{1}{3} \qquad P(B \mid A^c) = \frac{P(A^c \cap B)}{P(A^c)} = \frac{0}{\frac{4}{5}} = 0$$

$$P(B \mid C) = \frac{P(B \cap C)}{P(C)} = \frac{\frac{1}{15}}{\frac{14}{15}} = \frac{1}{14}$$

$$P(A \mid C) = \frac{P(A \cap C)}{P(C)} = \frac{\frac{2}{15}}{\frac{14}{15}} = \frac{1}{7}$$

$$P(C \mid A^c) = \frac{P(A^c \cap C)}{P(A^c)} = \frac{\frac{4}{5}}{\frac{4}{5}} = 1$$

3.77   a.   Define the following events:
A: {adult owns at least one gun}
B: {adult owns a hand gun}

From the exercise, we know that $P(A) = .26$ and $P(B \mid A) = .05$. Thus, $P(A) = .26$.

b. $P(A \cap B) = P(B|A)P(A) = .26(.05) = .013$

3.79 Define the following events:

C: {Speeding is cause of fatal crash}
M: {Missing curve is cause of fatal crash}

From the problem, $P(C) = .3$ and $P(C \cap M) = .12$

$$P(M|C) = \frac{P(C \cap M)}{P(C)} = \frac{.12}{.30} = .40$$

3.81 a. Define the following events:
A: {response indicates a perceived unfairness}
B: {angry emotion}

$$P(A) = \frac{594}{10,797} = .055$$

b. $P(B|A) = \dfrac{127}{594} = .214$

3.83 The probability that at least one acquires gastroenteritis is equal to 1 minus the probability
that neither acquires gastroenteritis. We will assume that the events that you acquire
gastroenteritis and the event that your friend acquires gastroenteritis are independent.

$P(\text{at least one acquires gastroenteritis}) = 1 - P(\text{neither acquires gastroenteritis})$

$= 1 - P(\text{you do not acquire gastroenteritis})P(\text{friend does not acquire gastroenteritis})$

$= 1 - \left(\dfrac{994}{1000}\right)\left(\dfrac{993}{1000}\right) = 1 - .987 = .013$

3.85 Define the following event:

R: {Fish is red snapper}

From the problem, $P(R^c) = .77$

a. $P(R) = 1 - P(R^c) = 1 - .77 = .23$

b. $P(\text{At least one customer is served red snapper}) = 1 - P(\text{No customers are served red
snapper}) = 1 - P(R^c \cap R^c \cap R^c \cap R^c \cap R^c) = 1 - .77(.77)(.77)(.77)(.77) = 1 - .271 = .729$

3.87 a. $P(\text{Species extinct}) = 38/132 = .2879$

b.   If 9 species are chosen and all are extinct, then for the 10th pick, there are only $132 - 9 = 123$ species to pick from, of which $38 - 9 = 29$ are extinct. P(Species extinct on 10th selection | first 9 species are extinct) $= 29 / 123 = .236$.

3.89   a.   Define the following events:
A: {ambulance travels to location A under 8 minutes}
B: {ambulance travels to location B under 8 minutes}
C: {ambulance is busy}

From the exercise, we know that $P(A\,|\,C^c) = .58$, $P(B\,|\,C^c) = .42$, and $P(C) = .3$.

$$P(A) = P(A\,|\,C^c)P(C^c) = .58(1-.3) = .58(.7) = .406$$

b. $P(B) = P(B\,|\,C^c)P(C^c) = .42(1-.3) = .42(.7) = .294$

3.91 $P(A\,|\,B) = 0$

3.93   a.   The 8 outcomes are:

A and C win first round matches and A wins final (A defeats B, C defeats D, A defeats C)
A and C win first round matches and C wins final (A defeats B, C defeats D, C defeats A)
A and D win first round matches and A wins final (A defeats B, D defeats C, A defeats D)
A and D win first round matches and D wins final (A defeats B, D defeats C, D defeats A)
B and C win first round matches and B wins final (B defeats A, C defeats D, B defeats C)
B and C win first round matches and C wins final (B defeats A, C defeats D, C defeats B)
B and D win first round matches and B wins final (B defeats A, D defeats C, B defeats D)
B and D win first round matches and D wins final (B defeats A, D defeats C, D defeats B)

b.   If all the players are of equal ability, then each of the above outcomes are equally likely. Each outcome has a probability of 1/8 of occurring. P(A wins) $= 2/8 = .25$.

c.   There are 2 outcomes where A wins the tournament:

A and C win first round matches and A wins final (A defeats B, C defeats D, A defeats C)
A and D win first round matches and A wins final (A defeats B, D defeats C, A defeats D)

$$P(A \text{ wins}) = P(A \text{ defeats } B \cap C \text{ defeats } D \cap A \text{ defeats } C)$$
$$+ P(A \text{ defeats } B \cap D \text{ defeats } C \cap A \text{ defeats } D)$$

$$= P(A \text{ defeats } B)P(C \text{ defeats } D)P(A \text{ defeats } C)$$
$$+ P(A \text{ defeats } B)P(D \text{ defeats } C)P(A \text{ defeats } D)$$

$$= .9(.4)(.7) + .9(1 - .4)(.6) = .252 + .324 = .576$$

3.95   a.   Define the following event:

        A: {Psychic picks box with crystal}
        If psychic is just guessing, $P(A) = 1/10 = .1$

    b. P(Psychic guesses correct at least once in 7 trials)
        $= 1 - P$(Psychic does not guess correct in 7 trials)

$$= 1 - P(A^c \cap A^c \cap A^c \cap A^c \cap A^c \cap A^c \cap A^c) = 1 - .9(.9)(.9)(.9)(.9)(.9)$$
$$= 1 - .478 = .522$$

        (This assumes that the trials are independent)

    c.   If the psychic is just guessing, then P(Psychic does not guess correct in 7 trials) $= .478$.

        Thus, if the person was not a psychic (and was merely guessing) we would expect the person to guess wrong in all seven trials about half the time. This would not be a rare event. Thus, this outcome would support the notion that the person is guessing.

3.97   a.   The total number of outcomes is a combination of 14 numbers taken 4 at a time.

$$\binom{14}{4} = \frac{14!}{4!(14-4)!} = \frac{14!}{4!10!} = \frac{14 \cdot 13 \cdot 12 \cdots 2 \cdot 1}{4 \cdot 3 \cdot 2 \cdot 1 \cdot 10 \cdot 9 \cdots 2 \cdot 1} = 1,001$$

    b.   The probability that any team obtains the first pick in the draft is that team's number of combinations divided by 1000. The probabilities are:

NBA Lottery Team	Number of Combinations	Probability
Worst Record	250	.250
2nd worst record	200	.200
3rd worst record	157	.157
4th worst record	120	.120
5th worst record	89	.089
6th worst record	64	.064
7th worst record	44	.044
8th worst record	29	.029
9th worst record	18	.018
10th worst record	11	.011
11th worst record	7	.007
12th worst record	6	.006
13th worst record	5	.005
Total	1000	1.000

    c.   Define the following events:

        W_{i1}: {i^{th} worst team gets 1st pick}
        W_{12}: {Worst team gets 2nd pick}

        If the 2nd worst team obtains the number 1 pick, then that team is removed from the group that could obtain the number 2 pick. Thus, the 200 combinations assigned to the team with the 2nd worst record are removed from the available combinations. Now, there are only 800 possible combinations. The probability that the team with the worst record will get the 2nd pick in the draft given that the team with the 2nd worst record gets the first pick would be $P(W_{12} \mid W_{21}) = 250 / 800 = .313$.

d.   If the 3rd worst team obtains the number 1 pick, then that team is removed from the group that could obtain the number 2 pick. Thus, the 157 combinations assigned to the team with the 3rd worst record are removed from the available combinations. Now, there are only $1000 - 157 = 843$ possible combinations. The probability that the team with the worst record will get the 2nd pick in the draft given that the team with the 3rd worst record gets the first pick would be $P(W_{12} \mid W_{31}) = 250 / 843 = .297$.

e.   We need to find $P(W_{12} \mid W_{11}^c) = \dfrac{P(W_{12} \cap W_{11}^c)}{P(W_{11}^c)} = \dfrac{P(W_{12})}{1 - P(W_{11})}$

Now, $P(W_{11}^c) = 1 - P(W_{11}) = 1 - .250 = .750$

$$P(W_{12}) = P(W_{12} \mid W_{21})P(W_{21}) + P(W_{12} \mid W_{31})P(W_{31})$$
$$+ P(W_{12} \mid W_{41})P(W_{41}) + \cdots + + P(W_{12} \mid W_{13\,1})P(W_{13\,1})$$

To aid in the computation, we will use the following table:

NBA Lottery Team	Number of Combinations	Probability $P(W_{i1})$	$P(W_{12} \mid W_{i1})$	$P(W_{12} \mid W_{i1})P(W_{i1})$
Worst Record	250	.250		
2nd worst record	200	.200	.313	.0626
3rd worst record	157	.157	.297	.0466
4th worst record	120	.120	.284	.0341
5th worst record	89	.089	.274	.0244
6th worst record	64	.064	.267	.0171
7th worst record	44	.044	.262	.0115
8th worst record	29	.029	.257	.0075
9th worst record	18	.018	.255	.0046
10th worst record	11	.011	.253	.0028
11th worst record	7	.007	.252	.0018
12th worst record	6	.006	.252	.0015
13th worst record	5	.005	.251	.0013
Total	1000	1.000		.2158

Thus,
$$P(W_{12}) = P(W_{12} \mid W_{21})P(W_{21}) + P(W_{12} \mid W_{31})P(W_{31})$$
$$+ P(W_{12} \mid W_{41})P(W_{41}) + \cdots + + P(W_{12} \mid W_{13\,1})P(W_{13\,1}) = .2158$$

Finally, $P(W_{12} \mid W_{11}^c) = \dfrac{P(W_{12} \cap W_{11}^c)}{P(W_{11}^c)} = \dfrac{P(W_{12})}{1 - P(W_{11})} = \dfrac{.2158}{.750} = .288$

3.99   A representative sample exhibits characteristics typical of those possessed by the population of interest. The most common way to obtain a representative sample is to use a random sample.

3.101   A blind study is one where the participants in the study do not know what treatment they are receiving.

3.103   a. $\dbinom{600}{3} = \dfrac{600!}{(3!)(597!)} = \dfrac{(600)(599)(598)}{(1)(2)(3)} = 35,820,200$

b.   All sets are equally likely in random sampling, so that the probability of any particular sample is:

$$\frac{1}{35,820,200}$$

c.   Answers will vary. Suppose we use Table I and start in row 18, column 5, the first 3 digits and go down. We have to ignore any numbers that are greater than 600. We will list all the numbers and highlight the numbers that are ignored for each sample of size 3:

949	496	144	763	470
585	781	982	078	133
099	812	789	061	587
143	642	826	277	197
741	827	533	988	248
242	464	278	188	469
873	672	742	174	846
073	073	101	530	444
964	299	954	709	267
264	319	142	496	422
664	253	417	889	863
264	707	967	482	189
943	383	897	772	679
773	533	337	774	308
561	919	512	893	040
552	876	819	312	200
886	493	816	232	023
129		309	426	846
301			091	396
491				016
				344
				232

Notice that no sample of size 3 is repeated. Since the probability in part b was so small, we would not expect any duplicate samples.

d.   Using MINITAB, the generated sample of size 3 is 36, 362 and 76.

3.105 a. Decide on a starting point on the random number table. Then take the first *n* numbers reading down, and this would be the sample. Group the digits on the random number table into groups of 7 (for part **b**) or groups of 4 (for part **c**). Eliminate any duplicates and numbers that begin with zero since they are not valid telephone numbers.

    b. Starting in Row 6, column 5, take the first 10 seven-digit numbers reading down. The telephone numbers are:

$$
\begin{array}{c}
277\text{-}5653 \\
988\text{-}7231 \\
188\text{-}7620 \\
174\text{-}5318 \\
530\text{-}6059 \\
709\text{-}9779 \\
496\text{-}2669 \\
889\text{-}7433 \\
482\text{-}3752 \\
772\text{-}3313
\end{array}
$$

    c. Starting in Row 10, column 7, take the first 5 four-digit numbers reading down. The 5 telephone numbers are:

$$
\begin{array}{c}
373\text{-}3886 \\
373\text{-}5686 \\
373\text{-}1866 \\
373\text{-}3632 \\
373\text{-}6768
\end{array}
$$

3.107 Using MINITAB, we will generate 50 random numbers between 1 and 100. These 50 women will receive the kiwifruit breakfast. The others will receive the banana breakfast.

The answers can vary. One possible solution follows. Seventy random numbers between 1 and 100 were generated using MINITAB to account for duplicate numbers. After deleting all the duplicate numbers, there were 56 random numbers remaining. The first fifty numbers were kept. They were:

  1 4 7 9 11 12 17 18 19 20 23 26 28 29 32 33 34 36 38 39 40 43 46 51 56
57 58 61 64 67 69 70 71 72 73 74 75 80 81 82 84 85 86 89 90 91 94 95 96 97

Thus, the women numbered above will receive the kiwifruit breakfast. The others will receive the banana breakfast.

3.109 Answers can vary. One possible solution follows. First, we will number the students from 1 to 120. Then we will generate 90 random numbers. The first 30 generated numbers will be assigned to group 1. The next 30 generated numbers will be assigned to group 2. The next 30 generated numbers will be assigned to group 3. The remaining 30 students will be assigned to group 4. Using MINITAB, we will generate 150 random numbers between 1 and 120. After deleting all the duplicate numbers, there were 93 random numbers remaining. The first 90 numbers were kept. The first 30 numbers sorted were:

1 2 4 6 18 20 25 33 36 41 46 52 61 66 67 70 78 84 86 90 92 94 103 104 106 107 110 112 113 120

Students assigned to these numbers were assigned to group 1.

The next 30 numbers sorted were:

5 7 8 11 12 22 23 29 40 43 47 48 49 51 54 58 59 60 63 65 75 79 80 82 85 89 93 100 101 109

Students assigned to these numbers were assigned to group 2.

The next 30 numbers sorted were:

3 10 13 15 17 26 27 31 35 37 39 42 44 55 56 62 68 74 77 87 95 96 97 99 102 105 114 117 118 119

Students assigned to these numbers were assigned to group 3. The remaining 30 students were assigned to group 4.

3.111   First, number the intersections from 1 to 5000. Using the random number table, select a starting point that contains 4 digits. Following either the row or column, select successive 4 digit numbers until 50 different 4 digit numbers between 0001 and 5000 are selected. The intersections corresponding to these 50 four digit numbers are selected for sampling.

The second method requires the rows to be numbered from 00 to 99 and the columns to be numbered from 1 to 50. Using the random number table, select a starting point that contains 2 digits. Following either the row or column, select successive 2 digit numbers in sets of 2. The first 2 digit number will correspond to the row number. The second 2 digit number must be between 1 and 50 and corresponds to the column number. This procedure is followed until 50 pairs of 2 digit numbers are selected that correspond to 50 different intersections.

3.113   a. $A \cup B$

b. B^c

c. $A \cap B$

d. $A^c \mid B$

3.115   a.   Since events A and B are mutually exclusive, $P(A \cap B) = 0$.

$$P(A \mid B) = \frac{P(A \cap B)}{P(B)} = \frac{0}{.3} = 0$$

b.   If Events A and B are independent, then $P(A \mid B) = P(A)$. From part **a**, we know $P(A \mid B) = 0$. However, we also know $P(A) = .2$. Thus, events A and B are not independent.

3.117 We know $P(A|B) = \dfrac{P(A \cap B)}{P(B)}$. Thus, $P(B) = \dfrac{P(A \cap B)}{P(A|B)} = \dfrac{.4}{.8} = .5$.

3.119 a.

$(1, H)$	$(2, 1)$	$(3, H)$	$(4, 1)$	$(5, H)$	$(6, 1)$
$(1, T)$	$(2, 2)$	$(3, T)$	$(4, 2)$	$(5, T)$	$(6, 2)$
	$(2, 3)$		$(4, 3)$		$(6, 3)$
	$(2, 4)$		$(4, 4)$		$(6, 4)$
	$(2, 5)$		$(4, 5)$		$(6, 5)$
	$(2, 6)$		$(4, 6)$		$(6, 6)$

b.   Each simple event is an intersection of two independent events. Each simple event whose first element is 1, 3, or 5 has probability

$$\left(\frac{1}{6}\right)\left(\frac{1}{2}\right) = \frac{1}{12}$$

while each simple event whose first element is 2, 4, or 6 has probability

$$\left(\frac{1}{6}\right)\left(\frac{1}{6}\right) = \frac{1}{36}$$

c. $P(A) = P\{(1,H),(3,H),(5,H)\} = \dfrac{1}{12} + \dfrac{1}{12} + \dfrac{1}{12} = \dfrac{3}{12} = \dfrac{1}{4}$

$P(B) = P\{(1,H),(1,T),(3,H),(3,T),(5,H),(5,T)\} = \dfrac{6}{12} = \dfrac{1}{2}$

d. A^c: {all except $(1, H)$, $(3, H)$, and $(5, H)$}
B^c: $\{(2, 1), (2, 2), (2, 3), (2, 4), (2, 5), (2, 6), (4, 1), (4, 2), (4, 3), (4, 4), (4, 5), (4, 6),$
    $(6, 1), (6, 2), (6, 3), (6, 4), (6, 5), (6, 6)\}$
$A \cap B$: $\{(1, H), (3, H), (5, H)\}$
$A \cup B$: $\{(1, H), (1, T), (3, H), (3, T), (5, H), (5, T)\}$

e. $P(A^c) = 1 - P(A) = 1 - \dfrac{1}{4} = \dfrac{3}{4}$

$P(B^c) = 1 - P(B) = 1 - \dfrac{1}{2} = \dfrac{1}{2}$

$P(A \cap B) = P(A) = \dfrac{1}{4}$

$P(A \cup B) = P(A) + P(B) - P(A \cap B) = \dfrac{1}{4} + \dfrac{1}{2} - \dfrac{1}{4} = \dfrac{1}{2}$

$P(A|B) = \dfrac{P(A \cap B)}{P(B)} = \dfrac{1/4}{1/2} = \dfrac{1}{2}$

$P(B|A) = \dfrac{P(A \cap B)}{P(A)} = \dfrac{1/4}{1/4} = 1$

f. $P(A \cap B) \neq 0$, so that A and B are not mutually exclusive.
$P(A|B) \neq P(A)$, so that A and B are not independent.

3.121   a.   Because events A and B are independent, we have:

$$P(A \cap B) = P(A)P(B) = .3(.1) = .03$$

Thus, $P(A \cap B) \neq 0$, and the two events cannot be mutually exclusive.

b. $P(A\,|\,B) = \dfrac{P(A \cap B)}{P(B)} = \dfrac{.03}{.1} = .3$ $P(B\,|\,A) = \dfrac{P(A \cap B)}{P(A)} = \dfrac{.03}{.3} = .1$

c. $P(A \cup B) = P(A) + P(B) - P(A \cap B) = .3 + .1 - .03 = .37$

3.123   Use the relative frequency as an estimate for the probability. Thus,
$P(\text{physically assaulted}) = 600/12{,}000 = .05$.

3.125   From the problem, to find probabilities, we must convert the percents to proportions by
dividing by 100. Thus, $P(1 \text{ star}) = 0$, $P(2 \text{ stars}) = .0408$, $P(3 \text{ stars}) = .1735$, $P(4 \text{ stars}) = $
.6020, and $P(5 \text{ stars}) = .1837$.

a.   False. The probability of an event cannot be 4. The probability of an event must be
between 0 and 1.

b.   True. $P(4 \text{ or } 5 \text{ stars}) = P(4 \text{ stars}) + P(5 \text{ stars}) = .6020 + .1837 = .7857$.

c.   True. No cars have a rating of 1 star, thus $P(1 \text{ star}) = 0$.

d.   False. $P(2 \text{ stars}) = .0408$ and $P(5 \text{ stars}) = .1837$. Since .0408 us smaller than .1837,
the car has a better chance of having a 5-star rating than a 2-star rating.

3.127   Suppose we define the following event:

$A = \{\text{eighth-grader scores above 655 on mathematics assessment test}\}$

Then the probability that a randomly selected eighth-grader has a score of 655 or below on
the mathematics assessment test is:

$P(A^c) = 1 - P(A) = 1 - .05 = .95$.

3.129   a.   Define the following event:

A: {Beech tree is damaged by fungi}

$$P(A) = \frac{49}{188} = .261$$

b.   There would be 3 sample points for this experiment: trunk, leaves, and branch. We can convert the percentages to proportions and use these to estimate the probabilities.

Sample Points	Probabilities
Trunk	.85
Leaves	.10
Branch	.05
Total	**1.00**

3.131   a.   The sample space is the listing of all possible nearshore bar conditions. The sample space is {Single, shore parallel; Other; Planar}.

b.   Define the following events:

SP: {Single, shore parallel}
P: {Planar}
O: {Other}

Assuming that all the sample points are equally likely, $P(SP) = 2/6 = .333$.
$P(O) = 2/6 = .333$. $P(P) = 2/6 = .333$

c. $P(P \cup SP) = 4/6 = .667$

d.   The sample space is the listing of all possible beach conditions. The sample space is {No dunes/flat; Bluff/scarp; Single dune; Not observed}.

e.   Define the following events:

N: {No dunes/flat}
B: {Bluff/scarp}
D: {Single dune}
NO: {Not observed}

Assuming that all the sample points are equally likely, $P(N) = 2/6 = .333$.
$P(B) = 1/6 = .167$. $P(D) = 2/6 = .333$. $P(NO) = 1/6 = .167$.

f. $P(N^C) = 1 - P(NC) = 1 - .333 = .667$.

3.133   a.   The event $A \cap B$ is the event the outcome is black and odd. The event is $A \cap B$: {11, 13, 15, 17, 29, 31, 33, 35}

b.   The event $A \cup B$ is the event the outcome is black or odd or both. The event $A \cup B$ is {2, 4, 6, 8, 10, 11, 13, 15, 17, 20, 22, 24, 26, 28, 29, 31, 33, 35, 1, 3, 5, 7, 9, 19, 21, 23, 25, 27}

c.   Assuming all events are equally likely, each has a probability of 1/38.

$$P(A) = 18\left(\frac{1}{38}\right) = \frac{18}{38} = \frac{9}{19}$$

$$P(B) = 18\left(\frac{1}{38}\right) = \frac{18}{38} = \frac{9}{19}$$

$$P(A \cap B) = 8\left(\frac{1}{38}\right) = \frac{8}{38} = \frac{4}{19}$$

$$P(A \cup B) = 28\left(\frac{1}{38}\right) = \frac{28}{38} = \frac{14}{19}$$

$$P(C) = 18\left(\frac{1}{38}\right) = \frac{18}{38} = \frac{9}{19}$$

d.   The event $A \cap B \cap C$ is the event the outcome is odd and black and low. The event $A \cap B \cap C$ is {11, 13, 15, 17}.

e. $P(A \cup B) = P(A) + P(B) - P(A \cap B) = \dfrac{9}{19} + \dfrac{9}{19} - \dfrac{4}{19} = \dfrac{14}{19}$. Events A and B are not mutually exclusive because $P(A \cap B) \neq 0$.

f. $P(A \cap B \cap C) = 4\left(\dfrac{1}{38}\right) = \dfrac{4}{38} = \dfrac{2}{19}$

g.   The event $A \cup B \cup C$ is the event the outcome is odd or black or low. The event $A \cup B \cup C$ is:

   {1, 2, 3, ... , 29, 31, 33, 35}

   or

   {All simple events except 00, 0, 30, 32, 34, 36}

h. $P(A \cup B \cup C) = 32\left(\dfrac{1}{38}\right) = \dfrac{32}{38} = \dfrac{16}{19}$

3.135   a.   Define the following event:

   F_i: {Player makes a foul shot on i^{th} attempt} $P(F_i) = .8$

$$P(F_i^c) = 1 - P(F_i) = 1 - .8 = .2$$

   The event "the player scores on both shots" is $F_1 \cap F_2$. If the throws are independent, then:

$$P(F_1 \cap F_2) = P(F_1)P(F_2) = .8(.8) = .64$$

   The event "the player scores on exactly one shot" is $(F_1 \cap F_2^c) \cup (F_1^c \cap F_2)$

$$P\left[(F_1 \cap F_2^c) \cup (F_1^c \cap F_2)\right] = P(F_1 \cap F_2^c) + P(F_1^c \cap F_2)$$
$$= P(F_1)P(F_2^c) + P(F_1^c)P(F_2)$$
$$= .8(.2) + .2(.8) = .16 + .16 = .32$$

The event "the player scores on neither shot" is $(F_1^c \cap F_2^c)$.

$$P(F_1^c \cap F_2^c) = P(F_1^c)P(F_2^c) = .2(.2) = .04$$

b.   We know $P(F_1) = .8$, $P(F_2 \mid F_1) = .9$, and $P(F_2 \mid F_1^c) = .7$

The probability the player scores on both shots is:

$$P(F_1 \cap F_2) = P(F_2 \mid F_1)P(F_1) = .9(.8) = .72$$

The probability the player scores on exactly one shot is:

$$P(F_1 \cap F_2^c) + P(F_1^c \cap F_2) = P(F_2^c \mid F_1)P(F_1) + P(F_2 \mid F_1^c)P(F_1^c)$$
$$= \left[1 - P(F_2 \mid F_1)\right]P(F_1) + P(F_2 \mid F_1^c)P(F_1^c)$$
$$= (1 - .9)(.8) + .7(.2) = .08 + .14 = .22$$

The probability the player scores on neither shot is:

$$P(F_1^c \cap F_2^c) = P(F_2^c \mid F_1^c)P(F_1^c) = \left[1 - P(F_2 \mid F_1^c)\right]P(F_1^c) = (1 - .7)(.2) = .06$$

c.   Two consecutive foul shots are probably dependent. The outcome of the second shot probably depends on the outcome of the first.

3.137   Define the following events:

A: {Seed carries single spikelets}
B: {Seed carries paired spikelets}
C: {Seed produces ears with single spikelets}
D: {Seed produces ears with paired spikelets}

From the problem, $P(A) = .4$, $P(B) = .6$, $P(C \mid A) = .29$, $P(D \mid A) = .71$, $P(C \mid B) = .26$, and $P(D \mid B) = .74$.

a. $P(A \cap C) = P(C \mid A)P(A) = .29(.4) = .116$

b. $P(D) = P(A \cap D) + P(B \cap D) = P(D \mid A)P(A) + P(D \mid B)P(B) = .71(.4) + .74(.6)$
$$= .284 + .444 = .728$$

3.139   Define the following events:

A_1: {Component 1 in System A works properly}
A_2: {Component 2 in System A works properly}
A_3: {Component 3 in System A works properly}
B_1: {Component 1 in System B works properly}
B_2: {Component 2 in System B works properly}
B_3: {Component 3 in System B works properly}

B_4: {Component 4 in System B works properly}
A: {System A works properly}
B: {System B works properly}
C: {Subsystem C works properly}
D: {Subsystem D works properly}

$$P(A_1) = 1 - P(A_1^c) = 1 - .12 = .88 \qquad P(B_1) = 1 - P(B_1^c) = 1 - .1 = .9$$

$$P(A_2) = 1 - P(A_2^c) = 1 - .09 = .91 \qquad P(B_2) = 1 - P(B_2^c) = 1 - .1 = .9$$

$$P(A_3) = 1 - P(A_3^c) = 1 - .11 = .89 \qquad P(B_3) = 1 - P(B_3^c) = 1 - .1 = .9$$

$$P(B_4) = 1 - P(B_4^c) = 1 - .1 = .9$$

a. $P(A) = P(A_1 \cap A_2 \cap A_3) = P(A_1)P(A_2)P(A_3) = .88(.91)(.89) = .7127$
   (since the three components operate independently)

b. $P(A^c) = 1 - P(A) = 1 - .7127 - .2873$

c. Subsystem C works properly if both components B_1 and B_2 work properly.

$$P(C) = P(B_1 \cap B_2) = P(B_1)P(B_2) = .9(.9) = .81$$
   (since the 2 components operate independently)

Similarly, subsystem D works properly if both components B_3 and B_4 work properly.

$$P(D) = P(B_3 \cap B_4) = P(B_3)P(B_4) = .9(.9) = .81$$

System B operates properly if either subsystem C or subsystem D operates properly.

$$P(B) = P(C \cup D) = P(C) + P(D) - P(C \cap D) = P(C) + P(D) - P(C)P(D)$$
$$= .81 + .81 - .81(.81) = .9639$$

d. The probability that exactly one subsystem in System B fails is:

$$P(C \cap D^c) + P(C^c \cap D) = P(C)P(D^c) + P(C^c)P(D)$$
$$= .81(1 - .81) + (1 - .81)(.81) = .1539 + .1539 = .3078$$

e. System B fails to operate only if both subsystems C and D fail.

$$P(B^c) = P(C^c \cap D^c) = P(C^c)P(D^c) = (1 - .81)(1 - .81) = .0361$$

f. If System B operates correctly 99% of the time, then it fails 1% of the time. The probability that one of the subsystems fails is 1 - .81 = .19. The probability that n subsystems fail is $.19^n$. Thus, we must find n such that

$$.19^n \leq .01 \implies n \geq 3$$

3.141   Define the following events:

   A: {Man never smoked cigars}

   B: {Man formerly smoked cigars}

   C: {Man currently smokes cigars}

   D: {Man died from Cancer}

   a. $P(A \cap D) = \dfrac{782}{137,243} = .006$

   b. $P(B \cap D) = \dfrac{91}{137,243} = .0007$

   c. $P(C \cap D) = \dfrac{141}{137,243} = .001$

   d. $P(D \mid C) = \dfrac{P(D \cap C)}{P(C)} = \dfrac{141/137,243}{7,866/137,243} = \dfrac{141}{7,866} = .018$

   e. $P(D \mid A) = \dfrac{P(D \cap A)}{P(A)} = \dfrac{782/137,243}{121,529/137,243} = \dfrac{782}{121,529} = .006$

3.143   a.   The odds in favor of Smarty Jones are $\dfrac{1}{3}$ to $\left(1 - \dfrac{1}{3}\right)$ or $\dfrac{1}{3}$ to $\dfrac{2}{3}$ or 1 to 2

   b.   If the odds are 1 to 1, P(Smarty Jones will win) $= \dfrac{1}{1+1} = \dfrac{1}{2} = .5$

   c.   If the odds against Smarty Jones winning are 3 to 2, the odds for Smarty Jones winning are 2 to 3. The probability Smarty Jones will win is $\dfrac{2}{2+3} = \dfrac{2}{5} = .4$

3.145   Define the following event:

   A: {Antigens match}

   From the problem, $P(A) = .25$ and $P(A^c) = 1 - P(A) = 1 - .25 = .75$

   a.   The probability that one sibling has a match is $P(A) = .25$.

   b.   The probability that all three will match is:

   $$P(AAA) = P(A \cap A \cap A) = P(A)P(A)P(A) = .25^3 = .0156$$

c.   The probability that none of the three match is:

$$P(A^c A^c A^c) = P(A^c \cap A^c \cap A^c) = P(A^c)P(A^c)P(A^c) = .75^3 = .4219$$

d.   For this part, $P(A) = .001$ and $P(A^c) = 1 - P(A) = 1 - .001 = .999$

$$P(A) = .001$$

$$P(AAA) = P(A \cap A \cap A) = P(A)P(A)P(A) = .001^3 = .000000001$$

$$P(A^c A^c A^c) = P(A^c \cap A^c \cap A^c) = P(A^c)P(A^c)P(A^c) = .999^3 = .9970$$

3.147   We will work with the following events:

A: {Woman is pregnant}
B: {Pregnancy test is positive}

The given probabilities can be written as:

$$P(A) = .75, \quad P(B|A) = .99, \quad P(B|A^c) = .02$$

Then,

$$P(A|B) = \frac{P(A \cap B)}{P(B)} = \frac{P(B|A)P(A)}{P(A \cap B) + P(A^c \cap B)}$$

$$= \frac{P(B|A)P(A)}{P(B|A)P(A) + P(B|A^c)P(A^c)}$$

$$= \frac{(.99)(.75)}{(.99)(.75) + (.02)(.25)} = \frac{.7425}{.7475} \approx .993$$

3.149   If each of 3 players uses a fair coin, the sample space is:

$$HHH \quad HTT \quad HHT \quad THT \quad HTH \quad TTH \quad THH \quad TTT$$

Since each event is equally likely, each event will have probability 1/8.

The probability of odd man out on the first roll is:

$$P(HHT) + P(HTH) + P(THH) + P(HTT) + P(THT) + P(TTH)$$
$$= 1/8 + 1/8 + 1/8 + 1/8 + 1/8 + 1/8 = 6/8 = 3/4$$

If one player uses a two-headed coin, the sample space will be (assume 3rd player has the two-headed coin):

$$HHH \quad THH \quad HTH \quad TTH$$

Since each event is equally likely, each will have probability 1/4.

The probability the 3rd player is odd man out is $P(TTH) = 1/4$.

From the first part, the probability the 3rd player is odd man out is $P(HHT) + P(TTH)$ $= 1/8 + 1/8 = 1/4$. Thus, the two probabilities are the same.

3.151   First, we will list all possible sample points for placing a car (C) and 2 goats (G) behind doors #1, #2, and #3. If the first position corresponds to door #1, the second position corresponds to door #2, and the third position corresponds to door #3, the sample space is:

(C G G) (G C G) (G G C)

Now, suppose you pick door #1. Initially, the probability that you will win the car is 1/3 – only one of the sample points has a car behind door #1.

The host will now open a door behind which is a goat. If you pick door #1 in the first sample point (C G G), the host will open either door #2 or door #3. Suppose he opens door #3 (it really does not matter). If you pick door #1 in the second sample point (G C G), the host will open door #3. If you pick door #1 in the third sample point (G G C), the host will open door #2. Now, the new sample space will be:

(C G) (G C) (G C)

where the first position corresponds to door #1 (the one you chose) and the second position corresponds to the door that was not opened by the host.

Now, if you keep door #1, the probability that you win the car is 1/3. However, if you switch to the remaining door, the probability that you win the car is now 2/3. Based on these probabilities, it is to your advantage to switch doors.

The above could be repeated by selecting door #2 initially or door #3 initially. In either of these cases, again, the probability of winning the car is 1/3 if you do not switch and 2/3 if you switch. Thus, Marilyn was correct.

Discrete Random Variables

Chapter
4

4.1   A random variable is a rule that assigns one and only one value to each simple event of an experiment.

4.3   a.   Since we can count the number of words spelled correctly, the random variable is discrete.

   b.   Since we can assume values in an interval, the amount of water flowing through the Hoover Dam in a day is continuous.

   c.   Since time is measured on an interval, the random variable is continuous.

   d.   Since we can count the number of bacteria per cubic centimeter in drinking water, this random variable is discrete.

   e.   Since we cannot count the amount of carbon monoxide produced per gallon of unleaded gas, this random variable is continuous.

   f.   Since weight is measured on an interval, weight is a continuous random variable.

4.5   a.   The reaction time difference is continuous because it lies within an interval.

   b.   Since we can count the number of violent crimes, this random variable is discrete.

   c.   Since we can count the number of near misses in a month, this variable is discrete.

   d.   Since we can count the number of winners each week, this variable is discrete.

   e.   Since we can count the number of free throws made per game by a basketball team, this variable is discrete.

   f.   Since distance traveled by a school bus lies in some interval, this is a continuous random variable.

4.7   The values x can assume are 0, 1, 2, 3, 4, …. Thus, x is a discrete random variable.

4.9   Annual rainfall can take on any value in an interval. Thus, annual rainfall is a continuous random variable. The number of ant species can assume only a countable number of outcomes. Thus, number of ant species is a discrete random variable.

4.11   Answers will vary. An example of a discrete random variable of interest to a psychologist might be the number of times a person has violent dreams in a week.

4.13   Answers will vary. An example of a discrete random variable of interest to a hospital nurse might be the number of medications a patient receives in 24 hours.

4.15   The probability distribution of a discrete random variable can be represented by a table, graph, or formula that specifies the probability associated with each possible value the random variable can assume.

4.17   The expected value of a random variable, $E(x)$, does not always equal a specific value of the random variable, x. The $E(x)$ is the mean of the probability distribution and does not have to equal a specific value of x.

4.19   a. $p(22) = .25$

    b. $p(20) + p(24) = .15 + .20 = .35$

    c. $P(x \leq 23) = p(20) + p(21) + p(22) + p(23) = .15 + .10 + .25 + .30 = .80$

4.21   a. $P(x \leq 12) = P(x = 10) + P(x = 11) + P(x = 12) = .2 + .3 + .2 = .7$

    b. $P(x > 12) = 1 - P(x \leq 12) = 1 - .7 = .3$

    c. $P(x \leq 14) = P(x = 10) + P(x = 11) + P(x = 12) + P(x = 13) + P(x = 14)$
$$= .2 + .3 + .2 + .1 + .2 = 1$$

    d. $P(x = 14) = .2$

    e. $P(x \leq 11 \text{ or } x > 12) = P(x \leq 11) + P(x > 12)$
$$= P(x = 10) + P(x = 11) + P(x = 13) + P(x = 14)$$
$$= .2 + .3 + .1 + .2 = .8$$

4.23   a. $\mu = E(x) = \sum xp(x) = 1(.2) + 2(.4) + 4(.2) + 10(.2) = .2 + .8 + .8 + 2 = 3.8$

    b. $\sigma^2 = E[(x - \mu)^2] = \sum (x - \mu)^2 p(x)$
$$= (1 - 3.8)^2 (.2) + (2 - 3.8)^2 (.4) + (4 - 3.8)^2 (.2) + (10 - 3.8)^2 (.2)$$
$$= 1.568 + 1.296 + .008 + 7.688 = 10.56$$

    c. $\sigma = \sqrt{10.59} = 3.2496$

    d.   The average value of x over many trials is 3.8.

    e.   No. The random variable can only take on values 1, 2, 4, or 10.

    f.   Yes. It is possible that μ can be equal to an actual value of x.

4.25   a.   It would seem that the mean of both would be 1 since they both are symmetric distributions centered at 1.

    b.   The distribution of x seems more variable since there appears to be greater probability for the two extreme values of 0 and 2 than there is in the distribution of y.

c.   For x: $\mu = E(x) = \sum xp(x) = 0(.3) + 1(.4) + 2(.3) = 0 + .4 + .6 = 1$

$\sigma^2 = E[(x-\mu)^2] = \sum (x-\mu)^2 p(x)$

$= (0-1)^2(.3) + (1-1)^2(.4) + (2-1)^2(.3) = .3 + 0 + .3 = .6$

For y: $\mu = E(y) = \sum yp(y) = 0(.1) + 1(.8) + 2(.1) = 0 + .8 + .2 = 1$

$\sigma^2 = E[(y-\mu)^2] = \sum (y-\mu)^2 p(y)$

$= (0-1)^2(.1) + (1-1)^2(.8) + (2-1)^2(.1) = .1 + 0 + .1 = .2$

The variance for x is larger than that for y.

4.27   a. $p(0) + p(1) + p(2) + p(3) + p(4) = .09 + .30 + .37 + .20 + .04 = 1.00$

b. $P(x = 3 \text{ or } 4) = p(3) + p(4) = .20 + .04 = .24$

c. $P(x < 2) = p(0) + p(1) = .09 + .30 = .39$

d. $E(x) = \sum xp(x) = 0(.09) + 1(.30) + 2(.37) + 3(.20) + 4(.04) = 0 + .30 + .74 + .60 + .16 = 1.8$

In a random sample of 4 homes, the average number of homes with high dust mite levels is 1.8.

e. $\sigma^2 = E[(x-\mu)^2] = \sum (x-\mu)^2 p(x)$

$= (0-1.8)^2(.09) + (1-1.8)^2(.30) + (2-1.8)^2(.37) + (3-1.8)^2(.20) + (4-1.8)^2(.04)$

$= .2916 + .1920 + .0148 + .2880 + .1936 = .98$

$\sigma = \sqrt{.98} = .9899$

f. $\mu \pm 2\sigma \Rightarrow 1.8 \pm 2(.9899) \Rightarrow 1.8 \pm 1.9798 \Rightarrow (-.1798,\ 3.7798)$

$P(-.1798 < x < 3.7798) = p(0) + p(1) + p(2) + p(3) = .09 + .03 + .37 + .20 = .96$

Chebyshev's Theorem says that the interval $\mu \pm 2\sigma$ will contain at least .75 of the data. The Empirical Rule says that approximately .95 of the data will be contained in the interval. Both Chebyshev's Theorem and the Empirical Rule fit the distribution.

4.29   a.   There are 4 sample points: Boy-Boy, Boy-Girl, Girl-Boy, and Girl-Girl.

b.   If all the sample points are equally likely, then each would have a probability of ¼.

c. $P(x=0) = p(0) = \dfrac{1}{4}, \ P(x=1) = p(1) = \dfrac{1}{4} + \dfrac{1}{4} = \dfrac{2}{4} = \dfrac{1}{2}, \ P(x=2) = p(2) = \dfrac{1}{4}$

d.   Using the proportions in the table as approximate probabilities, the distribution of x is:
$P(x=0) = p(0) = .222, \ P(x=1) = p(1) = .259 + .254 = .513, \ P(x=2) = p(2) = .265$

e. $E(x) = \sum xp(x) = 0(.222) + 1(.513) + 2(.265) = 1.043$. Over all two-child families, the average number of boys is 1.043.

4.31   a. $p(1) = (.23)(.77)^{1-1} = .23(.77)^0 = .23$.   The probability that a contaminated cartridge is selected on the first sample is .23.

b. $p(5) = (.23)(.77)^{5-1} = .23(.77)^4 = .081$. The probability that the first contaminated cartridge will be selected on the 5th sample is .081.

c. $P(x \geq 2) = 1 - P(x \leq 1) = 1 - p(1) = 1 - .23 = .77$. The probability that the first contaminated cartridge will be selected on the second trial or later is .77.

4.33   a.   Define the following events:

M: {Miami Beach, FL}
C: {Coney Island, NY}
S: {Surfside, CA}
B: {Monmouth Beach, NJ}
O: {Ocean City, NJ}
L: {Spring Lake, NJ}

The possible pairs of beach hotspots that can be selected are:

MC, MS, MB, MO, ML, CS, CB, CO, CL, SB, SO, SL, BO, BL, OL

b.   If we are just selecting 2 hotspots from 6, each of the 15 combinations are equally likely. Thus, each would have a probability of 1/15.

c.   The value of x for each of the pairs is:

Pair	x		Pair	x
MC	0		CL	1
MS	0		SB	1
MB	1		SO	0
MO	0		SL	1
ML	1		BO	1
CS	0		BL	2
CB	1		OL	1
CO	0			

d.   The probability distribution for x is:

x	$p(x)$
0	6/15
1	8/15
2	1/15

e. $\mu = E(x) = \sum xp(x) = 0(6/15) + 1(8/15) + 2(1/15) = 10/15 = .667$

The average total number of hotspots with a planar nearshore bar condition in each pair is .667.

4.35   For a \$5 bet, you will either win \$5 or lose \$5 (−\$5). The probability distribution for the net winnings is:

x	$p(x)$
−5	20/38
5	18/38

$$\mu = E(x) = \sum xp(x) = -5\left(\frac{20}{38}\right) + 5\left(\frac{18}{38}\right) = -.263$$

Over a large number of trials, the average winning for a \$5 bet on red is -\$0.263.

4.37   a. x is the sum of the grill numbers. To find the probability distribution for x, we divide the number of students choosing the particular grill display combination divided by the total number of students or 124. The table below shows the probability distribution for x.

Grill Display Combination	Number of Students	x	$p(x)$
1-2-3	35	6	35/124 = .282
1-2-4	8	7	8/124 = .065
1-2-5	42	8	42/124 = .339
2-3-4	4	9	4/124 = .032
2-3-5	1	10	1/124 = .008
2-4-5	34	11	34/124 = .274
Total	**124**		**1.00**

b. $P(x > 10) = P(x = 11) = .274$.

4.39   Let x = bookie's earnings per dollar wagered. Then x can take on values \$1 (you lose) and \$−5 (you win). The only way you win is if you pick 3 winners in 3 games. If the probability of picking 1 winner in 1 game is .5, then $P(www) = p(w)p(w)p(w) = .5(.5)(.5) = .125$ (assuming games are independent).

Thus, the probability distribution for x is:

x	$p(x)$
\$1	.875
\$−5	.125

$$E(x) = E(x) = \sum xp(x) = 1(.875) - 5(.125) = .875 - .625 = \$.25$$

On the average, the bookie will earn \$.25 for each bet made.

4.41   a.   Each point in the system can have one of 2 status levels, "free" or "obstacle". Define the following events:

A_F: {Point A is free} A_O: {Point A is obstacle}
B_F: {Point B is free} B_O: {Point B is obstacle}
C_F: {Point C is free} C_O: {Point C is obstacle}

Thus, the sample points for the space are:

$A_FB_FC_F$, $A_FB_FC_O$, $A_FB_OC_F$, $A_FB_OC_O$, $A_OB_FC_F$, $A_OB_FC_O$, $A_OB_OC_F$, $A_OB_OC_O$

The variable x is the total number of links that are "free". Thus, x can have values 0, 1, or 2.

b.   Since it is stated that the probability of any point in the system having a "free" status is .5, the probability of any point having an "obstacle" status is also .5, Thus, the probability of each of the sample points above is $P(A_iB_iC_i) = .5(.5)(.5) = .125$.

The values of x, the number of free links in the system, for each sample point are listed below. A link is free if both the points are free. Thus, a link from A to B is free if A is free and B is free. A link from B to C is free if B is free and C is free.

Sample point	x	Probability
$A_FB_FC_F$	2	.125
$A_FB_FC_O$	1	.125
$A_FB_OC_F$	0	.125
$A_FB_OC_O$	0	.125
$A_OB_FC_F$	1	.125
$A_OB_FC_O$	0	.125
$A_OB_OC_F$	0	.125
$A_OB_OC_O$	0	.125

The probability distribution for x is:

x	Probability
0	.625
1	.250
2	.125

4.43   The formula for $p(x)$ for a binomial random variable with $n = 7$ and $p = .2$ is:

$$p(x) = \binom{n}{x} p^x q^{n-x} = \binom{7}{x} .2^x .8^{7-x} \qquad (x = 0,1,2,...,7)$$

4.45   a. $\dfrac{6!}{2!(6-2)!} = \dfrac{6!}{2!4!} = \dfrac{6 \cdot 5 \cdot 4 \cdot 3 \cdot 2 \cdot 1}{(2 \cdot 1)(4 \cdot 3 \cdot 2 \cdot 1)} = 15$

   b. $\dbinom{5}{2} = \dfrac{5!}{2!(5-2)!} = \dfrac{5!}{2!3!} = \dfrac{5 \cdot 4 \cdot 3 \cdot 2 \cdot 1}{(2 \cdot 1)(3 \cdot 2 \cdot 1)} = 10$

   c. $\dbinom{7}{0} = \dfrac{7!}{0!(7-0)!} = \dfrac{7!}{0!7!} = \dfrac{7 \cdot 6 \cdot 5 \cdot 4 \cdot 3 \cdot 2 \cdot 1}{(1)(7 \cdot 6 \cdot 5 \cdot 4 \cdot 3 \cdot 2 \cdot 1)} = 1$

   (Note: $0! = 1$)

   d. $\dbinom{6}{6} = \dfrac{6!}{6!(6-6)!} = \dfrac{6!}{6!0!} = \dfrac{6 \cdot 5 \cdot 4 \cdot 3 \cdot 2 \cdot 1}{(6 \cdot 5 \cdot 4 \cdot 3 \cdot 2 \cdot 1)(1)} = 1$

   e. $\dbinom{4}{3} = \dfrac{4!}{3!(4-3)!} = \dfrac{4!}{3!1!} = \dfrac{4 \cdot 3 \cdot 2 \cdot 1}{(3 \cdot 2 \cdot 1)(1)} = 4$

4.47   a. $\mu = np = 25(.5) = 12.5$

   $\sigma^2 = np(1-p) = 25(.5)(.5) = 6.25$

   $\sigma = \sqrt{\sigma^2} = \sqrt{6.25} = 2.5$

   b. $\mu = np = 80(.2) = 16$

   $\sigma^2 = np(1-p) = 80(.2)(.8) = 12.8$

   $\sigma = \sqrt{\sigma^2} = \sqrt{12.8} = 3.578$

   c. $\mu = np = 100(.6) = 60$

   $\sigma^2 = np(1-p) = 100(.6)(.4) = 24$

   $\sigma = \sqrt{\sigma^2} = \sqrt{24} = 4.899$

   d. $\mu = np = 70(.9) = 63$

   $\sigma^2 = np(1-p) = 70(.9)(.1) = 6.3$

   $\sigma = \sqrt{\sigma^2} = \sqrt{6.3} = 2.510$

e. $\mu = np = 60(.8) = 48$

$\sigma^2 = np(1-p) = 60(.8)(.2) = 9.6$

$\sigma = \sqrt{\sigma^2} = \sqrt{9.6} = 3.098$

f. $\mu = np = 1,000(.04) = 40$

$\sigma^2 = np(1-p) = 1,000(.04)(.96) = 38.4$

$\sigma = \sqrt{\sigma^2} = \sqrt{38.4} = 6.197$

4.49   a. $P(x < 10) = P(x \le 9) = 0$

b. $P(x \ge 10) = 1 - P(x \le 9) = 1 - .002 = .998$

c. $P(x = 2) = P(x \le 2) - P(x \le 1) = .206 - .069 = .137$

4.51 x is a binomial random variable with $n = 4$.

a.   If the probability distribution of x is symmetric, $p(0) = p(4)$ and $p(1) = p(3)$.

Since $p(x) = \binom{n}{x} p^x q^{n-x} \quad x = 0, 1, \ldots , n,$

When $n = 4$,

$$\binom{4}{0} p^0 q^4 = \binom{4}{4} p^4 q^0 \Rightarrow \frac{4!}{0!4!} p^0 q^4 = \frac{4!}{4!0!} p^4 q^0 \Rightarrow q^4 = p^4 \Rightarrow p = q$$

Since $p + q = 1$, $p = .5$

Therefore, the probability distribution of x is symmetric when $p = .5$.

b.   If the probability distribution of x is skewed to the right, then the mean is greater than the median. Therefore, there are more small values in the distribution $(0, 1)$ than large values $(3, 4)$. Therefore, p must be smaller than .5. Let $p = .2$ and the probability distribution of x will be skewed to the right.

c.   If the probability distribution of x is skewed to the left, then the mean is smaller than the median. Therefore, there are more large values in the distribution $(3, 4)$ than small values $(0, 1)$. Therefore, p must be larger than .5. Let $p = .8$ and the probability distribution of x will be skewed to the left.

d.   In part **a**, x is a binomial random variable with $n = 4$ and $p = .5$.

$$p(x) = \binom{4}{x} .5^x .5^{4-x} \quad x = 0, 1, 2, 3, 4$$

$$p(0) = \binom{4}{0} .5^0 .5^4 = \frac{4!}{0!4!} .5^4 = 1(.5)^4 = .0625$$

$$p(1) = \binom{4}{1}.5^1.5^3 = \frac{4!}{1!3!}.5^4 = 4(.5)^4 = .25$$

$$p(2) = \binom{4}{2}.5^2.5^2 = \frac{4!}{2!2!}.5^4 = 6(.5)^4 = .375$$

$$p(3) = p(1) = .25 \text{ (since the distribution is symmetric)}$$

$$p(4) = p(0) = .0625$$

The probability distribution of x in tabular form is:

x	0	1	2	3	4
$p(x)$	.0625	.25	.375	.25	.0625

$$\mu = np = 4(.5) = 2$$

The graph of the probability distribution of x when $n = 4$ and $p = .5$ is as follows.

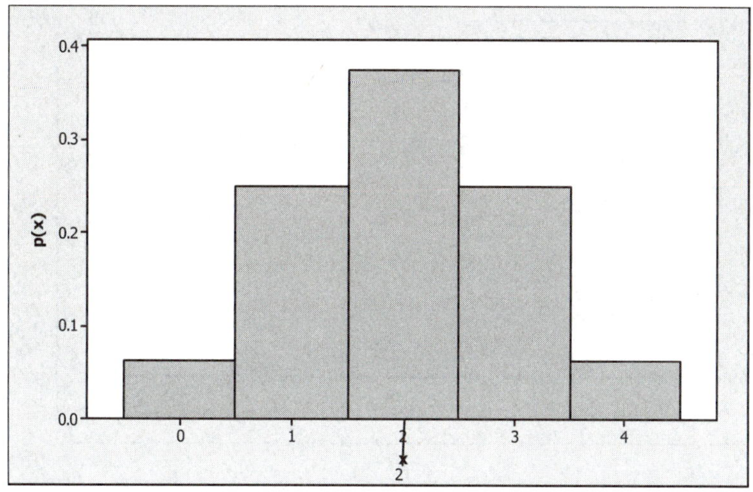

In part **b**, x is a binomial random variable with $n = 4$ and $p = .2$.

$$p(x) = \binom{4}{x}.2^x.8^{4-x} \qquad x = 0, 1, 2, 3, 4$$

$$p(0) = \binom{4}{0}.2^0.8^4 = 1(1).8^4 = .4096$$

$$p(1) = \binom{4}{1}.2^1.8^3 = 4(.2)(.8)^3 = .4096$$

$$p(2) = \binom{4}{2}.2^2.8^2 = 6(.2)^2(.8)^2 = .1536$$

$$p(3) = \binom{4}{3}.2^3.8^1 = 4(.2)^3(.8) = .0256$$

$$p(4) = \binom{4}{4}.2^4.8^0 = 1(.2)^4(1) = .0016$$

The probability distribution of x in tabular form is:

x	0	1	2	3	4
$p(x)$	.4096	.4096	.1536	.0256	.0016

$$\mu = np = 4(.2) = .8$$

The graph of the probability distribution of x when $n = 4$ and $p = .2$ is as follows:

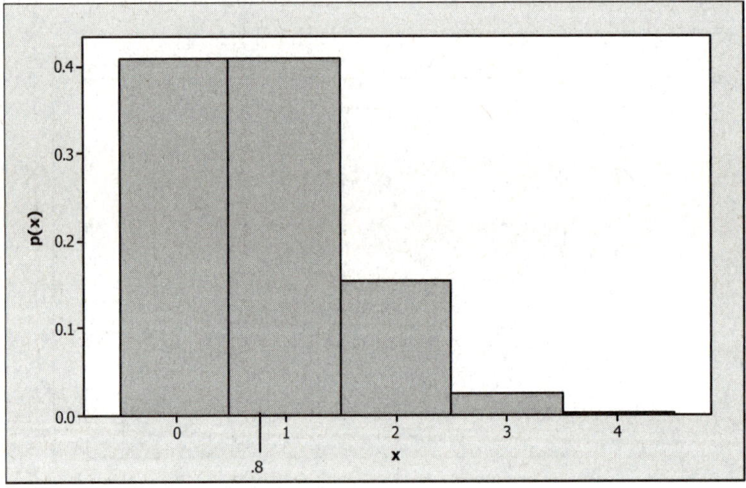

In part **c**, x is a binomial random variable with $n = 4$ and $p = .8$.

$$p(x) = \binom{4}{x}.8^x.2^{4-x} \quad x = 0, 1, 2, 3, 4$$

$$p(0) = \binom{4}{0}.8^0.2^4 = 1(1).2^4 = .0016$$

$$p(1) = \binom{4}{1}.8^1.2^3 = 4(.8)(.2)^3 = .0256$$

$$p(2) = \binom{4}{2}.8^2.2^2 = 6(.8)^2(.2)^2 = .1536$$

$$p(3) = \binom{4}{3}.8^3.2^1 = 4(.8)^3(.2) = .4096$$

$$p(4) = \binom{4}{4}.8^4.2^0 = 1(.8)^4(1) = .4096$$

The probability distribution of x in tabular form is:

x	0	1	2	3	4
$p(x)$	.0016	.0256	.1536	.4096	.4096

Note: The distribution of x when $n = 4$ and $p = .2$ is the reverse of the distribution of x when $n = 4$ and $p = .8$.

$$\mu = np = 4(.8) = 3.2$$

The graph of the probability distribution of x when $n = 4$ and $p = .8$ is as follows:

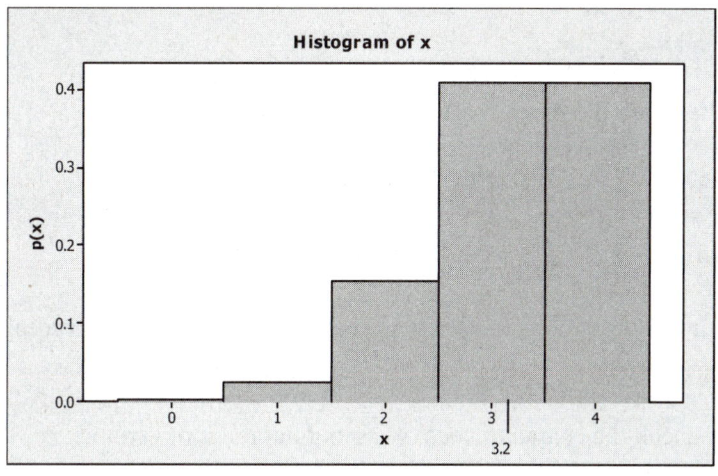

e.   In general, when $p = .5$, a binomial distribution will be symmetric regardless of the value of n. When p is less than .5, the binomial distribution will be skewed to the right; and when p is greater than .5, it will be skewed to the left. (Refer to parts **a**, **b**, and **c**.)

4.53   a.   For this experiment, there are $n = 100$ identical trials. Each trial can result in 2 possible outcomes (mouse responds positively or mouse does not respond positively). Let $S =$ mouse responds positively and $F =$ mouse does not respond positively. Then, $P(S) = p = .4$ and $P(F) = q = 1 - .4 = .6$. We can assume the trials are independent. Thus, the conditions of a binomial random variable hold.

b. $E(x) = \mu = np = 100(.4) = 40$

c. $V(x) = \sigma^2 = npq = 100(.4)(.6) = 24$

d.   Since $p = .4$ is close to .5, the distribution should be fairly symmetric. Thus, we would expect most of the observations to fall within 2 standard deviation of the mean. The standard deviation is $\sigma = \sqrt{24} = 4.899$. The interval is

$$\mu \pm 2\sigma \Rightarrow 40 \pm 2(4.899) \Rightarrow 40 \pm 9.798 \Rightarrow (30.202, 49.798)$$

4.55   a.   We will check the characteristics of a binomial random variable:

1.   This experiment consists of $n = 20$ identical trials.

2.   There are only 2 possible outcomes for each trial. SBIRS detects intruding object (S) or not (F).

3.   The probability of S remains the same from trial to trial. In this case, $p = P(S) = .8$ for each trial.

4.   The trials are independent. Whether SBIRS detects one intruding object should not effect whether it detects another.

5. x = number of intruding objects detected by SBIRS in 20 trials.

b. $n = 20$ and $p = P(S) = .8$

c.   Using the Table II, Appendix A with $n = 20$ and $p = .8$,

$$P(x = 15) = 1 - P(x \le 15) - P(x \le 14) = .370 - .196 = .174.$$

d. $P(x \ge 15) = 1 - P(x \le 14) = 1 - .196 = .804$

e. $E(x) = \mu = E(x) = np = 20(.8) = 16$. The average number of intruding objects detected in 20 trials is 16.

4.57   a.   We will check the characteristics of a binomial random variable:

1.   This experiment consists of $n = 15$ identical trials.

2.   There are only 2 possible outcomes for each trial. Hotel guests are aware and participate in the hotel's conservation efforts (S) or not (F).

3.   The probability of S remains the same from trial to trial. In this case, $p = P(S) = .66(.72) = .4752$ for each trial.

4.   The trials are independent. Whether one hotel guest is aware and participates in the hotel's conservation efforts should not effect whether another guest is aware and participates.

5. x = number of hotel guests who are aware and participate in the hotel's conservation efforts in 15 trials.

The characteristics of the binomial random variable are met.

b.   Define the following events:

G: {hotel guest is aware of the hotel's "green" conservation program}
P: {hotel guest participates in the conservation program}

We are given $P(G) = .66$ and $P(P|G) = .72$.

$$P(S) = P(P \cap G) = P(P|G)P(G) = .66(.72) = .4752.$$

c. $P(x \geq 10) = p(10) + p(11) + p(12) + p(13) + p(14) + p(15)$

$$= \binom{15}{10}.45^{10}(.55)^{15-10} + \binom{15}{11}.45^{11}(.55)^{15-11} + \binom{15}{12}.45^{12}(.55)^{15-12}$$

$$+ \binom{15}{13}.45^{13}(.55)^{15-13} + \binom{15}{14}.45^{14}(.55)^{15-14} + \binom{15}{15}.45^{15}(.55)^{15-10}$$

$$= \frac{15!}{10!5!}.45^{10}(.55)^{5} + \frac{15!}{11!4!}.45^{11}(.55)^{4} + \frac{15!}{12!3!}.45^{12}(.55)^{3}$$

$$+ \frac{15!}{13!2!}.45^{13}(.55)^{2} + \frac{15!}{14!1!}.45^{14}(.55)^{1} + \frac{15!}{15!0!}.45^{15}(.55)^{0}$$

$$= .0515 + .0191 + .0052 + .0010 + .0001 + .0000 = .0769$$

4.59   a.   Let x = number of commissioners favoring an issue in 4 trials. Then x is a binomial random variable with $n = 4$ and $p = .5$.

$$P(x = 2) = \binom{4}{2}.5^{2}.5^{4-2} = \frac{4!}{2!(4-2)!}.5^{2}.5^{2} = \frac{4(3)(2)(1)}{2(1)(2)(1)}.25(.25) = .375$$

b.   Let x = number of commissioners favoring an issue in 2 trials. Then x is a binomial random variable with $n = 2$ and $p = .5$.

$$P(x = 1) = \binom{2}{1}.5^{1}.5^{2-1} = \frac{2!}{1!(2-1)!}.5^{1}.5^{1} = \frac{2(1)}{1(1)}.5(.5) = .5$$

4.61   a.   Let x = number of women out of 15 that have been abused. Then x is a binomial random variable with $n = 15$, S = woman has been abused, F = woman has not been abused, $P(S) = p = \dfrac{1}{3}$ and $q = 1 - p = 1 - \dfrac{1}{3} = \dfrac{2}{3}$.

$$P(x \geq 4) = 1 - P(x < 4) = 1 - P(x = 0) - P(x = 1) - P(x = 2) - P(x = 3)$$

$$= 1 - \frac{15!}{0!15!}\left(\frac{1}{3}\right)^0\left(\frac{2}{3}\right)^{15} - \frac{15!}{1!14!}\left(\frac{1}{3}\right)^1\left(\frac{2}{3}\right)^{14} - \frac{15!}{2!13!}\left(\frac{1}{3}\right)^2\left(\frac{2}{3}\right)^{13} - \frac{15!}{3!12!}\left(\frac{1}{3}\right)^3\left(\frac{2}{3}\right)^{12}$$

$$= 1 - .0023 - .0171 - .0599 - .1299 = .7908$$

b.   Using Table II, Appendix A, with $n = 15$ and $p = .1$,

$$P(x \geq 4) = 1 - P(x \leq 3) = 1 - .944 = .056$$

c.   We sampled 15 women and actually found that 4 had been abused. If $p = 1/3$, the probability of observing 4 or more abused women is .7908. If $p = .1$, the probability of observing 4 or more abused women is only .056. If $p = .1$, we would have seen a very unusual event because the probability is so small (.056). If $p = 1/3$, we would have seen an event that was very common because the probability was very large (.7908). Since we normally do not see rare events, the true probability of abuse is probably close to 1/3.

4.63   a.   If the psychic is just guessing, she has 1 chance out of 10 of guessing the correct decision. Thus, $p = 1/10 = .1$.

b.   Let $x =$ number of correct decisions in 7 trials. Then x is a binomial random variable with $n = 7$ and $p = .1$. The expected number of correct decisions in 7 trials is

$$E(x) = np = 7(.1) = .7 \ .$$

Thus, if the psychic is just guessing, we would expect her to make less than 1 correct decision in 7 trials.

c. $P(x = 0) = \binom{7}{0}.1^0.9^7 = .4783$

d.   Now, let $p = .5$. $P(x = 0) = \binom{7}{0}.5^0.5^7 = .0078$

e.   Yes. If the psychic is just guessing, the probability that she makes no correct decisions is .4783. Thus, if the psychic, in fact, made no correct decisions, this would not be an unusual event if she is guessing. However, on the other hand, if she does have ESP ($p = .5$) and made no correct decisions, this would be a very unusual event ($P(x = 0) = .0078$). The evidence indicates the psychic probably is just guessing and does not have ESP.

4.65   a.   We must assume that the probability that a specific type of ball meets the requirements is always the same from trial to trial and the trials are independent. To use the binomial probability distribution, we need to know the probability that a specific type of golf ball meets the requirements.

b.   For a binomial distribution,

$$\mu = np$$
$$\sigma = \sqrt{npq}$$

In this example, n = two dozen = $2 \cdot 12 = 24$.

$p = .10$ (Success here means the golf ball *does not* meet standards.)
$q = .90$
$\mu = \mu = np = 24(.10) = 2.4$
$\sigma = \sqrt{npq} = \sqrt{24(.10)(.90)} = 1.47$

c.   In this situation,

p = Probability of success
  = Probability golf ball *does* meet standards
  = .90
$q = 1 - .90 = .10$
$n = 24$
$E(y) = \mu = np = 24(.90) = 21.6$
$\sigma = \sqrt{npq} = \sqrt{24(.10)(.90)} = 1.47$ (Note that this is the same as in part **b**.)

4.67   A normal distribution is a bell-shaped curve with the center of the distribution at μ.

4.69   A normal distribution with $\mu = 0$ and $\sigma = 1$ is a standard normal distribution.

4.71   a. $P(0 < z < 2.00) = .4772$
       (from Table III, Appendix A)

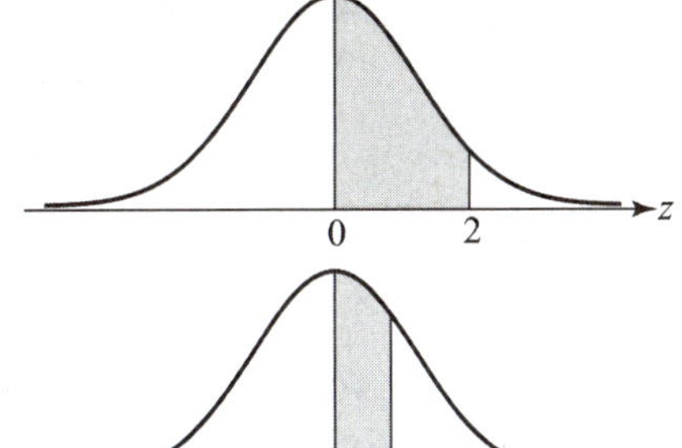

b. $P(0 < z < 1.00) = .3413$
     (from Table III, Appendix A)

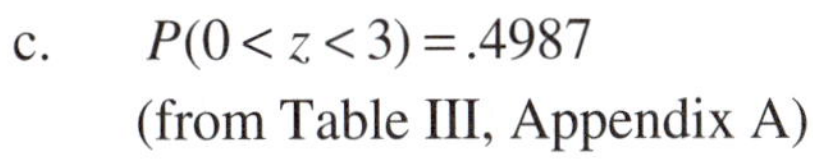

c. $P(0 < z < 3) = .4987$
     (from Table III, Appendix A)

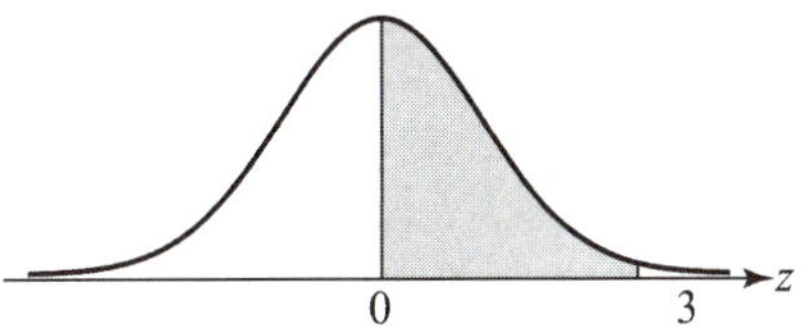

d. $P(0 < z < .58) = .2190$
     (from Table III, Appendix A)

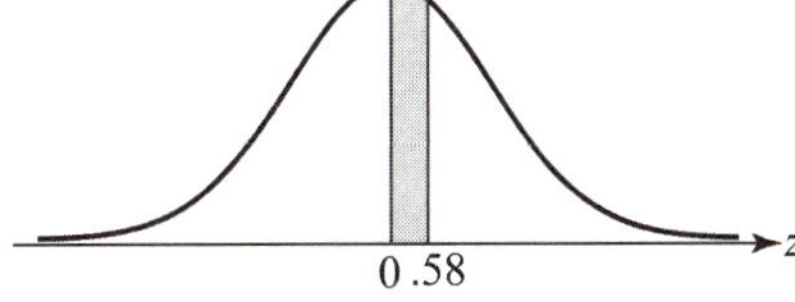

4.73   a. $P(z=1)=0$, since a single point does not have an area.

b. $P(z \leq 1) = P(z \leq 0) + P(0 < z \leq 1)$
$= A_1 + A_2$
$= .5 + .3413 = .8413$
(Table III, Appendix A)

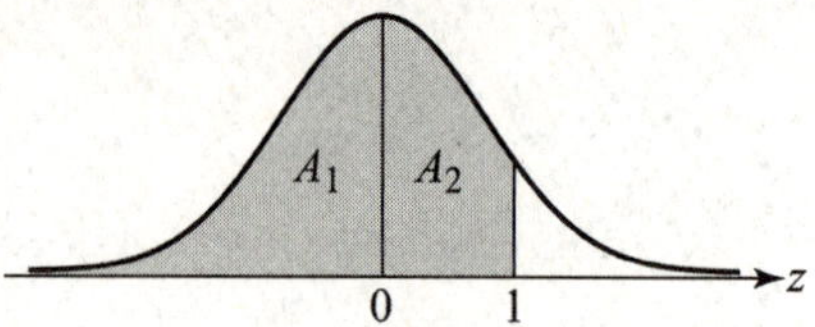

c. $P(z < 1) = P(z \leq 1) = .8413$ (Refer to part **b**.)

d. $P(z > 1) = 1 - P(z \leq 1) = 1 - .8413 = .1587$ (Refer to part **b**.)

e. $P(-1 \leq z \leq 1) = P(-1 \leq z \leq 0) + P(0 \leq z \leq 1)$
$= A_1 + A_2$
$= .3413 + .3413 = .6826$

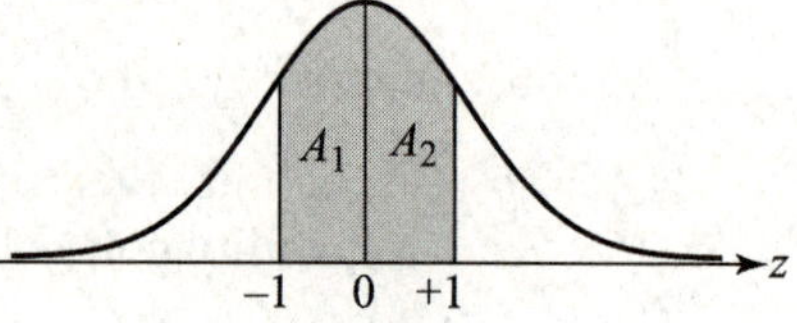

f. $P(-2 \leq z \leq 2) = P(-2 \leq z \leq 0) + P(0 \leq z \leq 2)$
$= A_1 + A_2$
$= .4772 + .4772 = .9544$

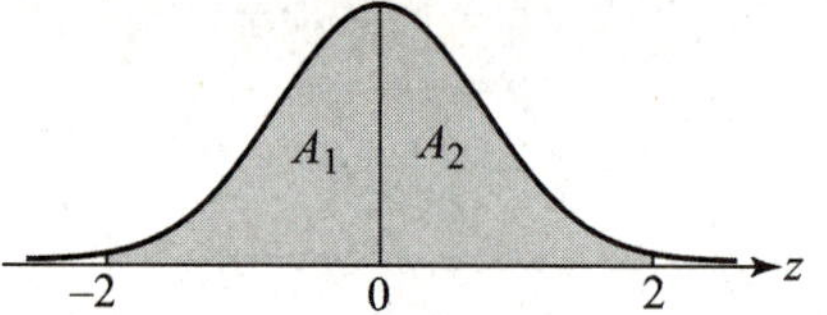

g. $P(-2.16 \leq z \leq 0.55) = P(-2.16 \leq z \leq 0) + P(0 \leq z \leq 0.55)$
$= A_1 + A_2$
$= .4846 + .2088 = .6934$

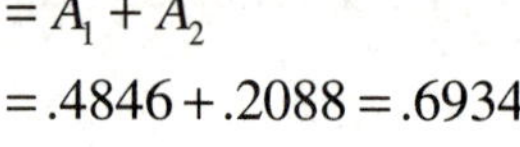
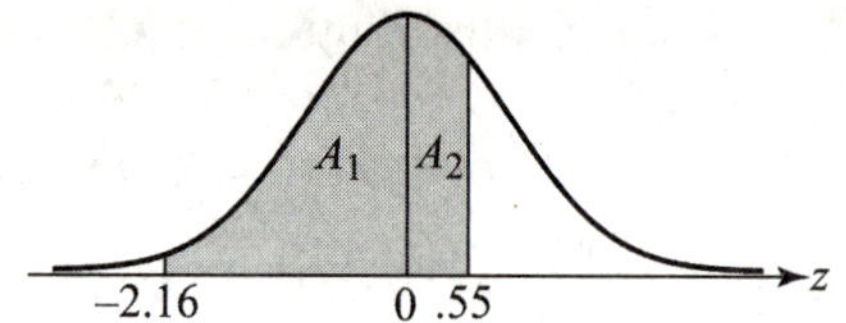

h. $P(-.42 < z < 1.96) = P(-.42 \leq z \leq 0) + P(0 \leq z \leq 1.96)$
$= A_1 + A_2$
$= .1628 + .4750 = .6378$

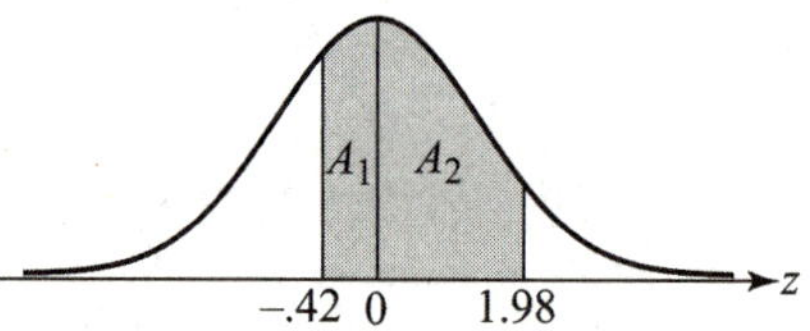

4.75   a. $z = 1$

b. $z = -1$

c. $z = 0$

d. $z = -2.5$

e. $z = 3$

4.77   Using Table III, Appendix A:

a. $P(z \geq z_0) = .05$

$A_1 = .5 - .05 = .4500$

Looking up the area .4500 in Table III gives z_0 = 1.645.

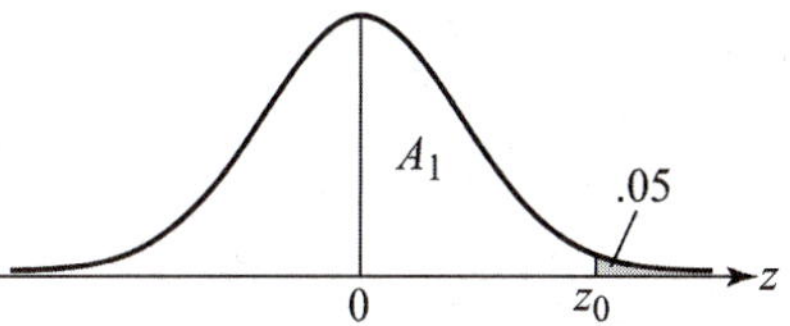

b. $P(z \geq z_0) = .025$

$A_1 = .5 - .025 = .4750$

Looking up the area .4750 in Table III gives $z_0 = 1.96$.

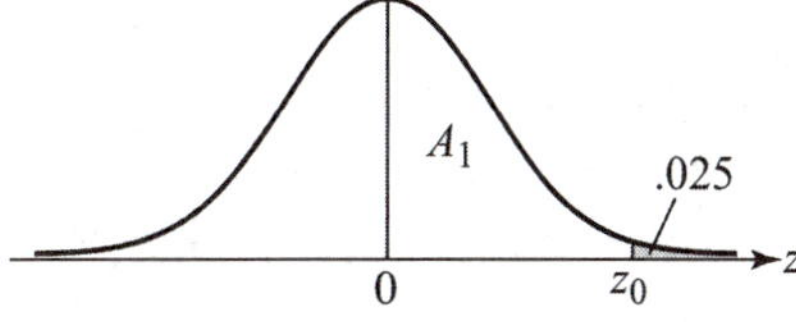

c. $P(z \leq z_0) = .025$

$A_1 = .5 - .025 = .4750$

Looking up the area .4750 in Table III gives $z = 1.96$. Since z_0 is to the left of 0, $z_0 = -1.96$.

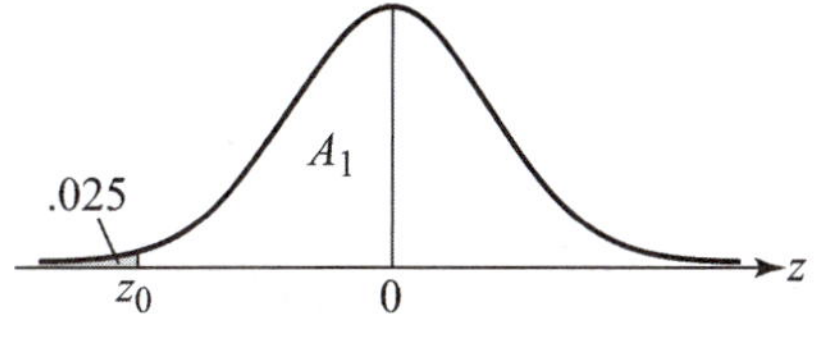

d. $P(z \geq z_0) = .10$

$A_1 = .5 - .10 = .4000$

Looking up the area .4000 in Table III gives $z_0 = 1.28$.

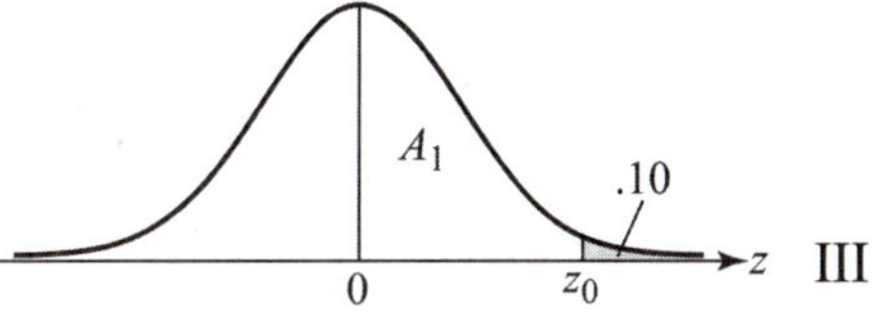

e. $P(z > z_0) = .10$

$A_1 = .5 - .10 = .4000$

$z_0 = 1.28$ (same as in **d**)

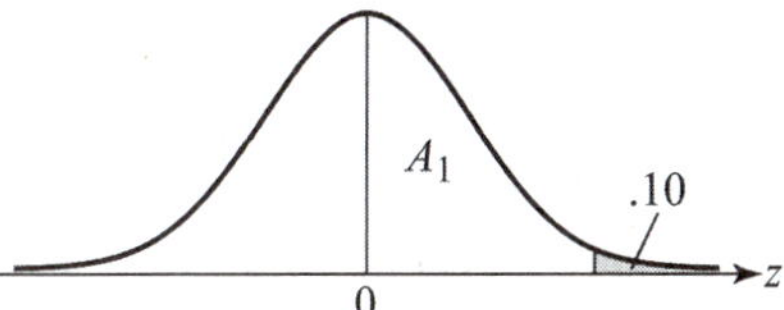

4.79   a. $P(x \geq x_0) = .5 \Rightarrow P\left(z \geq \dfrac{x_0 - 30}{8}\right)$

$= P(z \geq z_0) = .5$

$\Rightarrow z_0 = 0 = \dfrac{x_0 - 30}{8}$

$\Rightarrow x_0 = 8(0) + 30 = 30$

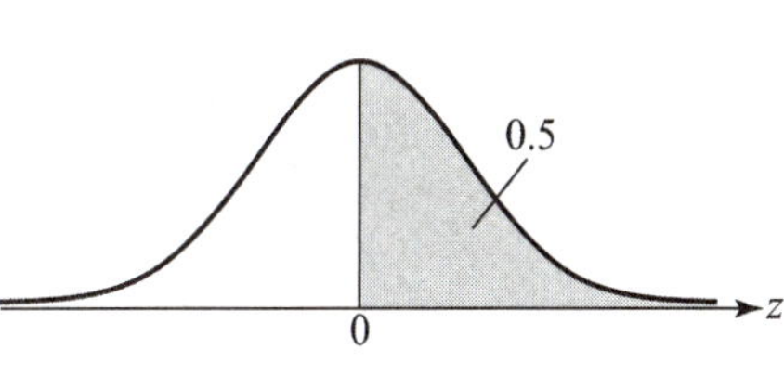

b. $P(x < x_0) = .025 \Rightarrow P\left(z < \dfrac{x_0 - 30}{8}\right)$

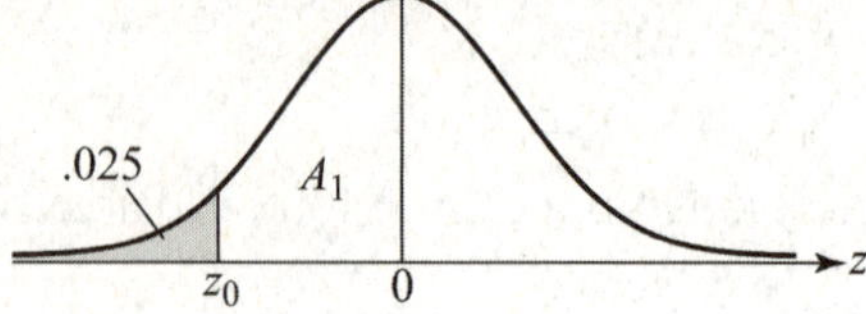

$$= P(z < z_0) = .025$$

$A_1 = .5 - .025 = .4750$

Looking up the area .4750 in Table III gives $z_0 = 1.96$.
Since z_0 is to the left of 0, $z_0 = -1.96$.

$$z_0 = -1.96 = \dfrac{x_0 - 30}{8} \Rightarrow x_0 = 8(-1.96) + 30 = 14.32$$

c. $P(x > x_0) = .10 \Rightarrow P\left(z > \dfrac{x_0 - 30}{8}\right)$

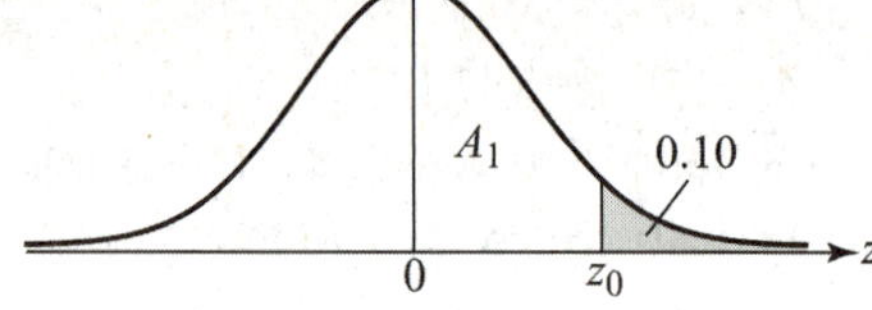

$$= P(z > z_0) = .10$$

$A_1 = .5 - .10 = .4000$

Looking up the area .4000 in Table III gives $z_0 = 1.28$.

$$z_0 = 1.28 = \dfrac{x_0 - 30}{8} \Rightarrow x_0 = 8(1.28) + 30 = 40.24$$

d. $P(x > x_0) = .95 \Rightarrow P\left(z > \dfrac{x_0 - 30}{8}\right)$

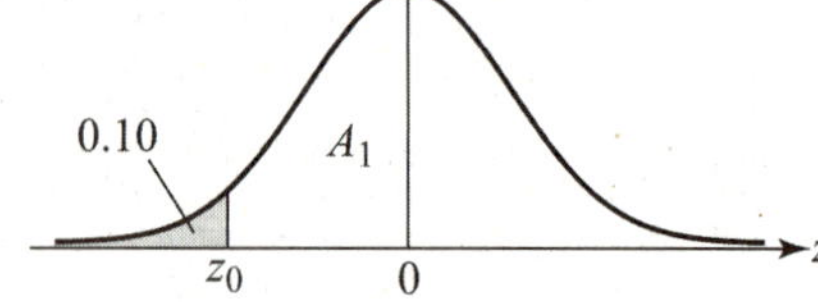

$$= P(z > z_0) = .95$$

$A_1 = .95 - .50 = .4500$

Looking up the area .4500 in Table III gives $z_0 = 1.645$.
Since z_0 is to the left of 0, $z_0 = -1.645$.

$$z_0 = -1.645 = \dfrac{x_0 - 30}{8} \Rightarrow x_0 = 8(-1.645) + 30 = 16.84$$

e. $P\left(x < x_0\right) = .10 \Rightarrow P\left(z < \dfrac{x_0 - 30}{8}\right)$

$$P(z < z_0) = .10$$

$A_1 = .5 - .10 = .4000$

Looking up the area .4000 in Table III gives $z_0 = 1.28$. Since z_0 is to the left of 0,
$z_0 = -1.28$.

$$z_0 = -1.28 = \dfrac{x_0 - 30}{8} \Rightarrow x_0 = 8(-1.28) + 30 = 19.76$$

f. $P(x < x_0) = .80 \Rightarrow P\left(z < \dfrac{x_0 - 30}{8}\right)$

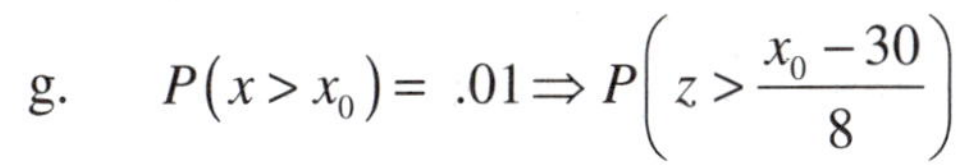

$$P(z < z_0) = .80$$

$A_1 = .80 - .5 = .3000$

Looking up the area .3000 in Table III gives $z_0 = 0.84$.

$$z_0 = 0.84 = \frac{x_0 - 30}{8} \Rightarrow x_0 = 8(.84) + 30 = 36.72$$

g. $P(x > x_0) = .01 \Rightarrow P\left(z > \dfrac{x_0 - 30}{8}\right)$

$$P(z > z_0) = .01$$

$A_1 = .5 - .01 = .4900$

Looking up the area .4900 in Table III gives $z_0 = 2.33$.

$$z_0 = 2.33 = \frac{x_0 - 30}{8} \Rightarrow x_0 = 8(2.33) + 30 = 48.64$$

4.81   a. $z = \dfrac{x - \mu}{\sigma} = \dfrac{10 - 11}{4} = \dfrac{-1}{4} = -.25$

b. $P(10 < x < 15) = P\left(\dfrac{10 - 11}{4} < z < \dfrac{15 - 11}{4}\right) = P(-.25 < z < 1)$

$$= P(-.25 < z < 0) + P(0 < z < 1)$$
$$= .0987 + .3413 = .4400$$

(Using Table III, Appendix A)

c. $P(x > 20) = P\left(z > \dfrac{20 - 11}{4}\right) = P(z > 2.25)$

$$= .5 - P(0 < z < 2.25) = .5 - .4878 = .0122$$

(Using Table III, Appendix A)

4.83   a.   Let $x =$ age of powerful woman. Using Table III, Appendix A,

$$P(55 < x < 60) = P\left(\frac{55 - 50}{6.4} < z < \frac{60 - 50}{6.4}\right) = P(.78 < z < 1.56)$$
$$= .4406 - .2823 = .1583$$

b.
$$P(48 < x < 52) = P\left(\frac{48-50}{6.4} < z < \frac{52-50}{6.4}\right) = P(-.31 < z < .31)$$
$$= .1217 + .1217 = .2434$$

c.
$$P(x < 35) = P\left(z < \frac{35-50}{6.4}\right) = P(z < -2.34) = .5000 - .4904 = .0096$$

d.
$$P(x > 40) = P\left(z > \frac{40-50}{6.4}\right) = P(z > -1.56) = .5000 + .4406 = .9406$$

e. We need to find the probability that a randomly selected woman is less than or equal to 25. $P(x \le 25) = P\left(z \le \frac{25-50}{6.4}\right) = P(z \le -3.91) \approx .5 - .5 = 0$. Since this observation would be very rare (probability almost 0), it is very unlikely that the woman is one of the 50 most powerful women.

4.85   a.   Let x = transmission delay of an RSVP linked wireless device. Using Table III, Appendix A,

$$P(x < 57) = P\left(z \le \frac{57-48.5}{8.5}\right) = P(z \le 1.00) = .5000 + .3413 = .8413$$

b.
$$P(40 < x < 60) = P\left(\frac{40-48.5}{8.5} < z < \frac{60-48.5}{8.5}\right) = P(-1.00 < z < 1.35)$$
$$= .3413 + .4115 = .7528$$

4.87   a.   Let x = driver's head injury rating. Using Table III, Appendix A,

$$P(500 < x < 700) = P\left(\frac{500-605}{185} < z < \frac{700-605}{185}\right) = P(-.57 < z < .51)$$
$$= .2157 + .1950 = .4107$$

b.
$$P(400 < x < 500) = P\left(\frac{400-605}{185} < z < \frac{500-605}{185}\right) = P(-1.11 < z < -.57)$$
$$= .3665 - .2157 = .1508$$

c.
$$P(x < 850) = P\left(z < \frac{850-605}{185}\right) = P(z < 1.32) = .5 + .4066 = .9066$$

d.
$$P(x > 1,000) = P\left(z > \frac{1,000-605}{185}\right) = P(z > 2.14) = .5 - .4838 = .0162$$

e. $\quad P(x > x_o) = .10 \Rightarrow P\left(z > \dfrac{x_o - 605}{185}\right) = P(z > z_0)$

$A_1 = .5 - .1 = .4000$

Looking up the area .4000 in Table IV gives $z_0 = 1.28$.

$z_0 = 1.28 = \dfrac{x_o - 605}{185} \Rightarrow x_o = 185(1.28) + 605 = 841.8$

4.89   a.   Let x = ambulance response time at Station A. Then x has a normal distribution with $\mu = 7.5$ and $\sigma = 2.5$. Using Table III, Appendix A,

$P(x \le 9) = P\left(z \le \dfrac{9 - 7.5}{2.5}\right) = P(z \le .60) = .5 + .2257 = .7257$. Since this probability is

less than the requirement of .9, the regulations are not met at EMS station A.

     b. $P(x \le 2) = P\left(z \le \dfrac{2 - 7.5}{2.5}\right) = P(z \le -2.20) = .5 - .4861 = .0139$. If the call was serviced

by Station A, the probability of a response time of 2 minutes or less would be extremely unusual. Thus, we would conclude that the call was probably not serviced by Station A.

4.91   a.   Let x = height of women over 20 years old. We are given that $\mu = 64$ and that the distribution of x is normal. However, we are given no information about the standard deviation. The answers to parts a through e will depend on what each person uses as the standard deviation.

     f-a.   Find x_L and x_U such that $P(x_L < x < x_U) = .50$.

First, we need to find z_0 such that $P(-z_0 < z < z_0) = .50$. Since we know that the center of the z distribution is 0, half of the area or .50/2 = .25 will be between $-z_0$ and 0 and half will be between 0 and z_0.

Look up .25 in the body of Table III, Appendix A to find $z_0 = .67$.

We know $\quad P(x_L < x < x_U) = P\left(\dfrac{x_L - 64}{2.6} < z < \dfrac{x_U - 64}{2.6}\right) = P(-.67 < z < .67) = .50$

Thus,

$\dfrac{x_L - 64}{2.6} = -.67 \quad$ and $\quad \dfrac{x_U - 64}{2.6} = .67$

$\Rightarrow x_L = 64 - .67(2.6) = 64 - 1.74 = 62.26$

$\Rightarrow x_U = 64 + .67(2.6) = 64 + 1.74 = 65.74$

     f-b.   Find x_L and x_U such that $P(x_L < x < x_U) = .75$.

First, we need to find z_0 such that $P(-z_0 < z < z_0) = .75$. Since we know that the center of the z distribution is 0, half of the area or $.75/2 = .375$ will be between $-z_0$ and 0 and half will be between 0 and z_0.

Look up .375 in the body of Table III, Appendix A to find $z_0 = 1.15$.

We know

$$P(x_L < x < x_U) = P\left(\frac{x_L - 64}{2.6} < z < \frac{x_U - 64}{2.6}\right) = P(-1.15 < z < 1.15) = .75$$

Thus,

$$\frac{x_L - 64}{2.6} = -1.15 \quad \text{and} \quad \frac{x_U - 64}{2.6} = 1.15$$

$$\Rightarrow x_L = 64 - 1.15(2.6) = 64 - 2.99 = 61.01$$

$$\Rightarrow x_U = 64 + 1.15(2.6) = 64 + 2.99 = 66.99$$

f-c.   Find x_L and x_U such that $P(x_L < x < x_U) = .90$.

First, we need to find z_0 such that $P(-z_0 < z < z_0) = .90$. Since we know that the center of the z distribution is 0, half of the area or $.90/2 = .45$ will be between $-z_0$ and 0 and half will be between 0 and z_0.

Look up .45 in the body of Table III, Appendix A to find $z_0 = 1.645$.

We know

$$P(x_L < x < x_U) = P\left(\frac{x_L - 64}{2.6} < z < \frac{x_U - 64}{2.6}\right) = P(-1.645 < z < 1.645) = .90$$

Thus,

$$\frac{x_L - 64}{2.6} = -1.645 \quad \text{and} \quad \frac{x_U - 64}{2.6} = 1.645$$

$$\Rightarrow x_L = 64 - 1.645(2.6) = 64 - 4.28 = 59.72$$

$$\Rightarrow x_U = 64 + 1.645(2.6) = 64 + 4.28 = 68.28$$

f-d.   Find x_L and x_U such that $P(x_L < x < x_U) = .95$.

First, we need to find z_0 such that $P(-z_0 < z < z_0) = .95$. Since we know that the center of the z distribution is 0, half of the area or $.95/2 = .475$ will be between $-z_0$ and 0 and half will be between 0 and z_0.

Look up .475 in the body of Table III, Appendix A to find $z_0 = 1.96$.

We know

$$P(x_L < x < x_U) = P\left(\frac{x_L - 64}{2.6} < z < \frac{x_U - 64}{2.6}\right) = P(-1.96 < z < 1.96) = .95$$

Thus,

$$\frac{x_L - 64}{2.6} = -1.96 \quad \text{and} \quad \frac{x_U - 64}{2.6} = 1.96$$

$$\Rightarrow x_L = 64 - 1.96(2.6) = 64 - 5.10 = 58.90$$

$$\Rightarrow x_U = 64 + 1.96(2.6) = 64 + 5.10 = 69.10$$

f-e.   Find x_L and x_U such that $P(x_L < x < x_U) = .99$.

First, we need to find z_0 such that $P(-z_0 < z < z_0) = .99$. Since we know that the center of the z distribution is 0, half of the area or $.99/2 = .495$ will be between $-z_0$ and 0 and half will be between 0 and z_0.

We look up .495 in the body of Table III, Appendix A to find $z_0 = 2.575$.

We know $P(x_L < x < x_U) = P\left(\frac{x_L - 64}{2.6} < z < \frac{x_U - 64}{2.6}\right) = P(-2.575 < z < 2.575) = .99$

Thus,

$$\frac{x_L - 64}{2.6} = -2.575 \quad \text{and} \quad \frac{x_U - 64}{2.6} = 2.575$$

$$\Rightarrow x_L = 64 - 2.575(2.6) = 64 - 6.70 = 57.30$$

$$\Rightarrow x_U = 64 + 2.575(2.6) = 64 + 6.70 = 70.70$$

4.93   Let x = number of additional Electoral College votes. Then x has a normal distribution with $\mu = 241.5$ and $\sigma = 49.8$. If the candidate wins the popular vote in California, then he/she needs only $270 - 55 = 215$ additional votes.

$$P(x \geq 215) = P\left(z \geq \frac{215 - 241.5}{49.8}\right) = P(z \geq -.53) = .5 + .2019 = .7019$$

4.95   a.   If z is a standard normal random variable,

$Q_L = z_L$ is the value of the standard normal distribution which has 25% of the data to the left and 75% to the right.

Find z_L such that $P(z < z_L) = .25$

$A_1 = .50 - .25 = .25$.

Look up the area $A_1 = .25$ in the body of Table III of Appendix A; $z_L = -.67$ (taking the closest value). If interpolation is used, $-.675$ would be obtained.

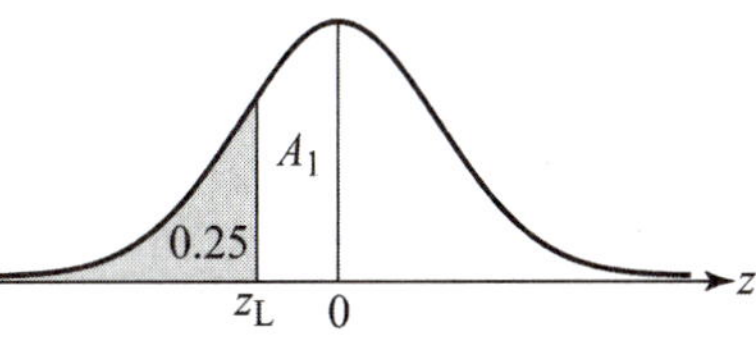

$Q_U = z_U$ is the value of the standard normal distribution which has 75% of the data to the left and 25% to the right.

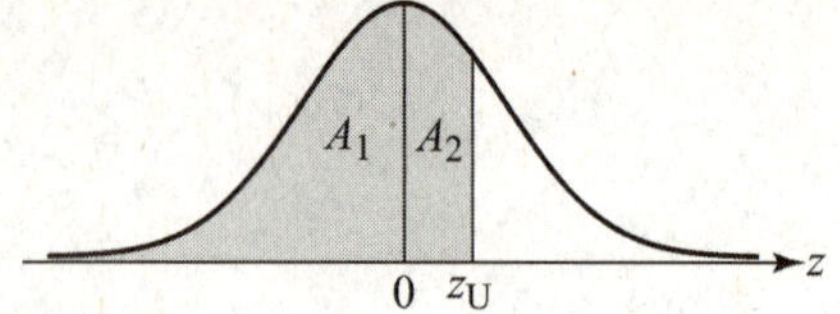

Find z_U such that $P(z < z_U) = .75$

$$A_1 + A_2 = P(z \leq 0) + P(0 \leq z \leq z_U)$$
$$= .5 + P(0 \leq z \leq z_U) = .75$$

Therefore, $P(0 \leq z \leq z_U) = .25$.

Look up the area .25 in the body of Table III of Appendix A; $z_U = .67$ (taking the closest value). If interpolation is used, .675 would be obtained.

b.   Recall that the inner fences of a box plot are located $1.5(Q_U - Q_L)$ outside the hinges (Q_L and Q_U).

To find the lower inner fence,

$$Q_L - 1.5(Q_U - Q_L) = -.67 - 1.5[.67 - (-.67)]$$
$$= -.67 - 1.5(1.34)$$
$$= -2.68 \quad (-2.70 \text{ if } z_L = -.675 \text{ and } z_U = .675)$$

The upper inner fence is:

$$Q_U + 1.5(Q_U - Q_L) = .67 + 1.5[.67 - (-.67)]$$
$$= .67 + 1.5(1.34)$$
$$= 2.68 \quad (2.70 \text{ if } z_L = -.675 \text{ and } z_U = .675)$$

c.   Recall that the outer fences of a box plot are located $3(Q_U - Q_L)$ outside the hinges (Q_L and Q_U).

To find the lower outer fence,

$$Q_L - 3(Q_U - Q_L) = -.67 - 3[.67 - (-.67)]$$
$$= -.67 - 3(1.34)$$
$$= -4.69 \quad (-4.725 \text{ if } z_L = -.675 \text{ and } z_U = .675)$$

The upper outer fence is:

$$Q_U + 3(Q_U - Q_L) = .67 + 3[.67 - (-.67)]$$
$$= .67 + 3(1.34)$$
$$= 4.69 \quad (4.725 \text{ if } z_L = -.675 \text{ and } z_U = .675)$$

d. $P(z < -2.68) + P(z > 2.68)$

$= 2P(z > 2.68)$

$= 2(.5000 - .4963)$

$= 2(.0037) = .0074$

(Table III, Appendix A)

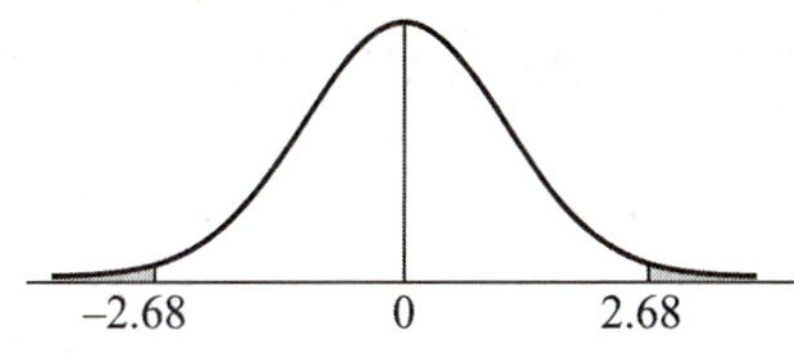

(or $2(.5000 - .4965) = .0070$ if -2.70 and 2.70 are used)

$P(z < -4.69) + P(z > 4.69)$

$= 2P(z > 4.69)$

$\approx 2(.5000 - .5000) \approx 0$

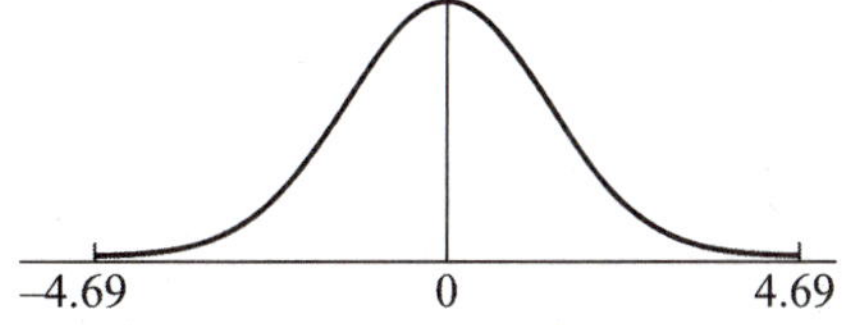

e.   In a normal probability distribution, the probability of an observation being beyond the inner fences is only .0074 and the probability of an observation being beyond the outer fences is approximately zero. Since the probabilities are so small, there should not be any observations beyond the inner and outer fences. Therefore, they are probably outliers.

4.97   Several statistical techniques are based on the assumption that the population is approximately normally distributed. Thus, it is important to determine whether the sample data come from a normal population so that the techniques can be properly applied.

4.99   a.   The proportion of measurements that one would expect to fall in the interval $\mu \pm \sigma$ is about .68.

b.   The proportion of measurements that one would expect to fall in the interval $\mu \pm 2\sigma$ is about .95.

c.   The proportion of measurements that one would expect to fall in the interval $\mu \pm 3\sigma$ is about .997.

4.101   If the data are normally distributed, then the normal probability plot should be an approximate straight line. Of the three plots, only plot c implies that the data are normally distributed. The data points in plot c form an approximately straight line. In both plots a and b, the plots of the data points do not form a straight line.

4.103 a.   Using MINITAB, the stem-and-leaf display is:

Stem-and-Leaf Display: Data

```
Stem-and-leaf of Data   N  = 40
Leaf Unit = 10

   14   0   00012223333344
  (10)  0   5556666889
   16   1   001334
   10   1   5689
    6   2   01
    4   2
    4   3   000
    1   3
    1   4   0
```

The distribution from the stem-and-leaf display looks like it is skewed to the right. Thus, it does not appear that the data are normally distributed.

b.   Using MINITAB, the descriptive statistics are:

Descriptive Statistics: Data

```
Variable   N  Mean  StDev  Minimum    Q1  Median     Q3  Maximum
Data      40 105.1   95.8     1.00  37.5    65.0  152.8    401.0
```

The lower quartile is $Q_L = 37.5$, the upper quartile is $Q_U = 152.8$, and $s = 95.8$.

c.   The interquartile range is $IQR = Q_U - Q_L = 152.8 - 37.5 = 115.3$.
$IQR / s = 115.3 / 95.8 = 1.204$. If the data are approximately normal, then this ratio should be 1.3. Since the actual ratio is 1.204 which is close to 1.3, it appears that the data could be approximately normal.

d.   Using MINITAB, the normal probability plot is:

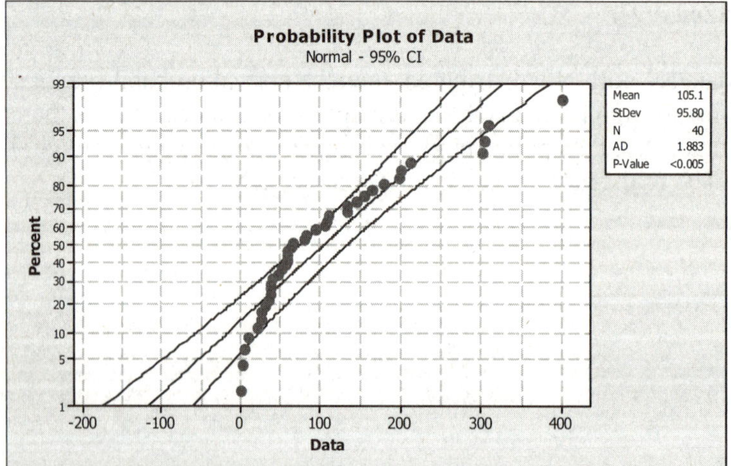

Since the data points do not fall along the straight line, the data do not appear to be normally distributed.

4.105   a. $\text{IQR} = Q_U - Q_L = 54 - 47 = 7$.

       b.   From the printout, $s = 6.444$.

       c.   If the data are approximately normal, then $\text{IQR}/s \approx 1.3$. For this data,
           $\text{IQR}/s = 7/6.444 = 1.09$. This is somewhat close to 1.3, so the data may be normal.

       d.   Using MINITAB, a relative frequency histogram of the data is:

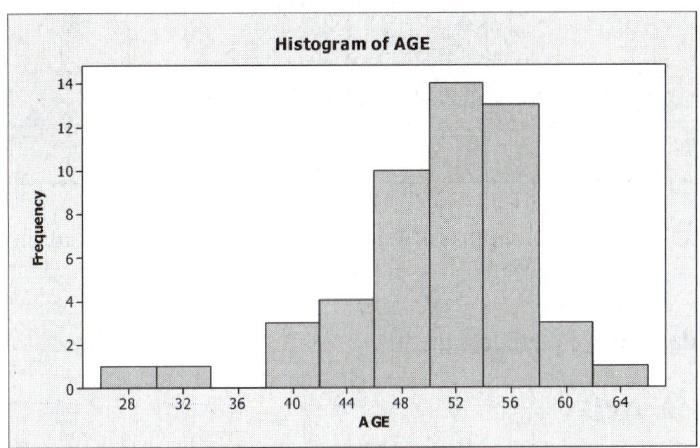

         From the histogram, the data appear to be close to mound-shaped (slightly skewed to the left), so the data may be normally distributed.

4.107   The interquartile range is $\text{IQR} = Q_U - Q_L = 6 - 1 = 5$. $\text{IQR}/s = 5/6.09 = .82$. This is much less than the 1.3 that we would expect if the data were normal. It appears that these data are not approximately normal. In addition, $\bar{x} = 4.71$ and $s = 6.09$. Since the number of updates cannot be negative, the data must be skewed to the right since one standard deviation below the mean would be a negative number which is impossible.

4.109   a.   From the graph, the histogram of the difference between the elevation estimates is too peaked. This means that there are too many observations in the very center of the distribution and not enough observations in the tails for the distribution to be normal.

       b.   The interval $\mu \pm 2\sigma$ would be $.28 \pm 2(1.6) \Rightarrow .28 \pm 3.2 \Rightarrow (-2.92, 3.48)$. From the graph, almost all the observations are between are between -2 and 2. Thus, the interval $(-2.92, 3.48)$ will contain more that 95% of the 400 elevation differences.

4.111   **American League**

       To determine if the distribution of batting averages is approximately normal, we will run through the tests. Using MINITAB, a histogram of the data with a normal curve included is:

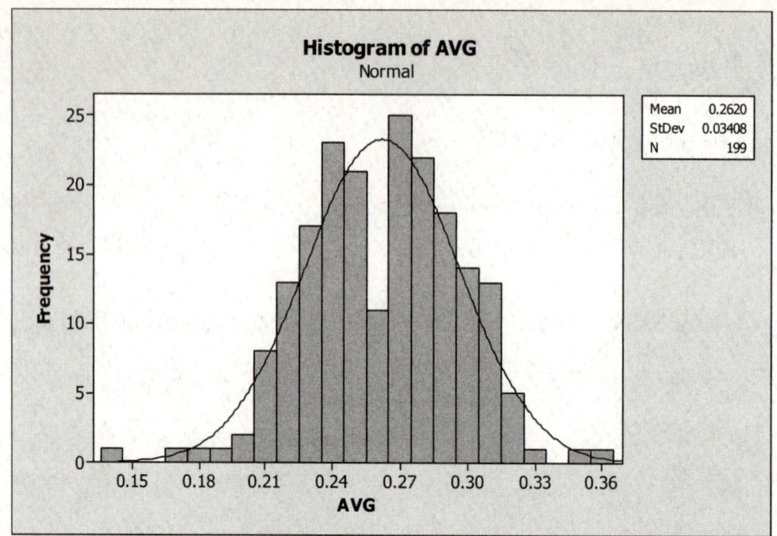

From the graph, the data appear to be approximately mound-shaped. The data may be normal.

Using MINITAB, the descriptive statistics are:

Descriptive Statistics: AVG

```
Variable     N     Mean     StDev  Minimum        Q1    Median        Q3  Maximum
AVG        199  0.26198  0.03408  0.14400   0.23600   0.26500   0.28600  0.36500
```

The interval $\bar{x} \pm s \Rightarrow .26198 \pm .03408 \Rightarrow (.22790, .29606)$ contains 133 of the 199 observations. The proportion is $133/199 = .668$. This is very close to the .68 from the Empirical Rule.

The interval $\bar{x} \pm 2s \Rightarrow .26198 \pm 2(.03408) \Rightarrow .26198 \pm .06816 \Rightarrow (.19382, .33014)$ contains 193 of the 199 observations. The proportion is $193/199 = .970$. This is somewhat larger than the .95 from the Empirical Rule.

The interval $\bar{x} \pm 3s \Rightarrow .26198 \pm 3(.03408) \Rightarrow .26198 \pm .10224 \Rightarrow (.15974, .36422)$ contains 197 of the 199 observations. The proportion is $197/199 = .990$. This is very close to the .997 from the Empirical Rule. Thus, it appears that the data may be normal.

The lower quartile is $Q_L = .236$ and the upper quartile is $Q_U = .286$. The interquartile range is $IQR = Q_U - Q_L = .286 - .236 = .050$. From the printout, $s = .03408$.
$IQR / s = .050 / .03408 = 1.467$. This is a fairly close to the 1.3 that we would expect if the data were normal. Thus, there is evidence that the data may be normal.

Using MINITAB, the normal probability plot is:

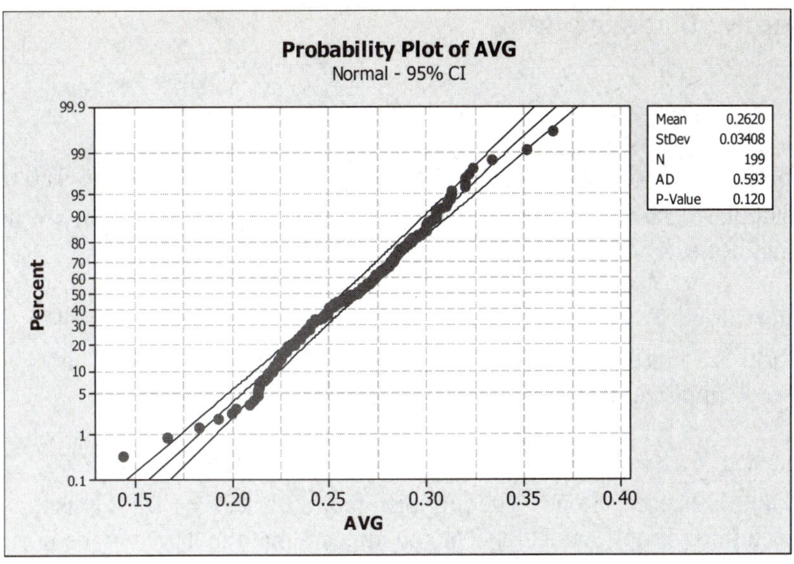

The data are very close to a straight line. Thus, it appears that the data may be normal.

From all of the indicators, it appears that the batting averages from the American League come from an approximate normal distribution.

National League

To determine if the distribution of batting averages is approximately normal, we will run through the tests. Using MINITAB, a histogram of the data with a normal curve included is:

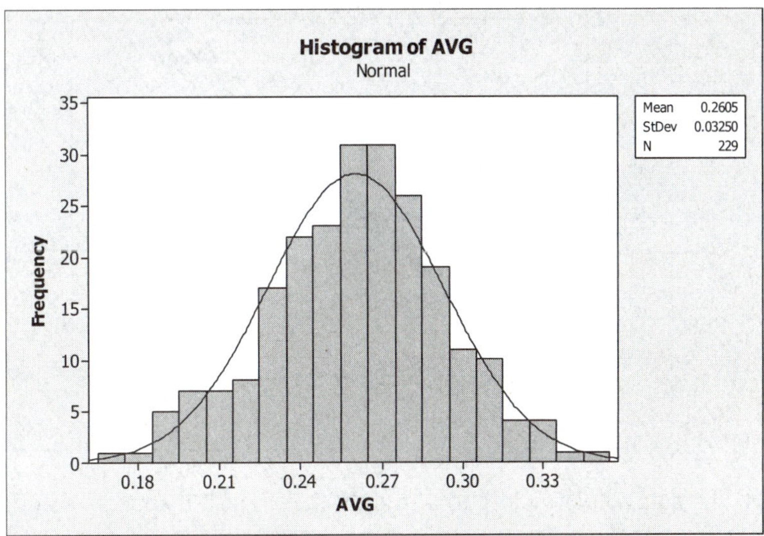

From the graph, the data appear to be approximately mound-shaped. The data may be normal.

Using MINITAB, the descriptive statistics are:

Descriptive Statistics: AVG

```
Variable    N     Mean    StDev  Minimum        Q1    Median        Q3  Maximum
AVG       229  0.26047  0.03250  0.17100   0.23850   0.26100   0.28200  0.35300
```

The interval $\bar{x} \pm s \Rightarrow .26047 \pm (.03250) \Rightarrow (.22797, .29297)$ contains 160 of the 229 observations.   The proportion is $160 / 229 = .699$. This is very close to the .68 from the Empirical Rule.

The interval $\bar{x} \pm 2s \Rightarrow .26047 \pm 2(.03250) \Rightarrow .26047 \pm .06500 \Rightarrow (.19547, .32547)$ contains 216 of the 229 observations. The proportion is $216 / 229 = .943$. This is very close to the .95 from the   Empirical Rule.

The interval $\bar{x} \pm 3s \Rightarrow .26047 \pm 3(.03250) \Rightarrow .26047 \pm .09750 \Rightarrow (.16297, .35797)$ contains 229 of the 229 observations. The proportion is $229 / 229 = 1.00$. This is very close to the .997 from the   Empirical Rule. Thus, it appears that the data may be normal.

The lower quartile is $Q_L = .2385$ and the upper quartile is $Q_U = .282$. The interquartile range is $\text{IQR} = Q_U - Q_L = .282 - .2385 = .0435$. From the printout, $s = .0325$. $\text{IQR} / s = .0435 / .0325 = 1.338$. This is a very close to the 1.3 that we would expect if the data were normal. Thus, there is evidence that the data may be normal.

Using MINITAB, the normal probability plot is:

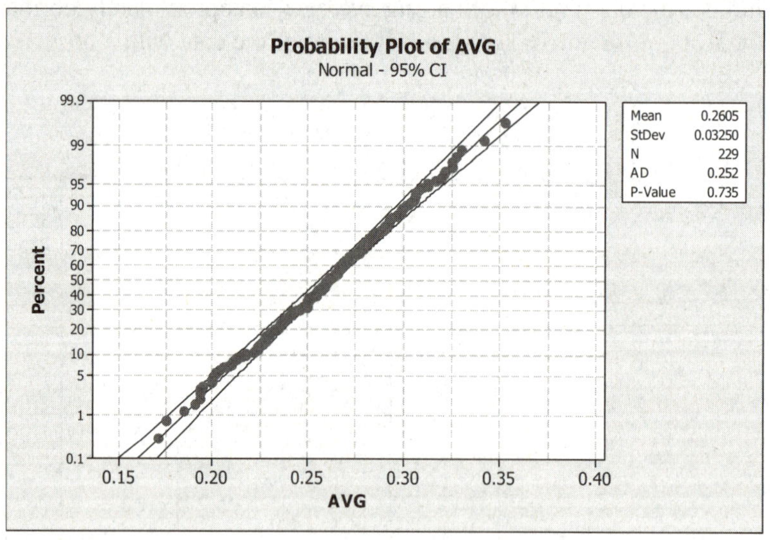

The data are very close to a straight line. Thus, it appears that the data may be normal.

From all of the indicators, it appears that the batting averages from the National League come from an approximate normal distribution.

4.113   We will look at the 4 methods for determining if the data are normal. First, we will look at a histogram of the data. Using MINITAB, the histogram of the driver's head injury ratings is:

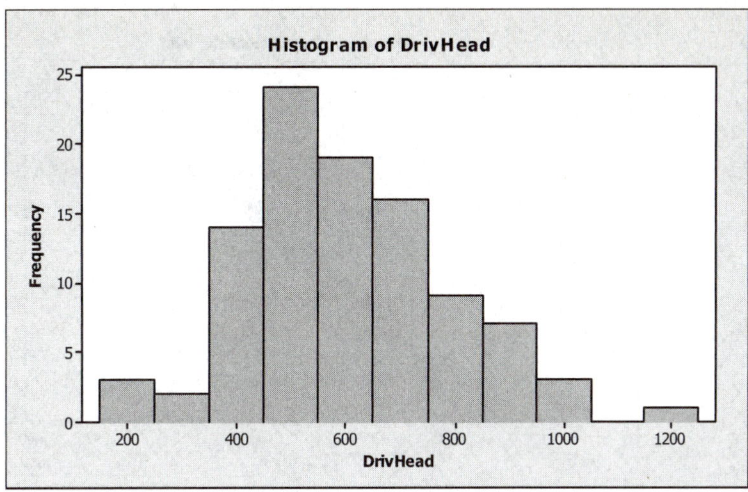

From the histogram, the data appear to be somewhat mound-shaped, but also somewhat skewed to the right. This indicates that the data may be normal.

Next, we look at the intervals $\bar{x} \pm s$, $\bar{x} \pm 2s$, $\bar{x} \pm 3s$. If the proportions of observations falling in each interval are approximately .68, .95, and 1.00, then the data are approximately normal.

Using MINITAB, the summary statistics are:

Descriptive Statistics: DrivHead

```
Variable   N   Mean    StDev  Minimum      Q1  Median      Q3  Maximum
DrivHead  98  603.7    185.4    216.0   475.0   605.0   724.3   1240.0
```

$\bar{x} \pm s \Rightarrow 603.7 \pm 185.4 \Rightarrow (418.3,\ 789.1)$ 68 of the 98 values fall in this interval. The proportion is .69. This is very close to the .68 we would expect if the data were normal.

$\bar{x} \pm 2s \Rightarrow 603.7 \pm 2(185.4) \Rightarrow 603.7 \pm 370.8 \Rightarrow (232.9,\ 974.5)$ 96 of the 98 values fall in this interval. The proportion is .98. This is a fair amount larger than the .95 we would expect if the data were normal.

$\bar{x} \pm 3s \Rightarrow 603.7 \pm 3(185.4) \Rightarrow 603.7 \pm 556.2 \Rightarrow (47.5,\ 1,159.9)$ 97 of the 98 values fall in this interval. The proportion is .99. This is fairly close to the .997 we would expect if the data were normal.

From this method, it appears that the data may be normal.

Next, we look at the ratio of the IQR to s. $IQR = Q_U - Q_L = 724.3 - 475 = 249.3$.

$\dfrac{IQR}{s} = \dfrac{249.3}{185.4} = 1.3$ This is equal to the 1.3 we would expect if the data were normal. This method indicates the data may be normal.

Finally, using MINITAB, the normal probability plot is:

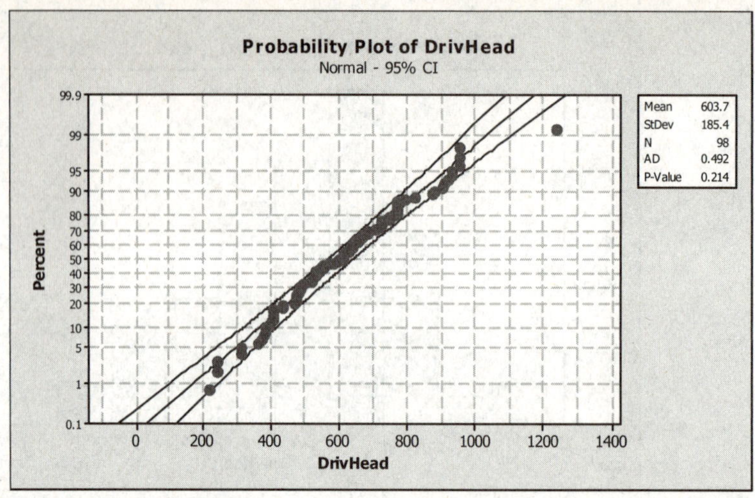

Since the data form a fairly straight line, the data may be normal.

From the 4 different methods, all indications are that the driver's head injury rating data are normal.

4.115   We are given that the mean is 13.2 and the standard deviation is 19.5. We know that the number of minutes per day each student used their laptop for taking notes cannot be negative. Since the standard deviation is larger than the mean, we know that no observations can be more than 1 standard deviation below the mean. Since the Empirical Rule indicates that approximately 16% of the observations will be more than 1 standard deviation below the mean, we know that the distribution of the number of minutes per day each student used their laptop for taking notes is unlikely to be normal.

4.117   When n is large, there often are not Binomial Tables for finding probabilities. If n is large, then finding the probabilities of events by using the formulas can be quite tedious and time consuming. The normal approximation gives very good estimates of the true probabilities.

4.119   a.   In order to approximate the binomial distribution with the normal distribution, the interval $\mu \pm 3\sigma \Rightarrow np \pm 3\sqrt{npq}$ should lie in the range 0 to n.

When $n = 25$ and $p = .4$,
$$np \pm 3\sqrt{npq} \Rightarrow 25(.4) \pm 3\sqrt{25(.4)(1-.4)}$$
$$\Rightarrow 10 \pm 3\sqrt{6} \Rightarrow 10 \pm 7.3485 \Rightarrow (2.6515, 17.3485)$$

Since the interval calculated does lie in the range 0 to 25, we can use the normal approximation.

b. $\mu = np = 25(.4) = 10 \qquad \sigma^2 = npq = 25(.4)(.6) = 6$

c. $P(x \geq 9) = 1 - P(x \leq 8) = 1 - .274 = .726$ (Table II, Appendix A)

d. $P(x \geq 9) \approx P\left(z \geq \dfrac{(9-.5)-10}{\sqrt{6}}\right)$

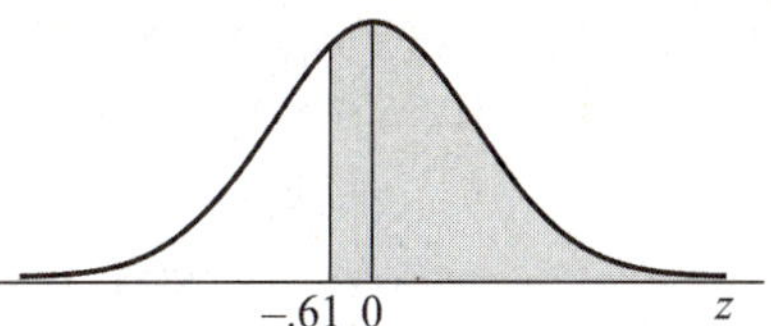

$= P(z \geq -.61) = .5000 + .2291 = .7291$

(Using Table III in Appendix A.)

4.121 a. Using Table II, $P(x \leq 11) = .345$

$\mu = np = 25(.5) = 12.5$, $\sigma = \sqrt{npq} = \sqrt{25(.5)(.5)} = 2.5$

Using the normal approximation,

$P(x \leq 11) \approx P\left(z \leq \dfrac{(11+.5)-12.5}{2.5}\right) = P(z \leq -.40) = .5 - .1554 = .3446$

(from Table III, Appendix A)

b. Using Table II, $P(x \geq 16) = 1 - P(x \leq 15) = 1 - .885 = .115$

Using the normal approximation,

$P(x \geq 16) \approx P\left(z \geq \dfrac{(16-.5)-12.5}{2.5}\right) = P(z \geq 1.2) = .5 - .3849 = .1151$

(from Table III, Appendix A)

c. Using Table II, $P(8 \leq x \leq 16) = P(x \leq 16) - P(x \leq 7) = .946 - .022 = .924$

Using the normal approximation,

$P(8 \leq x \leq 16) \approx P\left(\dfrac{(8-.5)-12.5}{2.5} \leq z \leq \dfrac{(16+.5)-12.5}{2.5}\right)$

$= P(-2.0 \leq z \leq 1.6) = .4772 + .4452 = .9224$

(from Table III, Appendix A)

4.123 x is a binomial random variable with $n = 100$ and $p = .4$.

$\mu \pm 3\sigma \Rightarrow np \pm 3\sqrt{npq} \Rightarrow 100(.4) \pm 3\sqrt{100(.4)(1-.4)}$

$\Rightarrow 40 \pm 3(4.8990) \Rightarrow (25.303, 54.697)$

Since the interval lies in the range 0 to 100, we can use the normal approximation to approximate the probabilities.

a. $P(x \leq 35) \approx P\left(z \leq \dfrac{(35+.5)-40}{4.899}\right)$

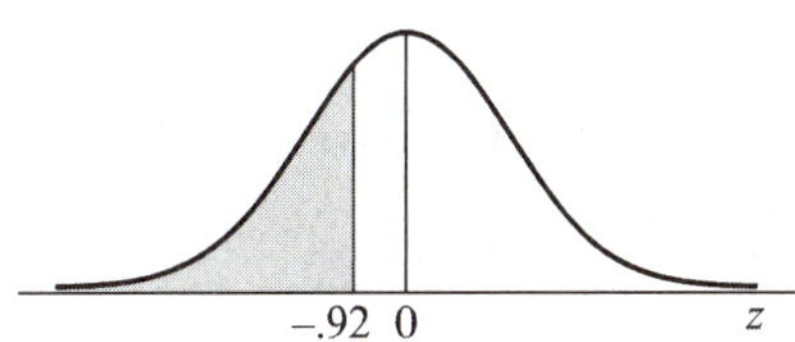

$= P(z \leq -.92)$

$= .5000 - .3212 = .1788$

(Using Table III in Appendix A.)

b. $P(40 \le x \le 50)$

$$\approx P\left(\frac{(40-.5)-40}{4.899} \le z \le \frac{(50+.5)-40}{4.899}\right)$$

$= P(-.10 \le z \le 2.14)$

$= P(-.10 \le z \le 0) + P(0 \le z \le 2.14)$

$= .0398 + .4838 = .5236$

(Using Table III in Appendix A.)

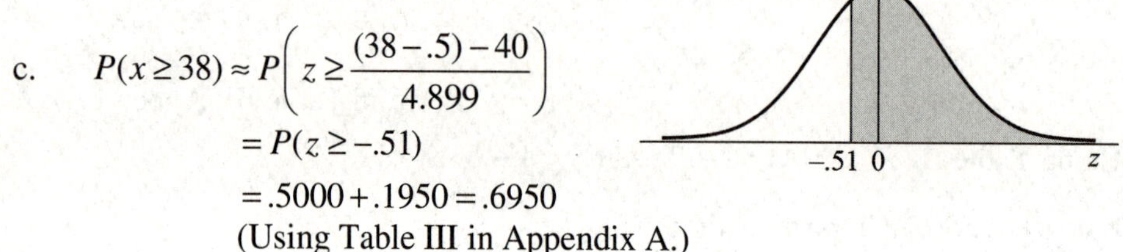

c. $P(x \ge 38) \approx P\left(z \ge \frac{(38-.5)-40}{4.899}\right)$

$= P(z \ge -.51)$

$= .5000 + .1950 = .6950$

(Using Table III in Appendix A.)

4.125   a.   Let x = number of bottles that contain tap water in 65 trials. Then x is a binomial random variable with $n = 65$ and $p = .25$.

$E(x) = \mu = np = 65(.25) = 16.25$

b. $\sigma = \sqrt{npq} = \sqrt{65(.25)(.75)} = \sqrt{12.1875} = 3.4911$

c. $z = \dfrac{x - \mu}{\sigma} = \dfrac{20 - 16.25}{3.4911} = 1.07$

d. $\mu \pm 3\sigma \Rightarrow 16.25 \pm 3(3.4911) \Rightarrow 16.25 \pm 10.4733 \Rightarrow (5.7767,\ 26.7233)$ Since the interval lies in the range 0 to 65, we can use the normal approximation to approximate the binomial probability.

$$P(x \ge 20) = P\left(z \ge \frac{(20-.5)-16.25}{3.4911}\right) = P(z \ge .93) = .5 - .3238 = .1762$$

(Using Table III, Appendix A)

4.127   a. $n = 10,000$ and $p = .15$

$E(x) = \mu = np = 10,000(.15) = 1,500$

$\sigma^2 = npq = 10,000(.15)(.85) = 1,275$

b. $\sigma = \sqrt{1,275} = 35.707$

Using the normal approximation to the binomial,

$$P(x > 1,600) \approx P\left(z > \frac{(1,600+.5)-1,500}{35.707} \right) = P(z > 2.81) = .5 - .4975 = .0025$$

(from Table III, Appendix A)

c. Using the normal approximation to the binomial,

$$P(x > 6,500) \approx P\left(z > \frac{(6,500+.5)-1,500}{35.707} \right) = P(z > 140.04) \approx .5 - .5 = 0$$

(from Table III, Appendix A)

Since this probability is essentially 0, we would not expect to see more than 6,500 deaths in any one year.

4.129 Let x = number of guppies that survive in 300 trials. Then x is a binomial random variable with $n = 300$ and $p = .6$.

$$\mu \pm 3\sigma \Rightarrow np \pm 3\sqrt{npq} \Rightarrow 300(.6) \pm 3\sqrt{300(.6)(1-.6)} \Rightarrow 180 \pm 3\sqrt{72} \Rightarrow 180 \pm 25.456$$
$$\Rightarrow (154.544,\ 205.456)$$

Since the interval lies in the range 0 to 300, we can use the normal approximation to approximate the binomial probability.

$$P(x < 100) = P\left(z < \frac{(100-.5)-180}{\sqrt{72}} \right) = P(z < -9.49) \approx .5 - .5 = 0$$

(Using Table III, Appendix A)

4.131 a. For $n = 100$ and $p = .01$:

$$\mu \pm 3\sigma \Rightarrow np \pm 3\sqrt{npq} \Rightarrow 100(.01) \pm 3\sqrt{100(.01)(.99)}$$
$$\Rightarrow 1 \pm 3(.995) \Rightarrow 1 \pm 2.985 \Rightarrow (-1.985,\ 3.985)$$

Since the interval does not lie in the range 0 to 100, we cannot use the normal approximation to approximate the probabilities.

b. For $n = 100$ and $p = .5$:

$$\mu \pm 3\sigma \Rightarrow np \pm 3\sqrt{npq} \Rightarrow 100(.5) \pm 3\sqrt{100(.5)(.5)}$$
$$\Rightarrow 50 \pm 3(5) \Rightarrow 50 \pm 15 \Rightarrow (35,\ 65)$$

Since the interval lies in the range 0 to 100, we can use the normal approximation to approximate the probabilities.

c. For $n = 100$ and $p = .9$:

$$\mu \pm 3\sigma \Rightarrow np \pm 3\sqrt{npq} \Rightarrow 100(.9) \pm 3\sqrt{100(.9)(.1)}$$
$$\Rightarrow 90 \pm 3(3) \Rightarrow 90 \pm 9 \Rightarrow (81, 99)$$

Since the interval lies in the range 0 to 100, we can use the normal approximation to approximate the probabilities.

4.133   Let x = number of guests who participate in the conservation program. Then x is a binomial random variable with $n = 200$ and $p = .45$.

$$\mu = np = 200(.45) = 90 \quad \text{and} \quad \sigma = \sqrt{npq} = \sqrt{200(.45)(.55)} = \sqrt{49.5} = 7.0356$$

$$\mu \pm 3\sigma \Rightarrow 90 \pm 3(7.0356) \Rightarrow 90 \pm 21.1068 \Rightarrow (68.8932, 111.1068)$$

Since the above is completely contained in the interval 0 to 200, the normal approximation is valid.

$$P(x > 110) = P\left(z > \frac{(110 + .5) - 90}{7.0356}\right) = P(z > 2.91) = .5 - .4982 = .0018$$

Since this probability is so small, it is very unlikely that the director's claim is correct.

4.135   a.   The expected number of passengers detained will be:

$$E(x) = \mu = np = 1,500(.2) = 300$$

b.   For $n = 4,000$, $E(x) = \mu = np = 4,000(.2) = 800$

c. $$P(x > 600) \approx P\left(z > \frac{(600 + .5) - 800}{\sqrt{4000(.2)(.8)}}\right) = P(z > -7.89) = .5 + .5 = 1.0$$

4.137   A population parameter is a numerical descriptive measure of a population. Because it is based on the observations in a population, its value is almost always unknown.

A sample statistic is a numerical descriptive measure of a sample. It is calculated from the observations in the sample.

4.139   a–b. The different samples of $n = 2$ with replacement and their means are:

Possible Samples	$\bar{x}$	Possible Samples	$\bar{x}$
0, 0	0	4, 0	2
0, 2	1	4, 2	3
0, 4	2	4, 4	4
0, 6	3	4, 6	5
2, 0	1	6, 0	3
2, 2	2	6, 2	4
2, 4	3	6, 4	5
2, 6	4	6, 6	6

c.   Since each sample is equally likely, the probability of any 1 being selected is

$$\frac{1}{4}\left(\frac{1}{4}\right) = \frac{1}{16}$$

d. $P(\overline{x} = 0) = \dfrac{1}{16}$

$P(\overline{x} = 1) = \dfrac{1}{16} + \dfrac{1}{16} = \dfrac{2}{16}$

$P(\overline{x} = 2) = \dfrac{1}{16} + \dfrac{1}{16} + \dfrac{1}{16} = \dfrac{3}{16}$

$P(\overline{x} = 3) = \dfrac{1}{16} + \dfrac{1}{16} + \dfrac{1}{16} + \dfrac{1}{16} = \dfrac{4}{16}$

$P(\overline{x} = 4) = \dfrac{1}{16} + \dfrac{1}{16} + \dfrac{1}{16} = \dfrac{3}{16}$

$P(\overline{x} = 5) = \dfrac{1}{16} + \dfrac{1}{16} = \dfrac{2}{16}$

$P(\overline{x} = 6) = \dfrac{1}{16}$

$\overline{x}$	$p(\overline{x})$
0	1/16
1	2/16
2	3/16
3	4/16
4	3/16
5	2/16
6	1/16

e.

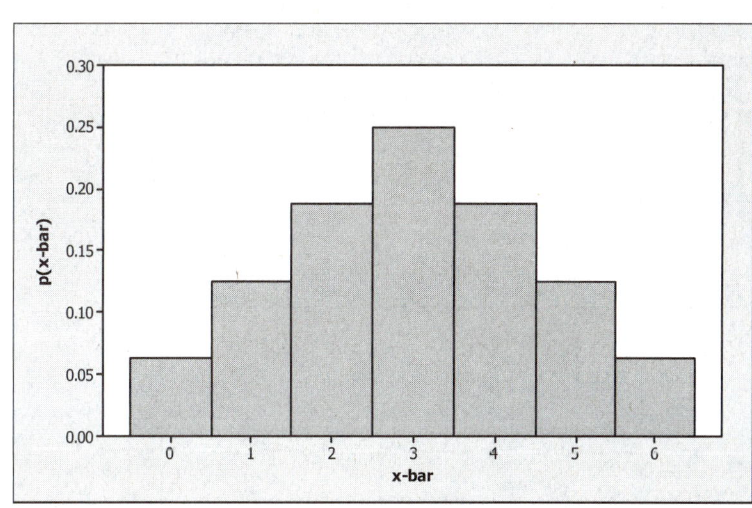

4.141   If the observations are independent of each other, then

$P(1, 1) = p(1)p(1) = .2(.2) = .04$
$P(1, 2) = p(1)p(2) = .2(.3) = .06$
$P(1, 3) = p(1)p(3) = .2(.2) = .04$
        etc.

a.

Possible Samples	$\bar{x}$	$p(\bar{x})$	Possible Samples	$\bar{x}$	$p(\bar{x})$
1, 1	1	.04	3, 4	3.5	.04
1, 2	1.5	.06	3, 5	4	.02
1, 3	2	.04	4, 1	2.5	.04
1, 4	2.5	.04	4, 2	3	.06
1, 5	3	.02	4, 3	3.5	.04
2, 1	1.5	.06	4, 4	4	.04
2, 2	2	.09	4, 5	4.5	.02
2, 3	2.5	.06	5, 1	3	.02
2, 4	3	.06	5, 2	3.5	.03
2, 5	3.5	.03	5, 3	4	.02
3, 1	2	.04	5, 4	4.5	.02
3, 2	2.5	.06	5, 5	5	.01
3, 3	3	.04			

Summing the probabilities, the probability distribution of $\bar{x}$ is:

$\bar{x}$	$p(\bar{x})$
1	.04
1.5	.12
2	.17
2.5	.20
3	.20
3.5	.14
4	.08
4.5	.04
5	.01

b.

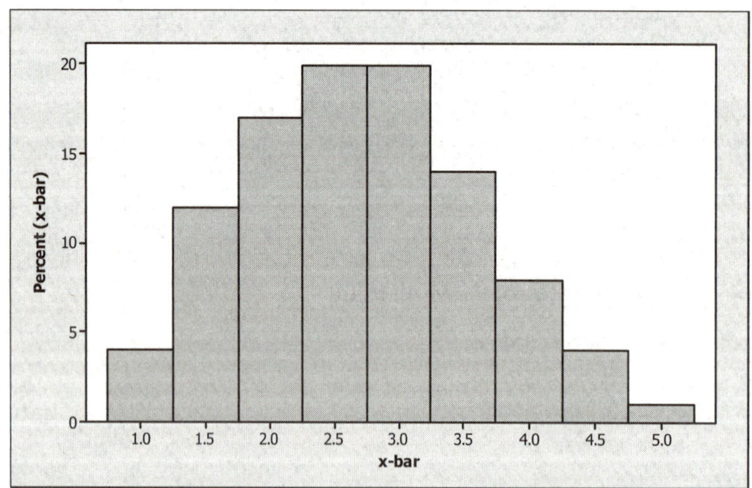

c. $P(\bar{x} \geq 4.5) = .04 + .01 = .05$

d.   No. The probability of observing $\bar{x} = 4.5$ or larger is small (.05).

4.143   a.   For a sample of size $n = 2$, the sample mean and sample median are exactly the same. Thus, the sampling distribution of the sample median is the same as that for the sample mean (see Exercise 4.141**a**).

   b.   The probability histogram for the sample median is identical to that for the sample mean (see Exercise 4.141**b**).

4.145   a.   Answers will vary. MINITAB was used to generate 500 samples of size $n = 25$ observations from a uniform population from 00 to 99. The first 10 samples along with the sample means are shown in the table below:

Sample	Observations	Mean	Var
1	29 57 34 57 82 63 57 94 59 28 29 48 29 12 88 7 93 25 67 69 84 59 28 90 48	53.44	679.76
2	83 27 46 42 18 40 52 5 81 38 58 28 30 16 57 60 45 92 82 19 30 12 42 16 13	41.28	604.79
3	68 15 75 26 46 69 46 81 97 23 38 17 7 47 32 85 98 20 69 8 28 33 7 34 61	45.2	809.75
4	60 38 42 88 85 1 58 5 74 36 4 73 68 99 17 65 13 41 49 97 15 6 95 65 9	48.12	1066.11
5	62 76 37 6 70 14 95 94 12 18 63 65 33 37 44 86 71 67 35 10 17 53 21 27 65	47.12	761.44
6	53 56 92 5 57 13 57 13 7 91 2 76 98 21 93 44 91 74 30 17 99 77 49 86 5	52.24	1178.44
7	38 54 18 62 31 70 72 85 50 57 8 44 48 8 21 50 90 75 2 86 50 1 87 43 49	47.96	741.54
8	54 29 55 23 50 90 66 46 9 30 39 29 33 67 67 93 60 46 28 2 11 63 96 61 46	47.72	631.63
9	32 30 33 92 57 53 40 88 18 60 9 3 7 34 98 62 25 13 67 3 98 96 40 66 25	45.96	976.29
10	7 24 53 4 5 48 10 21 52 23 82 51 55 76 56 44 81 32 57 27 36 84 37 22 49	41.44	582.84

Using MINTAB, the histogram of the sample means is:

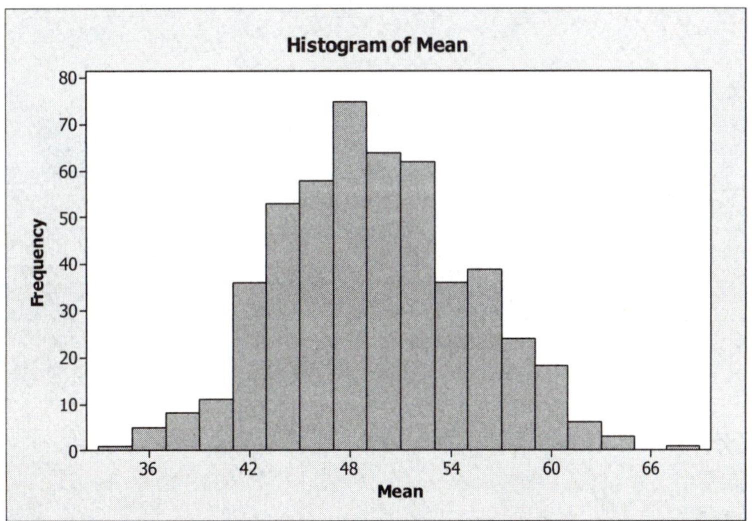

b.   The sample variances for the first 10 samples are also shown in the table in part a. Using MINITAB, the histogram of the 500 sample variances is:

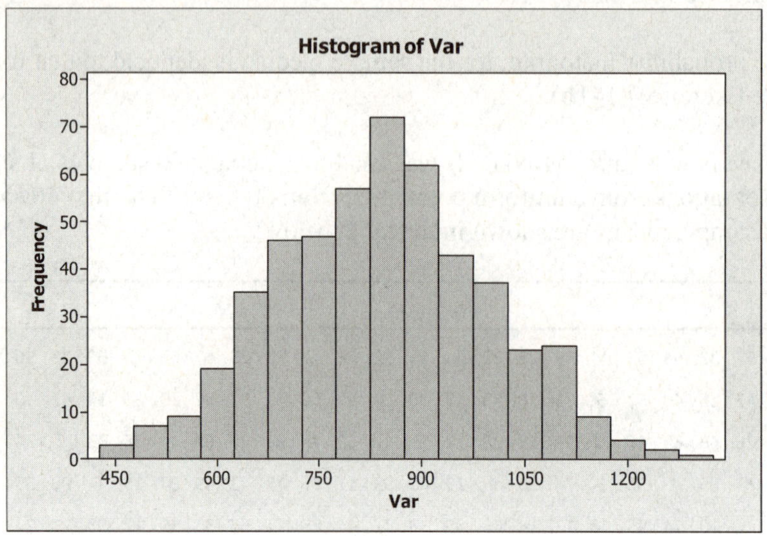

4.147   The mean of the sampling distribution of $\bar{x}$, $\mu_{\bar{x}}$, is the same as the mean of the population from which the sample is selected.

4.149   The Central Limit Theorem states: Consider a random sample of n observations selected from a population (*any* population) with mean μ and standard deviation σ. Then, when n is sufficiently large, the sampling distribution of $\bar{x}$ will be approximately a normal distribution with mean $\mu_{\bar{x}} = \mu$ and standard deviation $\sigma_{\bar{x}} = \sigma / \sqrt{n}$. The larger the sample size, the better will be the normal approximation to the sampling distribution of $\bar{x}$.

4.151   a. $\mu_{\bar{x}} = \mu = 100, \sigma_{\bar{x}} = \dfrac{\sigma}{\sqrt{n}} = \dfrac{\sqrt{100}}{\sqrt{4}} = 5$

b. $\mu_{\bar{x}} = \mu = 100, \sigma_{\bar{x}} = \dfrac{\sigma}{\sqrt{n}} = \dfrac{\sqrt{100}}{\sqrt{25}} = 2$

c. $\mu_{\bar{x}} = \mu = 100, \sigma_{\bar{x}} = \dfrac{\sigma}{\sqrt{n}} = \dfrac{\sqrt{100}}{\sqrt{100}} = 1$

d. $\mu_{\bar{x}} = \mu = 100, \sigma_{\bar{x}} = \dfrac{\sigma}{\sqrt{n}} = \dfrac{\sqrt{100}}{\sqrt{50}} = 1.414$

e. $\mu_{\bar{x}} = \mu = 100, \sigma_{\bar{x}} = \dfrac{\sigma}{\sqrt{n}} = \dfrac{\sqrt{100}}{\sqrt{500}} = .447$

f. $\mu_{\bar{x}} = \mu = 100, \sigma_{\bar{x}} = \dfrac{\sigma}{\sqrt{n}} = \dfrac{\sqrt{100}}{\sqrt{1000}} = .316$

4.153   a. $\mu = \sum xp(x) = 1(.1) + 2(.4) + 3(.4) + 8(.1) = 2.9$

$\sigma^2 = \sum (x - \mu)^2 p(x) = (1 - 2.9)^2(.1) + (2 - 2.9)^2(.4) + (3 - 2.9)^2(.4) + (8 - 2.9)^2(.1)$
$= .361 + .324 + .004 + 2.601 = 3.29$

$\sigma = \sqrt{3.29} = 1.814$

b.   The possible samples, values of $\bar{x}$, and associated probabilities are listed:

Possible Samples	$\bar{x}$	$p(\bar{x})$	Possible Samples	$\bar{x}$	$p(\bar{x})$
1, 1	1	.01	3, 1	2	.04
1, 2	1.5	.04	3, 2	2.5	.16
1, 3	2	.04	3, 3	3	.16
1, 8	4.5	.01	3, 8	5.5	.04
2, 1	1.5	.04	8, 1	4.5	.01
2, 2	2	.16	8, 2	5	.04
2, 3	2.5	.16	8, 3	5.5	.04
2, 8	5	.04	8, 8	8	.01

$P(1, 1) = p(1)p(1) = .1(.1) = .01$
$P(1, 3) = p(1)p(2) = .1(.4) = .04$
$P(1, 3) = p(1)p(3) = .1(.4) = .04$
etc.

The sampling distribution of $\bar{x}$ is:

$\bar{x}$	$p(\bar{x})$
1	.01
1.5	.08
2	.24
2.5	.32
3	.16
4.5	.02
5	.08
5.5	.08
8	.01
	1.00

c. $\mu_{\bar{x}} = E(\bar{x}) = \sum \bar{x}p(\bar{x}) = 1(.01) + 1.5(.08) + 2(.24) + 2.5(.32) + 3(.16) + 4.5(.02)$
$5(.08) + 5.5(.08) + 8(.01) = 2.9 = \mu$

$$\sigma_{\bar{x}}^2 = \sum (\bar{x} - \mu_{\bar{x}})^2\, p(\bar{x}) = (1 - 2.9)^2(.01) + (1.5 - 2.9)^2(.08) + (2 - 2.9)^2(.24)$$
$$+ (2.5 - 2.9)^2(.32) + (3 - 2.9)^2(.16) + (4.5 - 2.9)^2(.02)$$
$$+ (5 - 2.9)^2(.08) + (5.5 - 2.9)^2(.08) + (8 - 2.9)^2(.01)$$
$$= .0361 + .1568 + .1944 + .0512 + .0016 + .0512 + .3528$$
$$+ .5408 + .2601 = 1.645$$

$$\sigma_{\bar{x}} = \sqrt{1.645} = 1.283$$
$$\sigma_{\bar{x}} = \frac{\sigma}{\sqrt{n}} = \frac{1.814}{\sqrt{2}} = 1.283$$

4.155   a. $\mu_{\bar{x}} = \mu = 30, \quad \sigma_{\bar{x}} = \dfrac{\sigma}{\sqrt{n}} = \dfrac{16}{\sqrt{100}} = 1.6$

b.   By the Central Limit Theorem, the sampling distribution of $\bar{x}$ is approximately normal.

c. $P(\bar{x} \geq 28) = P\left(z \geq \dfrac{28 - 30}{1.6} \right) = P(z \geq -1.25) = .5 + .3944 = .8944$

d. $P(22.1 \leq \bar{x} \leq 26.8) = P\left(\dfrac{22.1 - 30}{1.6} \leq z \leq \dfrac{26.8 - 30}{1.6} \right) = P(-4.94 \leq z \leq -2)$
$$= .5 - .4772 = .0228$$

e. $P(\bar{x} \leq 28.2) = P\left(z \leq \dfrac{28.2 - 30}{1.6} \right) = P(z \leq -1.125) = .5 - \left(\dfrac{.3868 + .3708}{2} \right)$
$$= .5 - .3697 = .1303$$

f. $P(\bar{x} \geq 27.0) = P\left(z \geq \dfrac{27.0 - 30}{1.6} \right) = P(z \geq -1.88) = .5 + .4699 = .9699$

4.157   a.   By the Central Limit Theorem, the sampling distribution of $\bar{x}$ is approximately normal. The mean of the $\bar{x}$ distribution is $\mu_{\bar{x}} = \mu = 320$ and the standard deviation of the $\bar{x}$ distribution is $\sigma_{\bar{x}} = \dfrac{\sigma}{\sqrt{n}} = \dfrac{100}{\sqrt{100}} = \dfrac{100}{10} = 10$.

b. $P(300 < \bar{x} < 310) = P\left(\dfrac{300 - 320}{10} < z < \dfrac{310 - 320}{10} \right)$
$$= P(-2 < z < -1) = .4772 - .3413 = .1359$$
(Using Table III, Appendix A)

c. $P(\bar{x} > 360) = P\left(z > \dfrac{360 - 320}{10} \right) = P(z > 4) \approx .5 - .5 = 0$ (Using Table III, Appendix A)

4.159   a. $\mu_{\bar{x}}$ is the mean of the sampling distribution of $\bar{x}$. $\mu_{\bar{x}} = \mu = 79$.

b. $\sigma_{\bar{x}}$ is the standard deviation of the sampling distribution of $\bar{x}$. $\sigma_{\bar{x}} = \dfrac{\sigma}{\sqrt{n}} = \dfrac{23}{\sqrt{100}} = 2.3$

c. By the Central Limit Theorem, the sampling distribution of $\bar{x}$ is approximately normal.

d. $z = \dfrac{\bar{x} - \mu_{\bar{x}}}{\sigma_{\bar{x}}} = \dfrac{80 - 79}{2.3} = 0.43$

e. $P(\bar{x} > 80) = P(z > 0.43) = .5 - .1664 = .3336$ (Using Table III, Appendix A.)

4.161   Let $\bar{x}$ = sample mean amount of miraculin produced. By the Central Limit Theorem, the sampling distribution of $\bar{x}$ is approximately normal with $\mu_{\bar{x}} = \mu = 105.3$ and

$\sigma_{\bar{x}} = \dfrac{\sigma}{\sqrt{n}} = \dfrac{8.0}{\sqrt{64}} = 1$.

$P(\bar{x} < 103) = P\left(z < \dfrac{103 - 105.3}{1}\right) = P(z < -2.30) = .5 - .4893 = .0107$

(using Table III, Appendix A)

Since this probability is so small, we would not expect to observe a value of $\bar{x}$ less than 103 micro-grams per gram of fresh weight.

4.163   a. $E(\bar{x}) = \mu_{\bar{x}} = \mu = .10$ and $V(\bar{x}) = \sigma_{\bar{x}}^2 = \dfrac{\sigma^2}{n} = \dfrac{.10^2}{50} = .0002$

b. By the Central Limit Theorem, the sampling distribution of $\bar{x}$ is approximately normal if the sample size is sufficiently large. In this case, $n = 50$ is considered large.

c. $P(\bar{x} > .13) = P\left(z > \dfrac{.13 - .10}{\sqrt{.0002}}\right) = P(z > 2.12) = .5 - .4830 = .0170$

4.165   a. By the Central Limit Theorem, the sampling distribution of $\bar{x}$ is approximately normal if the sample size is sufficiently large. In this case, $n = 50$ is considered large.

$\mu_{\bar{x}} = \mu = .53$, $\sigma_{\bar{x}} = \dfrac{\sigma}{\sqrt{n}} = \dfrac{.193}{\sqrt{50}} = .0273$

b. $P(\bar{x} > .58) = P\left(z > \dfrac{.58 - .53}{.0273}\right) = P(z > 1.83) = .5 - .4664 = .0336$

(Using Table III, Appendix A)

c. $P(\bar{x} > .59 \mid \mu = .58) = P\left(z > \dfrac{.59 - .58}{.0273}\right) = P(z > .37) = .5 - .1443 = .3557$

(Using Table III, Appendix A) The above is the probability *after tensioning and loading*.

$$P(\bar{x} > .59 \mid \mu = .53) = P\left(z > \frac{.59 - .53}{.0273}\right) = P(z > 2.20) = .5 - .4772 = .0228$$

(Using Table III, Appendix A) The above is the probability *before tensioning*.

Since the probability *after tensioning* is not small, this would not be an unusual event. Since the probability before tensioning is very small, this would be an unusual event. The sample measurements were probably obtained after tensioning and loading.

4.167   a.   By the Central Limit Theorem, the sampling distribution of $\bar{x}$ is approximately normal if the sample size is sufficiently large. In this case, $n = 326$ is considered large.

$$\mu_{\bar{x}} = \mu = 6, \quad \sigma_{\bar{x}} = \frac{\sigma}{\sqrt{n}} = \frac{10}{\sqrt{326}} = .5538$$

$$P(\bar{x} > 7.5) = P\left(z > \frac{7.5 - 6}{.5538}\right) = P(z > 2.71) = .5 - .4966 = .0034$$

b. $$P(\bar{x} > 300) = P\left(z > \frac{300 - 6}{.5538}\right) = P(z > 530.88) \approx .5 - .5 = 0$$

Since it is essentially impossible to get a sample mean of 300 if the population mean was 6, we would conclude that the mean PFOA in people who live near DuPont's Teflon-making facility is not 6 ppb but something much larger.

4.169   **Handrubbing**

$$\mu_{\bar{x}} = \mu = 35, \quad \sigma_{\bar{x}} = \frac{\sigma}{\sqrt{n}} = \frac{59}{\sqrt{50}} = 8.344$$

$$P(\bar{x} < 30) = P\left(z < \frac{30 - 35}{8.344}\right) = P(z < -.60) = .5 - .2257 = .2743$$

(Using Table III, Appendix A.)

Handwashing

$$\mu_{\bar{x}} = \mu = 69, \quad \sigma_{\bar{x}} = \frac{\sigma}{\sqrt{n}} = \frac{106}{\sqrt{50}} = 14.991$$

$$P(\bar{x} < 30) = P\left(z < \frac{30 - 69}{14.991}\right) = P(z < -2.60) = .5 - .4953 = .0047$$

(Using Table III, Appendix A.)

This sample of workers probably came from the handrubbing group. The probability of seeing a mean of 30 or smaller for this group is .2743, while the probability of observing a mean of 30 or less for the handwashers is .0047. This would be an extremely rare event if the workers were handwashers because the probability is so small. It would not be a very unusual event for the handrubbers because the probability is not that small.

4.171   a.   This experiment consists of 100 trials. Each trial results in one of two outcomes: chip is defective or not defective. If the number of chips produced in one hour is much larger than 100, then we can assume the probability of a defective chip is the same on each trial and that the trials are independent. Thus, x is a binomial. If, however, the number of chips produced in an hour is not much larger than 100, the trials would not be independent. Then x would not be a binomial random variable.

   b.   This experiment consists of two trials. Each trial results in one of two outcomes: applicant qualified or not qualified. However, the trials are not independent. The probability of selecting a qualified applicant on the first trial is 3 out of 5. The probability of selecting a qualified applicant on the second trial depends on what happened on the first trial. Thus, x is not a binomial random variable. It is a hypergeometric random variable.

   c.   The number of trials is not a specified number in this experiment, thus x is not a binomial random variable. In this experiment, x is counting the number of calls received in a specific time frame. Thus, x is a Poisson random variable.

   d.   The number of trials in this experiment is 1000. Each trial can result in one of two outcomes: favor state income tax or not favor state income tax. Since 1000 is small compared to the number of registered voters in Florida, the probability of selecting a voter in favor of the state income tax is the same from trial to trial, and the trials are independent of each other. Thus, x is a binomial random variable.

4.173   The statement "The sample mean, $\bar{x}$, will always be equal to μ_x" is false. The sample mean is used to estimate the population mean, but the probability that it is equal to the population mean is 0.

4.175   From Table II, Appendix A:

   a. $P(x=14) = P(x \leq 14) - P(x \leq 13) = .584 - .392 = .192$

   b. $P(x \leq 12) = .228$

   c. $P(x > 12) = 1 - P(x \leq 12) = 1 - .228 = .772$

   d. $P(9 \leq x \leq 18) = P(x \leq 18) - P(x \leq 8) = .992 - .005 = .987$

   e. $P(8 < x < 18) = P(x \leq 17) - P(x \leq 8) = .965 - .005 = .960$

   f. $\mu = np = 20(.7) = 1.4$

   $\sigma^2 = npq = 20(.7)(.3) = 4.2$, $\sigma = \sqrt{4.2} = 2.049$

   g. $\mu \pm 2\sigma \Rightarrow 14 \pm 2(2.049) \Rightarrow 14 \pm 4.098 \Rightarrow (9.902, 18.098)$

$$P(9.902 < x < 18.098) = P(10 \leq x \leq 18) = P(x \leq 18) - P(x \leq 9)$$
$$= .992 - .017 = .975$$

4.177   a. $P(x \le 75) = P\left(z \le \dfrac{75-70}{10}\right)$

$= P(z \le .50)$

$= .5 + .1915 = .6915$

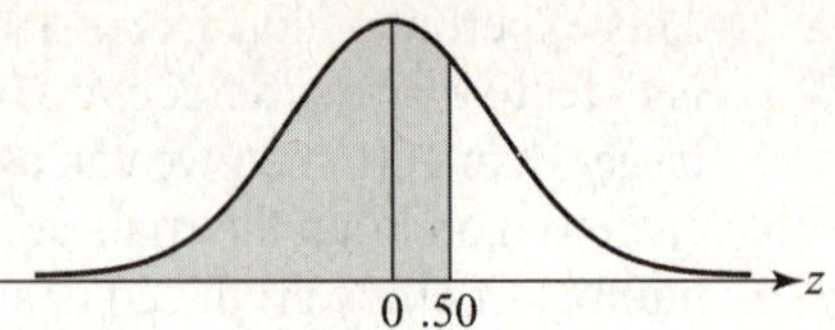

b. $P(x \ge 90) = P\left(z \ge \dfrac{90-70}{10}\right)$

$= P(z \ge 2.00)$

$= .5 - .4772 = .0228$

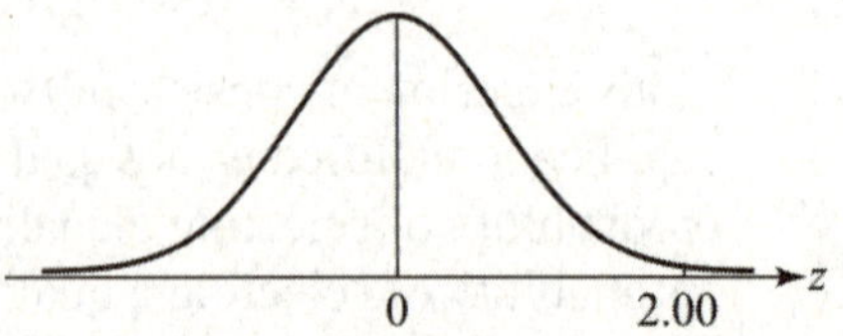

c. $P(60 \le x \le 75) = P\left(\dfrac{60-70}{10} \le z \le \dfrac{75-70}{10}\right)$

$= P(-1.00 \le z \le .50)$

$= .3413 + .1915 = .5328$

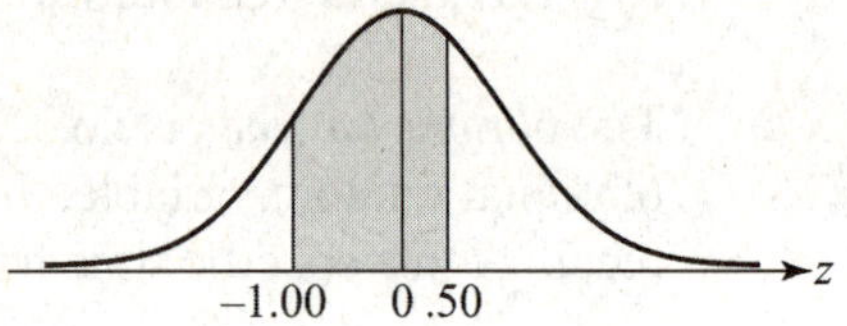

d.   This is the probability of the complement of the event in part **a**. Therefore:

$P(x > 75) = 1 - P(x \le 75) = 1 - .6915 = .3085$

e. $P(x = 75) = 0$ (True for any continuous random variable.)

f. $P(x \le 95) = P\left(z \le \dfrac{95-70}{10}\right)$

$= P(z \le 2.50)$

$= .5 + .4938 = .9938$

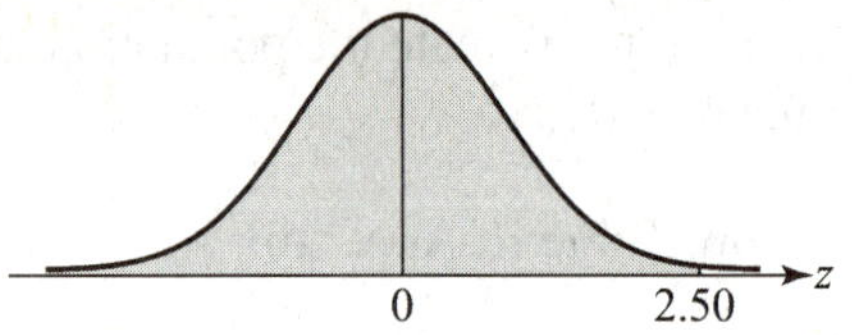

4.179 $\mu = np = 100(.5) = 50$, $\sigma = \sqrt{npq} = \sqrt{100(.5)(.5)} = 5$

a. $P(x \le 48) = P\left(z \le \dfrac{(48 + .5) - 50}{5}\right)$

$= P(z \le -.30)$

$= .5 - .1179 = .3821$

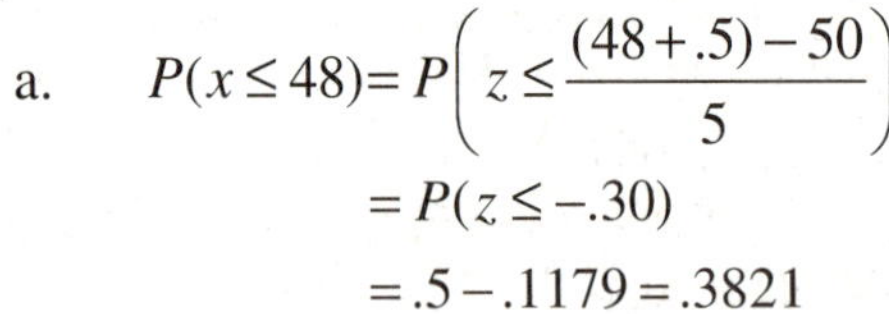

b. $P(50 \le x \le 65)$

$= P\left(\dfrac{(50 - .5) - 50}{5} \le z \le \dfrac{(65 + .5) - 50}{5}\right)$

$= P(-.10 \le z \le 3.10)$

$= .0398 + .5 = .5398$

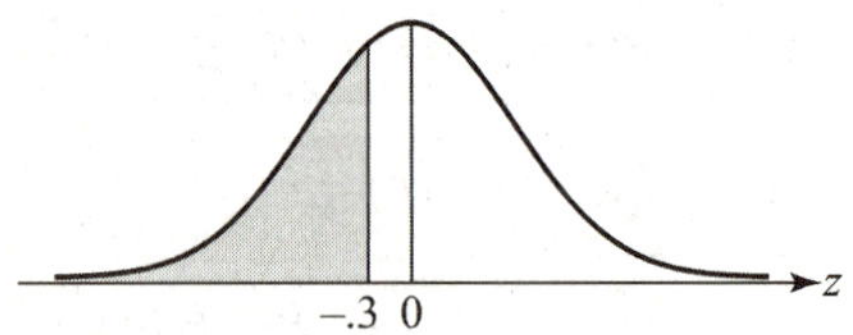

c. $P(x \geq 70) = P\left(z \geq \dfrac{(70-.5)-50}{5}\right)$

$= P(z \geq 3.90)$

$= .5 - .5 = 0$

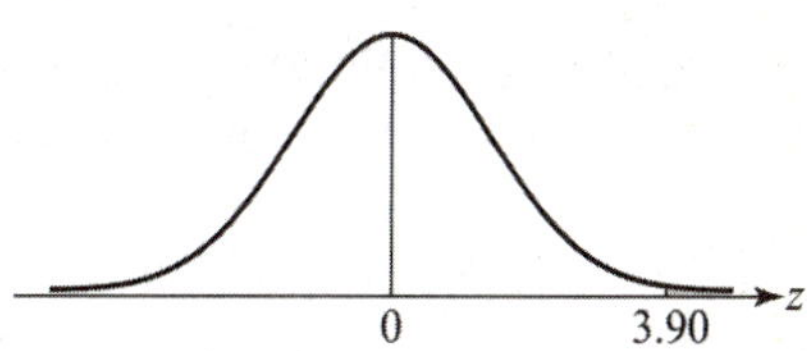

d. $P(55 \leq x \leq 58)$

$= P\left(\dfrac{(55-.5)-50}{5} \leq z \leq \dfrac{(58+.5)-50}{5}\right)$

$= P(.90 \leq z \leq 1.70)$

$= P(0 \leq z \leq 1.70) - P(0 \leq z \leq .90)$

$= .4554 - .3159 = .1395$

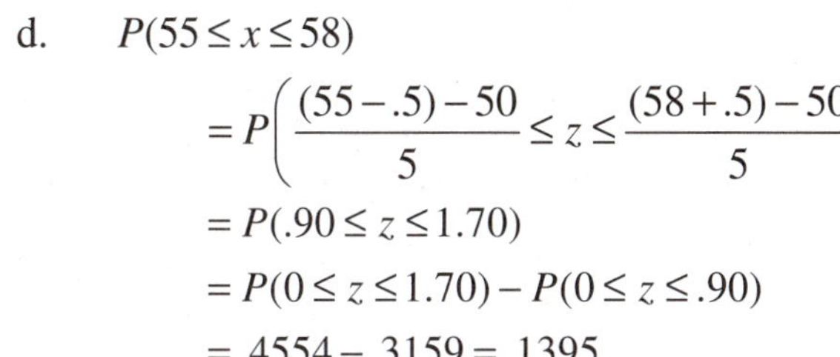

e. $P(x = 62)$

$= P\left(\dfrac{(62-.5)-50}{5} \leq z \leq \dfrac{(62+.5)-50}{5}\right)$

$= P(2.30 \leq z \leq 2.50)$

$= P(0 \leq z \leq 2.50) - P(0 \leq z \leq 2.30)$

$= .4938 - .4893 = .0045$

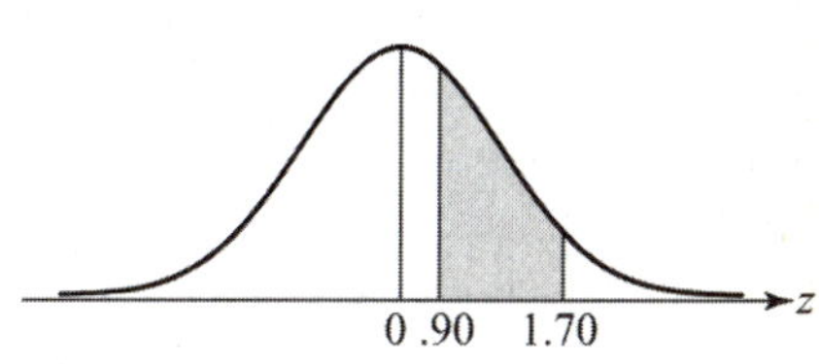

f. $P(x \leq 49 \text{ or } x \geq 72)$

$= P\left(z \leq \dfrac{(49+.5)-50}{5}\right) + P\left(z \geq \dfrac{(72-.5)-50}{5}\right)$

$= P(z \leq -.10) + P(z \geq 4.30)$

$= (.5 - .0398) + (.5 - .5) = .4602$

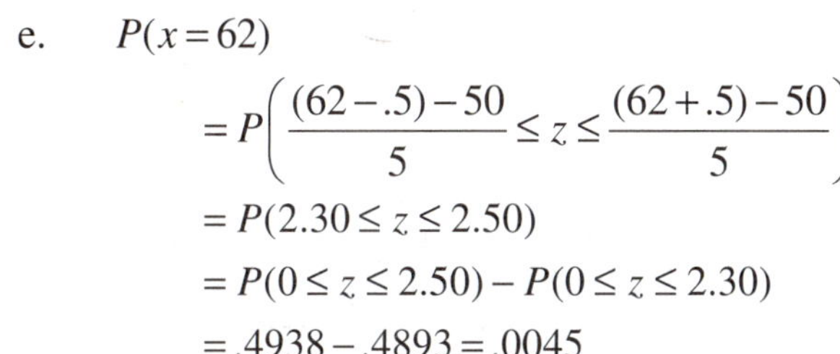

4.181   a. In order for this to be a valid probability distribution, $\sum p(x) = 1$ and $0 \leq p(x) \leq 1$ for all values of x.

In this distribution, $0 \leq p(x) \leq 1$ for all values of x and

$\sum p(x) = .051 + .099 + .093 + .635 + .122 = 1.000$. Thus, this is a valid probability distribution.

b. $P(x = 1) = .051$

c. $P(x \geq 4) = P(x = 4) + P(x = 5) = .635 + .122 = .757$

d. $P(x = 2 \text{ or } x = 3) = P(x = 2) + P(x = 3) = .099 + .093 = .192$

e. $E(x) = \mu = \sum xp(x) = 1(.051) + 2(.099) + 3(.093) + 4(.635) + 5(.122) = 3.678$

f. The average rating of all books rated is 3.678.

4.183   a.   For this experiment, there are $n = 200$ smokers (n identical trials). For each smoker, there are 2 possible outcomes: $S =$ smoker enters treatment program and $F =$ smoker does not enter treatment program. The probability of S (smoker enters treatment program) is the same from trial to trial. This probability is $P(S) = p = .05$. $P(F) = 1 - P(S) = 1 - .05 = .95$. The trials are independent and $x =$ number of smokers entering treatment program in 200 trials. Thus, x is a binomial random variable.

   b. $p = P(S) = .05$. Of all the smokers, only .05 or 5% enter treatment programs.

   c. $E(x) = np = 200(.05) = 10$. For all samples of 200 smokers, the average number of smokers who enter a treatment program will be 10.

4.185   Let $x =$ score for first time students on mathematics achievement test. Then x is a normal random variable with
$\mu = 77$ and $\sigma = 7.3$.

$$P(x \geq 70) = P\left(z \geq \frac{70 - 77}{7.3}\right)$$
$$= P(z \geq -.96)$$
$$= .3315 + .5 = .8315$$

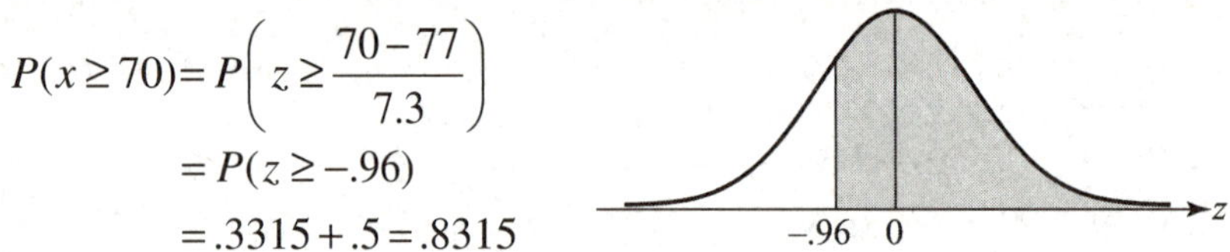

Thus, 83.15% of students will pass the test the first time.

4.187   a.   Let $x =$ alkalinity level of water specimens collected from the Han River.

   Using Table III, Appendix A,
$$P(x > 45) = P\left(z > \frac{45 - 50}{3.2}\right) = P(z > -1.56) = .5 + .4406 = .9406.$$

   b.   Using Table III, Appendix A,
$$P(x < 55) = P\left(z < \frac{55 - 50}{3.2}\right) = P(z < 1.56) = .5 + .4406 = .9406.$$

   c.   Using Table III, Appendix A,
$$P(51 < x < 52) = P\left(\frac{51 - 50}{3.2} < z < \frac{52 - 50}{3.2}\right) = P(.31 < z < .63) = .2357 - .1217 = .1140.$$

4.189   a. $\mu_{\bar{x}}$ is the mean of the sampling distribution of $\bar{x}$. $\mu_{\bar{x}} = \mu = 106$.

   b. $\sigma_{\bar{x}}$ is the standard deviation of the sampling distribution of $\bar{x}$. $\sigma_{\bar{x}} = \dfrac{\sigma}{\sqrt{n}} = \dfrac{16.4}{\sqrt{36}} = 2.73$

   c.   By the Central Limit Theorem, the sampling distribution of $\bar{x}$ is approximately normal.

   d. $z = \dfrac{\bar{x} - \mu_{\bar{x}}}{\sigma_{\bar{x}}} = \dfrac{100 - 106}{2.73} = -2.20$

e. $P(\bar{x} < 100) = P(z < -2.20) = .5 - .4864 = .0136$ (Using Table III, Appendix A.)

4.191 Let x = number of parents who yell at their child before, during, or after a meet in 20 trials. Then x has a binomial distribution with $n = 20$ and $p = .05$.

$P(x \geq 1) = 1 - P(x = 0) = 1 - .358 = .642$ (Using Table II, Appendix A)

4.193 a. Using Table III, Appendix A, with $\mu = 9.06$ and $\sigma = 2.11$,

$$P(x < 6) = \left(z < \frac{6 - 9.06}{2.11}\right) = P(z < -1.45) = .5 - P(-1.45 \leq z \leq 0)$$
$$= .5 - .4265 = .0735$$

b. $$P(8 < x < 10) = P\left(\frac{8 - 9.06}{2.11} < z < \frac{10 - 9.06}{2.11}\right) = P(-.50 < z < .45)$$
$$= P(-.50 < z < 0) + P(0 < z < .45) = .1915 + .1736 = .3651$$

c. $P(x < x_0) = .2000$. If $P(x < x_0) = .2000$, then $P\left(z < \frac{x_0 - 9.06}{2.11}\right) = P(z < z_0) = .2000$

If $P(z < z_0) = .2000$, then $P(z_0 < z < 0) = .3000$. Looking up .3000 in Table III, the z-score is .84. Since $z_0 < 0$, $z_0 = -.84$. Now, we must convert z_0 back to an x score.

$$z_0 = \frac{x_0 - 9.06}{2.11} \Rightarrow -.84 = \frac{x_0 - 9.06}{2.11} \Rightarrow x_0 = 2.11(-.84) + 9.06 = 7.29$$

4.195 Let x = number of beech trees damaged by fungi in 200 trees. Then x is a binomial random variable with $n = 200$ and $p = .25$.

$$\mu = np = 200(.25) = 50 \qquad \sigma = \sqrt{npq} = \sqrt{200(.25)(.75)} = \sqrt{37.5} = 6.124$$

$$P(x > 100) \approx P\left(z > \frac{(100 + .5) - 50}{6.124}\right) = P(z > 8.25) \approx 0$$

Since this probability is so small, it is almost impossible for more than half of the trees to have fungi damage.

4.197 For this problem, $\mu_{\bar{x}} = \mu = 4.59$ and $\sigma_{\bar{x}} = \dfrac{\sigma}{\sqrt{n}} = \dfrac{2.95}{\sqrt{50}} = .4172$

a. $P(\bar{x} \geq 6) = P\left(z \geq \dfrac{6 - 4.59}{.4172}\right) = P(z \geq 3.38) \approx .5 - .5 = 0$ (using Table III, Appendix A)

Since the probability of observing a sample mean CAHS score of 6 or higher is so small (p is essentially 0), we would not expect to see a sample mean of 6 or higher.

b. If a psychologist actually observed $x = 6.2$, we would conclude that he has seen an extremely unusual event, or the population from which the sample was drawn probably has a mean greater than 4.59.

4.199   a.   We must assume that the distribution of surface roughness scores is normally distributed because the sample size is only n = 20. Thus, the Central Limit Theorem will not be valid. If the distribution being sampled from is normal, then the sampling distribution of $\bar{x}$ is normal.

$$\mu_{\bar{x}} = \mu = 1.8, \quad \sigma_{\bar{x}} = \frac{\sigma}{\sqrt{n}} = \frac{.5}{\sqrt{20}} = .1118$$

$$P(\bar{x} > 1.85) = P\left(z > \frac{1.85 - 1.8}{.1118} \right) = P(z > .45) = .5 - .1736 = .3264$$

  b. $$\bar{x} = \frac{\sum x}{n} = \frac{37.62}{20} = 1.881$$

  c. $$P(\bar{x} > 1.881) = P\left(z > \frac{1.881 - 1.8}{.1118} \right) = P(z > .72) = .5 - .2642 = .2358$$

Since the probability of getting a sample mean of 1.881 or anything more unusual is not small (probability is .2358), the assumptions made in part **a** appear to be valid.

4.201   We know that $\mu = np = n(.4) = .4n$ and $\sigma = \sqrt{npq} = \sqrt{n(.4)(.6)} = \sqrt{.24n}$. Since $p = .4$ is close to .5, the distribution of x will be close to symmetrical. Therefore, we can use the Empirical Rule. We know that approximately 95% of all observations will fall within 2 standard deviation of the mean. Thus, approximately 97.5% of the observations will fall above two standard deviations below the mean. We will set 100 equal to the mean minus 2 standard deviations and solve for n:

$$.4n - 2\sqrt{.24n} = 100 \implies n - 5\sqrt{.24n} = 250 \implies n - 250 = 5\sqrt{.24n}$$

$$\implies n^2 - 500n + 62{,}500 = 6n \implies n^2 - 506n + 62{,}500 = 0$$

Now, using Pythagorean Theorem to solve for n, we get:

$$n = \frac{506 \pm \sqrt{506^2 - 4(62500)}}{2} = \frac{506 \pm 77.69}{2} \implies n = 214 \text{ or } n = 292$$

Since the mean is *np,* and for this exercise the mean must be greater than 100, we know that n cannot be 214. Thus, n has to be 292.

4.203   a.   Class I (very bright) consists of those with IQs above the 95th percentile. Thus, there would be .05 or 5% of the people in Class I.

Class II (bright) consists of those with IQs between the 75th and 95th percentile. Thus, there would be $.95 - .75 = .20$ or 20% of the people in Class II.

Class III (normal) consists of those with IQs between the 25th and 75th percentile. Thus, there would be $.75 - .25 = .50$ or 50% of the people in Class III.

Class IV (dull) consists of those with IQs between the 5th and 25th percentile. Thus, there would be $.25 - .05 = .20$ or 20% of the people in Class IV.

Class V (very dull) consists of those with IQs below the 5th percentile. Thus, there would be .05 or 5% of the people in Class V.

b.   Suppose we pick a pair of percentiles and compute the corresponding z-scores. Suppose we compute the z-scores for the 50th and 55th percentiles.

The z-score corresponding to the 50th percentile is $z = 0$. The z-score corresponding to the 55th percentile is: $P(z < z_0) = .55 \Rightarrow z_0 = .13$

Now, suppose we pick another pair of percentiles that are also 5 points apart. Suppose we compute the z-scores for the 94th and 99th percentiles.

The z-score corresponding to the 94th percentile is: $P(z < z_0) = .94 \Rightarrow z_0 = 1.555$

The z-score corresponding to the 99th percentile is: $P(z < z_0) = .99 \Rightarrow z_0 = 2.33$

The z-scores corresponding to the 50th and 55th percentiles are 0 and .13. The difference is $.13 - 0 = .13$. The z-scores corresponding to the 94th and 99th percentiles are 1.555 and 2.33. The difference is $2.33 - 1.555 = .775$. Even though the differences in the percentiles are the same, the differences in the z-scores are much different. The information in the z-scores is more informative.

c.   If the distribution of the IQs is skewed to the right, then the tail to the right is much longer than the tail to the left. If the distribution of the IQs is skewed to the left, then the tail to the left is much longer than the tail to the right. However, the proportions in the 5 cognitive classes would not differ, regardless of the shape of the distribution.

Class I (very bright) consists of those with IQs above the 95th percentile. Thus, there would be .05 or 5% of the people in Class I.

Class II (bright) consists of those with IQs between the 75th and 95th percentile. Thus, there would be $.95 - .75 = .20$ or 20% of the people in Class II.

Class III (normal) consists of those with IQs between the 25th and 75th percentile. Thus, there would be $.75 - .25 = .50$ or 50% of the people in Class III.

Class IV (dull) consists of those with IQs between the 5th and 25th percentile. Thus, there would be $.25 - .05 = .20$ or 20% of the people in Class IV.

Class V (very dull) consists of those with IQs below the 5th percentile. Thus, there would be .05 or 5% of the people in Class V.

Inferences Based on a Single Sample
Estimation with Confidence Intervals

Chapter **5**

5.1   The unknown population parameter (e.g., mean or proportion) that we are interested in estimating is called the target parameter.

5.3   An interval estimator estimates μ with a range of values, while a point estimator estimates μ with a single point.

5.5   Yes. As long as the sample size is sufficiently large, the Central Limit Theorem says the distribution of $\bar{x}$ is approximately normal regardless of the original distribution.

5.7   a.   For $\alpha = .10$, $\alpha / 2 = .10 / 2 = .05$. $z_{\alpha/2} = z_{.05}$ is the z-score with .05 of the area to the right of it. The area between 0 and $z_{.05}$ is $.5 - .05 = .4500$. Using Table III, Appendix A, $z_{.05} = 1.645$.

       b.   For $\alpha = .01$, $\alpha / 2 = .01 / 2 = .005$. $z_{\alpha/2} = z_{.005}$ is the z-score with .005 of the area to the right of it. The area between 0 and $z_{.005}$ is $.5 - .005 = .4950$. Using Table III, Appendix A, $z_{.005} = 2.58$.

       c.   For $\alpha = .05$, $\alpha / 2 = .05 / 2 = .025$. $z_{\alpha/2} = z_{.025}$ is the z-score with .025 of the area to the right of it. The area between 0 and $z_{.025}$ is $.5 - .025 = .4750$. Using Table III, Appendix A, $z_{.025} = 1.96$.

       d.   For $\alpha = .20$, $\alpha / 2 = .20 / 2 = .10$. $z_{\alpha/2} = z_{.10}$ is the z-score with .10 of the area to the right of it. The area between 0 and $z_{.10}$ is $.5 - .10 = .4000$. Using Table III, Appendix A, $z_{.10} = 1.28$.

5.9   a.   For confidence coefficient .95, $\alpha = .05$ and $\alpha / 2 = .05 / 2 = .025$. From Table III, Appendix A, $z_{.025} = 1.96$. The confidence interval is:

$$\bar{x} \pm z_{.025}\frac{\sigma}{\sqrt{n}} \Rightarrow 28 \pm 1.96\frac{20}{\sqrt{75}} \Rightarrow 28 \pm 4.53 \Rightarrow (23.47, 32.53)$$

       b. $$\bar{x} \pm z_{.025}\frac{\sigma}{\sqrt{n}} \Rightarrow 102 \pm 1.96\frac{20}{\sqrt{200}} \Rightarrow 102 \pm 2.77 \Rightarrow (99.23, 104.77)$$

       c. $$\bar{x} \pm z_{.025}\frac{\sigma}{\sqrt{n}} \Rightarrow 15 \pm 1.96\frac{20}{\sqrt{100}} \Rightarrow 15 \pm 3.92 \Rightarrow (11.08, 18.92)$$

       d. $$\bar{x} \pm z_{.025}\frac{\sigma}{\sqrt{n}} \Rightarrow 4.05 \pm 1.96\frac{20}{\sqrt{100}} \Rightarrow 4.05 \pm 3.92 \Rightarrow (0.13, 7.97)$$

       e.   No. Since the sample size in each part was large (n ranged from 75 to 200), the Central Limit Theorem indicates that the sampling distribution of $\bar{x}$ is approximately normal.

5.11   a.   For confidence coefficient .95, $\alpha = .05$ and $\alpha/2 = .05/2 = .025$. From Table III, Appendix A, $z_{.025} = 1.96$. The confidence interval is:

$$\bar{x} \pm z_{\alpha/2}\frac{s}{\sqrt{n}} \Rightarrow 83.2 \pm 1.96\frac{6.4}{\sqrt{100}} \Rightarrow 83.2 \pm 1.25 \Rightarrow (81.95,\ 84.45)$$

b.   The confidence coefficient of .95 means that in repeated sampling, 95% of all confidence intervals constructed will include μ.

c.   For confidence coefficient .99, $\alpha = .01$ and $\alpha/2 = .01/2 = .005$. From Table III, Appendix A, $z_{.005} = 2.58$. The confidence interval is:

$$\bar{x} \pm z_{\alpha/2}\frac{s}{\sqrt{n}} \Rightarrow 83.2 \pm 2.58\frac{6.4}{\sqrt{100}} \Rightarrow 83.2 \pm 1.65 \Rightarrow (81.55,\ 84.85)$$

d.   As the confidence coefficient increases, the width of the confidence interval also increases.

e.   Yes. Since the sample size is 100, the Central Limit Theorem applies. This ensures the distribution of $\bar{x}$ is normal, regardless of the original distribution.

5.13   a.   The point estimate for the average number of latex gloves used per week by all healthcare workers with a latex allergy is $\bar{x} = 19.3$.

b.   For confidence coefficient .95, $\alpha = .05$ and $\alpha/2 = .05/2 = .025$. From Table III, Appendix A, $z_{.025} = 1.96$. The 95% confidence interval is:

$$\bar{x} \pm z_{.025}\sigma_{\bar{x}} \Rightarrow \bar{x} \pm 1.96\frac{\sigma}{\sqrt{n}} \Rightarrow 19.3 \pm 1.96\frac{11.9}{\sqrt{46}} \Rightarrow 19.3 \pm 3.44 \Rightarrow (15.86,\ 22.74)$$

c.   We are 95% confident that the average number of latex gloves used per week by all healthcare workers with a latex allergy is between 15.86 and 22.74.

d.   We must assume that we have a random sample from the target population and that the sample size is sufficiently large.

5.15   For confidence coefficient .99, $\alpha = .01$ and $\alpha/2 = .01/2 = .005$. From Table III, Appendix A, $z_{.005} = 2.58$. The confidence interval is:

$$\bar{x} \pm z_{.005}\frac{s}{\sqrt{n}} \Rightarrow 39 \pm 2.58\frac{6}{\sqrt{100}} \Rightarrow 39 \pm 1.55 \Rightarrow (37.45,\ 40.55)$$

We are 99% confident that the true mean WR score for all convicted drug dealers is between 37.45 and 40.55.

5.17   a.   Some preliminary calculations are:

$$\bar{x} = \frac{\sum x}{n} = \frac{181.56}{504} = .3602$$

$$s^2 = \frac{\sum x^2 - \frac{\left(\sum x\right)^2}{n}}{n-1} = \frac{74.6546 - \frac{181.56^2}{504}}{504 - 1} = \frac{9.24977143}{503} = .018389$$

$$s = \sqrt{.018389} = .1356$$

For confidence coefficient .90, $\alpha = .10$ and $\alpha/2 = .10/2 = .05$. From Table III, Appendix A, $z_{.05} = 1.645$. The 90% confidence interval is:

$$\bar{x} \pm z_{.05}\sigma_{\bar{x}} \Rightarrow \bar{x} \pm 1.645\frac{\sigma}{\sqrt{n}} \Rightarrow .3602 \pm 1.645\frac{.1356}{\sqrt{504}} \Rightarrow .3602 \pm .0099 \Rightarrow (.3503, \ .3701)$$

b.   We are 90% confident that the true mean visible albedo value of all Canadian Arctic ice ponds is between .3503 and .3701.

   In repeated sampling 90% of all intervals constructed in the same manner will contain the true mean and 10% will not.

c.   **First-year Ice**:

   Some preliminary calculations are:

$$\bar{x} = \frac{\sum x}{n} = \frac{26.63}{88} = .3026$$

$$s^2 = \frac{\sum x^2 - \frac{\left(\sum x\right)^2}{n}}{n-1} = \frac{10.6923 - \frac{26.63^2}{88}}{88 - 1} = \frac{2.633698864}{87} = .03027$$

$$s = \sqrt{.03027} = .1740$$

The 90% confidence interval is:

$$\bar{x} \pm z_{.05}\sigma_{\bar{x}} \Rightarrow \bar{x} \pm 1.645\frac{\sigma}{\sqrt{n}} \Rightarrow .3026 \pm 1.645\frac{.1740}{\sqrt{88}} \Rightarrow .3026 \pm .0305 \Rightarrow (.2721, \ .3331)$$

We are 90% confident that the true mean visible albedo value of all First-year Canadian Arctic ice ponds is between .2721 and .3331.

Landfast Ice:

Some preliminary calculations are:

$$\bar{x} = \frac{\sum x}{n} = \frac{71.04}{196} = .3624$$

$$s^2 = \frac{\sum x^2 - \frac{\left(\sum x\right)^2}{n}}{n-1} = \frac{30.161 - \frac{71.04^2}{196}}{196-1} = \frac{4.41262449}{195} = .02263$$

$$s = \sqrt{.02263} = .1504$$

The 90% confidence interval is:

$$\bar{x} \pm z_{.05}\sigma_{\bar{x}} \Rightarrow \bar{x} \pm 1.645\frac{\sigma}{\sqrt{n}} \Rightarrow .3624 \pm 1.645\frac{.1504}{\sqrt{196}} \Rightarrow .3624 \pm .0177 \Rightarrow (.3447, \; .3801)$$

We are 90% confident that the true mean visible albedo value of all Landfast Canadian Arctic ice ponds is between .3447 and .3801.

Multi-year Ice:

Some preliminary calculations are:

$$\bar{x} = \frac{\sum x}{n} = \frac{83.89}{220} = .3813$$

$$s^2 = \frac{\sum x^2 - \frac{\left(\sum x\right)^2}{n}}{n-1} = \frac{33.8013 - \frac{83.89^2}{220}}{220-1} = \frac{1.81251773}{219} = .00827$$

$$s = \sqrt{.00827} = .0910$$

The 90% confidence interval is:

$$\bar{x} \pm z_{.05}\sigma_{\bar{x}} \Rightarrow \bar{x} \pm 1.645\frac{\sigma}{\sqrt{n}} \Rightarrow .3813 \pm 1.645\frac{.0910}{\sqrt{220}} \Rightarrow .3813 \pm .0101 \Rightarrow (.3712, \; .3914)$$

We are 90% confident that the true mean visible albedo value of all Multi-year Canadian Arctic ice ponds is between .3712 and .3914.

5.19 a. The parameter of interest is the mean effect size for all psychological studies of personality and aggressive behavior.

   b. No, the distribution does not look normal. The data appear to be skewed to the right. The shape of this distribution is not of interest because the sample size is large, $n = 109$. By the Central Limit Theorem, the sampling distribution of $\bar{x}$ is normal regardless of the original distribution.

   c. From the printout, the 95% confidence interval for μ is (0.4786, 0.8167). We are 95% confident that the true mean effect size is between 0.4786 and 0.8167.

d.   Yes. Since 0 is not contained in the 95% confidence interval, it is not a likely value for the true mean effect size. Since all the values in the 95% confident interval are above 0, the researchers are justified in concluding the true effect size mean is greater than 0 or that those who score high on a personality test are more aggressive than those who score low.

5.21   a.   For confidence coefficient .99, $\alpha = .01$ and $\alpha / 2 = .01 / 2 = .005$. From Table III, Appendix A, $z_{.005} = 2.58$. The confidence interval is:

$$\bar{x} \pm z_{\alpha/2} \frac{s}{\sqrt{n}} \Rightarrow 1.13 \pm 2.58 \frac{2.21}{\sqrt{72}} \Rightarrow 1.13 \pm .672 \Rightarrow (.458,\ 1.802)$$

We are 99% confident that the true mean number of pecks made by chickens pecking at blue string is between .458 and 1.802.

b.   Yes, there is evidence that chickens are more apt to peck at white string. The mean number of pecks for white string is 7.5. Since 7.5 is not in the 99% confidence interval for the mean number of pecks at blue string, it is not a likely value for the true mean for blue string.

5.23   a.   For confidence coefficient .95, $\alpha = .05$ and $\alpha / 2 = .05 / 2 = .025$. From Table III, Appendix A, $z_{.025} = 1.96$. The confidence interval is:

$$\bar{x} \pm z_{\alpha/2} \frac{s}{\sqrt{n}} \Rightarrow 19 \pm 1.96 \frac{65}{\sqrt{265}} \Rightarrow 19 \pm 7.826 \Rightarrow (11.174,\ 26.826)$$

b.   For confidence coefficient .95, $\alpha = .05$ and $\alpha / 2 = .05 / 2 = .025$. From Table III, Appendix A, $z_{.025} = 1.96$. The confidence interval is:

$$\bar{x} \pm z_{\alpha/2} \frac{s}{\sqrt{n}} \Rightarrow 7 \pm 1.96 \frac{49}{\sqrt{265}} \Rightarrow 7 \pm 5.90 \Rightarrow (1.10,\ 12.90)$$

c.   The SAT-Mathematics test would be more likely to have a mean change of 15 because 15 is in the 95% confidence interval for the mean change in SAT-Mathematics. Since 15 is in the confidence interval, it is a likely value. The value 15 is not in the 95% confidence interval for the SAT-Verbal. Thus, it is not a likely value.

5.25   a.   For confidence coefficient .95, $\alpha = .05$ and $\alpha / 2 = .05 / 2 = .025$. From Table III, Appendix A, $z_{.025} = 1.96$.

For Males, the 95% confidence interval is:

$$\bar{x} \pm z_{.025} \sigma_{\bar{x}} \Rightarrow \bar{x} \pm 1.96 \frac{\sigma}{\sqrt{n}} \Rightarrow 16.79 \pm 1.96 \frac{13.57}{\sqrt{128}} \Rightarrow 16.79 \pm 2.35 \Rightarrow (14.44,\ 19.14)$$

For Females, the 95% confidence interval is:

$$\bar{x} \pm z_{.025} \sigma_{\bar{x}} \Rightarrow \bar{x} \pm 1.96 \frac{\sigma}{\sqrt{n}} \Rightarrow 10.79 \pm 1.96 \frac{11.53}{\sqrt{184}} \Rightarrow 10.79 \pm 1.67 \Rightarrow (9.12,\ 12.46)$$

b.   Since the two intervals are independent, the probability that at least one of the 2 confidence intervals will not contain the population mean is equal to 1 minus the probability that neither of the 2 confidence intervals will not contain the population mean. Thus the probability is:

P(at least one interval will not contain mean)
$= 1 - P$(neither interval will not contain the mean)
$= 1 - P$(both will contain the mean) $= 1 - .95(.95) = 1 - .9025 = .0975$.

c.   The 95% confidence interval for the males is (14.44, 19.14). The 95% confidence interval for the females is (9.12, 12.46). Since all of the values in the male interval are larger than all the values in the female interval, we can infer that the males consume the most alcohol, on average, per week.

5.27   The two problems (and corresponding solutions) with using a small sample to estimate μ are:

1.   The shape of the sampling distribution of the sample mean $\bar{x}$ now depends on the shape of the population that is sampled. The Central Limit Theorem no longer applies since the sample size is small. However, if the sampled population is normal or approximately normal, then the sampling distribution of $\bar{x}$ is always normal.

2.   The population standard deviation σ is almost always unknown. Although it is still true that $\sigma_{\bar{x}} = \dfrac{\sigma}{\sqrt{n}}$, the sample standard deviation s may provide a poor approximation for σ when the sample size is small. Instead of using $z = \dfrac{\bar{x} - \mu}{\dfrac{\sigma}{\sqrt{n}}}$, we use $t = \dfrac{\bar{x} - \mu}{\dfrac{s}{\sqrt{n}}}$ as the statistic.

5.29   a.   If x is normally distributed, the sampling distribution of $\bar{x}$ is normal, regardless of the sample size. If n is large, then the test statistic is z. If n is small, then the test statistic would be t.

b.   If nothing is known about the distribution of x, the sampling distribution of $\bar{x}$ is approximately normal if n is sufficiently large. If n is not large, the distribution of $\bar{x}$ is unknown if the distribution of x is not known.

5.31   a. $P(t \geq t_0) = .025$ where df $= 10$
$t_0 = 2.228$

b. $P(t \geq t_0) = .01$ where df $= 17$
$t_0 = 2.567$

c. $P(t \le t_0) = .005$ where df $= 6$

Because of symmetry, the statement can be rewritten

$$P(t \ge -t_0) = .005 \text{ where df} = 6$$
$$t_0 = -3.707$$

d. $P(t \le t_0) = .05$ where df $= 13$
$$t_0 = -1.771$$

5.33   First, we must compute $\overline{x}$ and s.

$$\overline{x} = \frac{\sum x}{n} = \frac{30}{6} = 5 \qquad s^2 = \frac{\sum x^2 - \frac{\left(\sum x\right)^2}{n}}{n-1} = \frac{176 - \frac{(30)^2}{6}}{6-1} = \frac{26}{5} = 5.2$$

$$s = \sqrt{5.2} = 2.2804$$

a.   For confidence coefficient .90, $\alpha = .10$ and $\alpha / 2 = .10 / 2 = .05$. From Table IV, Appendix A, with df $= n - 1 = 6 - 1 = 5$, $t_{.05} = 2.015$. The 90% confidence interval is:

$$\overline{x} \pm t_{.05} \frac{s}{\sqrt{n}} \Rightarrow 5 \pm 2.015 \frac{2.2804}{\sqrt{6}} \Rightarrow 5 \pm 1.876 \Rightarrow (3.124,\ 6.876)$$

b.   For confidence coefficient .95, $\alpha = .05$ and $\alpha / 2 = .05 / 2 = .025$. From Table IV, Appendix A, with df $= n - 1 = 6 - 1 = 5$, $t_{.025} = 2.571$. The 95% confidence interval is:

$$\overline{x} \pm t_{.025} \frac{s}{\sqrt{n}} \Rightarrow 5 \pm 2.571 \frac{2.2804}{\sqrt{6}} \Rightarrow 5 \pm 2.394 \Rightarrow (2.606,\ 7.394)$$

c.   For confidence coefficient .99, $\alpha = .01$ and $\alpha / 2 = .01 / 2 = .005$. From Table IV, Appendix A, with df $= n - 1 = 6 - 1 = 5$, $t_{.005} = 4.032$. The 99% confidence interval is:

$$\overline{x} \pm t_{.005} \frac{s}{\sqrt{n}} \Rightarrow 5 \pm 4.032 \frac{2.2804}{\sqrt{6}} \Rightarrow 5 \pm 3.754 \Rightarrow (1.246,\ 8.754)$$

d.   a)   For confidence coefficient .90, $\alpha = .10$ and $\alpha / 2 = .10 / 2 = .05$. From Table IV, Appendix A, with df $= n - 1 = 25 - 1 = 24$, $t_{.05} = 1.711$. The 90% confidence interval is:

$$\overline{x} \pm t_{.05} \frac{s}{\sqrt{n}} \Rightarrow 5 \pm 1.711 \frac{2.2804}{\sqrt{25}} \Rightarrow 5 \pm .780 \Rightarrow (4.220,\ 5.780)$$

b)   For confidence coefficient .95, $\alpha = .05$ and $\alpha / 2 = .05 / 2 = .025$. From Table IV, Appendix A, with df $= n - 1 = 25 - 1 = 24$, $t_{.025} = 2.064$. The 95% confidence interval is:

$$\overline{x} \pm t_{.025} \frac{s}{\sqrt{n}} \Rightarrow 5 \pm 2.064 \frac{2.2804}{\sqrt{25}} \Rightarrow 5 \pm .941 \Rightarrow (4.059,\ 5.941)$$

c)   For confidence coefficient .99, $\alpha = .01$ and $\alpha / 2 = .01 / 2 = .005$. From Table IV, Appendix A, with df $= n - 1 = 25 - 1 = 24$, $t_{.005} = 2.797$. The 99% confidence interval is:

$$\overline{x} \pm t_{.005} \frac{s}{\sqrt{n}} \Rightarrow 5 \pm 2.797 \frac{2.2804}{\sqrt{25}} \Rightarrow 5 \pm 1.276 \Rightarrow (3.724,\ 6.276)$$

Increasing the sample size decreases the width of the confidence interval.

5.35   a.   The target parameter is $\mu =$ the mean trap spacing for the population of red spiny lobster fishermen fishing in Baja California Sur, Mexico.

b.   Using MINITAB, the descriptive statistics are:

Descriptive Statistics: Spacing

```
Variable   N    Mean   StDev   Minimum     Q1   Median     Q3   Maximum
Spacing    7   89.86   11.63     70.00   82.00   93.00   99.00    105.00
```

The point estimate for μ is $\overline{x} = 89.86$.

c.   We do not know what the distribution of the population of trap spacings is. The Central Limit Theorem also does not apply because the sample size ($n = 7$) is too small. The sample standard deviation, s, is no longer necessarily a good estimate for the population standard deviation σ.

d.   For confidence coefficient .95, $\alpha = .05$ and $\alpha / 2 = .05 / 2 = .025$. From Table IV, Appendix A, with df $= n - 1 = 7 - 1 = 6$, $t_{.025} = 2.447$. The confidence interval is:

$$\overline{x} \pm t_{.025,6} \frac{s}{\sqrt{n}} \Rightarrow 89.86 \pm 2.447 \frac{11.63}{\sqrt{7}} \Rightarrow 89.86 \pm 10.756 \Rightarrow (79.104,\ 100.616)$$

e.   We are 95% confident that the mean trap spacing for the population of red spiny lobster fishermen fishing in Baja California Sur, Mexico is between 79.104 and 100.616 meters.

f.   We must assume that the population of trap spacings is normal and that a random sample of trapping spacings was selected.

5.37   a.   From the printout, the 95% confidence interval is (652.76, 817.40).

b.   For confidence coefficient .95, $\alpha = .05$ and $\alpha / 2 = .05 / 2 = .025$. From Table IV, Appendix A, with df $= n - 1 = 12 - 1 = 11$, $t_{.025} = 2.201$. The confidence interval is:

$$\bar{x} \pm t_{.025,11}\frac{s}{\sqrt{n}} \Rightarrow 735.08 \pm 2.201\frac{129.565}{\sqrt{12}} \Rightarrow 735.08 \pm 82.32 \Rightarrow (652.76,\ 817.40)$$

    c.   We are 95% confident that the true mean number of minutes of daylight per day in Sharon, PA is between 652.76 and 817.40 minutes.

    d.   If a random sample of size 12 was taken from the population, then we could see most observations in one season, like summer or winter. Since the amount of daylight is greatly affected by the time of the year, this might not be the best plan. By sampling one observation from each month, we are guaranteed that the sample will be fairly representative. However, it will not be random.

5.39   a.   The point estimate for the mean amount of cesium in lichen specimens collected in Alaska is $\bar{x} = .009027$.

    b.   For confidence coefficient .95, $\alpha = .05$ and $\alpha/2 = .05/2 = .025$. From Table IV, Appendix A, with df $= n - 1 = 9 - 1 = 8$, $t_{.025} = 2.306$.

    c.   The 95% confidence interval is:

$$\bar{x} \pm t_{.025}\frac{s}{\sqrt{n}} \Rightarrow .009027 \pm 2.306\frac{.004854}{\sqrt{9}} \Rightarrow .009027 \pm .003731 \Rightarrow (.005296,\ .012758)$$

    d.   This interval is the same as that found on the printout.

    e.   We are 95% confident that the mean amount of cesium in lichen specimens collected in Alaska is between .005296 and .012758 microcuries per milliliter.

5.41   For confidence coefficient .90, $\alpha = .10$ and $\alpha/2 = .10/2 = .05$. From Table IV, Appendix A, with df $= n - 1 = 17 - 1 = 16$, $t_{.05} = 1.746$. The confidence interval is:

$$\bar{x} \pm t_{.05,16}\frac{s}{\sqrt{n}} \Rightarrow .11 \pm 1.746\frac{.19}{\sqrt{17}} \Rightarrow .11 \pm .08 \Rightarrow (.03,\ .19)$$

We are 90% confident that the true mean score difference for all amusiacs is between .03 and .19.

We must assume that the population of score differences is approximately normal and that the data were randomly selected.

5.43   a.   **Both Untreated**: For confidence coefficient .90, $\alpha = .10$ and $\alpha/2 = .10/2 = .05$. From Table IV, Appendix A, with df $= n - 1 = 29 - 1 = 28$, $t_{.05} = 1.701$. The 90% confidence interval is:

$$\bar{x} \pm t_{.05,28}\frac{s}{\sqrt{n}} \Rightarrow 20.9 \pm 1.701\frac{3.34}{\sqrt{29}} \Rightarrow 20.9 \pm 1.055 \Rightarrow (19.845,\ 21.955)$$

Male Treated: For confidence coefficient .90, $\alpha = .10$ and $\alpha/2 = .10/2 = .05$. From Table IV, Appendix A, with df $= n - 1 = 23 - 1 = 22$, $t_{.05} = 1.717$. The 90% confidence interval is:

$$\bar{x} \pm t_{.05,22} \frac{s}{\sqrt{n}} \Rightarrow 20.3 \pm 1.717 \frac{3.50}{\sqrt{23}} \Rightarrow 20.3 \pm 1.253 \Rightarrow (19.047,\ 21.553)$$

Female Treated: For confidence coefficient .90, $\alpha = .10$ and $\alpha/2 = .10/2 = .05$. From Table IV, Appendix A, with df $= n - 1 = 18 - 1 = 17$, $t_{.05} = 1.740$. The 90% confidence interval is:

$$\bar{x} \pm t_{.05,17} \frac{s}{\sqrt{n}} \Rightarrow 22.9 \pm 1.740 \frac{4.37}{\sqrt{18}} \Rightarrow 22.9 \pm 1.792 \Rightarrow (21.108,\ 24.692)$$

Both Treated: For confidence coefficient .90, $\alpha = .10$ and $\alpha/2 = .10/2 = .05$. From Table IV, Appendix A, with df $= n - 1 = 21 - 1 = 20$, $t_{.05} = 1.725$. The 90% confidence interval is:

$$\bar{x} \pm t_{.05,20} \frac{s}{\sqrt{n}} \Rightarrow 18.6 \pm 1.725 \frac{2.11}{\sqrt{21}} \Rightarrow 18.6 \pm 0.794 \Rightarrow (17.806,\ 19.394)$$

b.　The female/male pair that appears to produce the highest mean number of eggs is the **Female treated** pair. The 90% confidence interval for this pair is the highest. This interval does overlap with the **Both treated** pair and the **Males treated** pair, but just barely.

5.45　a.　For confidence coefficient .90, $\alpha = .10$ and $\alpha/2 = .10/2 = .05$. From Table IV, Appendix A, with df $= n - 1 = 4 - 1 = 3$, $t_{.05} = 2.353$. The confidence interval is:

$$\bar{x} \pm t_{.05,3} \frac{s}{\sqrt{n}} \Rightarrow 1.43 \pm 2.353 \frac{.13}{\sqrt{4}} \Rightarrow 1.43 \pm .153 \Rightarrow (1.277, 1.583)$$

b.　We are 90% confident that the true mean peptide score in alleles of the antigen-produced protein is between 1.277 and 1.583.

c.　"90% confidence" means that in repeated sampling, 90% of all intervals constructed in this manner will contain the true mean.

5.47　a.　For confidence coefficient .95, $\alpha = .05$ and $\alpha/2 = .05/2 = .025$. From Table IV, Appendix A, with df $= n - 1 = 15 - 1 = 14$, $t_{.025} = 2.145$. The confidence interval is:

$$\bar{x} \pm t_{.025} \frac{s}{\sqrt{n}} \Rightarrow 37.3 \pm 2.145 \frac{13.9}{\sqrt{15}} \Rightarrow 37.3 \pm 7.70 \Rightarrow (29.60, 45.00)$$

b.　We are 95% confident that the true mean adrenocorticotropin level in sleepers one-hour prior to anticipated waking is between 29.60 and 45.00

c.　Suppose we assume that the true mean adrenocorticotropin level of sleepers three hours before anticipated wake-up time is 25.5. Since the 95% confidence interval for the true mean adrenocorticotropin level in sleepers one-hour prior to anticipated waking (29.60, 45.00) does not contain 25.5, there is evidence to indicate that the true mean adrenocorticotropin level of sleepers one hour before anticipated wake-up time is greater than that of sleepers three hours prior to anticipated waking.

5.49   An unbiased estimator is one in which the mean of the sampling distribution is the parameter of interest, i.e., $E(\hat{p}) = p$.

5.51   a.   The sample size is large enough if both $n\hat{p} \geq 15$ and $n\hat{q} \geq 15$.

$n\hat{p} = 196(.64) = 125.44$ and $n\hat{q} = 196(.36) = 70.56$. Since both of these numbers are greater than or equal to 15, the sample size is sufficiently large to conclude the normal approximation is reasonable.

b.   For confidence coefficient .95, $\alpha = .05$ and $\alpha/2 = .05/2 = .025$. From Table III, Appendix A, $z_{.025} = 1.96$. The confidence interval is:

$$\hat{p} \pm z_{.025}\sqrt{\frac{\hat{p}\hat{q}}{n}} \Rightarrow .64 \pm 1.96\sqrt{\frac{.64(.36)}{196}} \Rightarrow .64 \pm .067 \Rightarrow (.537, .707)$$

c.   We are 95% confident the true value of p is between .372 and .468.

d.   "95% confidence" means that if repeated samples of size 196 were selected from the population and 95% confidence intervals formed, 95% of all confidence intervals will contain the true value of p.

5.53   The sample size is large enough if both $n\hat{p} \geq 15$ and $n\hat{q} \geq 15$.

a. $n\hat{p} = 500(.05) = 25$ and $n\hat{q} = 500(.95) = 475$. Since both of these numbers are greater than or equal to 15, the sample size is sufficiently large to conclude the normal approximation is reasonable.

b. $n\hat{p} = 100(.05) = 5$ and $n\hat{q} = 100(.95) = 95$. Since the first number is not greater than 15, the sample size is not sufficiently large to conclude the normal approximation is reasonable.

c. $n\hat{p} = 10(.5) = 5$ and $n\hat{q} = 10(.5) = 5$. Since neither of these numbers is greater than or equal to 15, the sample size is not sufficiently large to conclude the normal approximation is reasonable.

d. $n\hat{p} = 10(.3) = 3$ and $n\hat{q} = 10(.7) = 7$. Since neither of these numbers is greater than or equal to 15, the sample size is not sufficiently large to conclude the normal approximation is reasonable.

5.55   a.   The population of interest is all American adults.

b.   The sample in the study is the 1,000 American adults surveyed.

c. The parameter of interest is p = proportion of all adults who say Starbucks coffee is overpriced.

d. The sample size is large enough if both $n\hat{p} \geq 15$ and $n\hat{q} \geq 15$. For this problem, $\hat{p} = .73$. $n\hat{p} = 1000(.73) = 730$ and $n\hat{q} = 1000(.27) = 270$. Since both of these numbers are greater than or equal to 15, the sample size is sufficiently large to conclude the normal approximation is reasonable.
For confidence coefficient .95, $\alpha = .05$ and $\alpha/2 = .05/2 = .025$. From Table III, Appendix A, $z_{.025} = 1.96$. The 95% confidence interval is:

$$\hat{p} \pm z_{.025}\sqrt{\frac{pq}{n}} \approx \hat{p} \pm 1.96\sqrt{\frac{\hat{p}\hat{q}}{n}} \Rightarrow .73 \pm 1.96\sqrt{\frac{.73(.27)}{1000}} \Rightarrow .73 \pm .028 \Rightarrow (.702, .758)$$

We are 95% confident that the true proportion of American adults who think Starbucks coffee is overpriced is between .702 and .758.

5.57 a. The population of interest is the set of gun ownerships (Yes or No) of all adults in the U.S.

b. The parameter of interest is the true percentage or proportion, p, of all adults in the U.S. who own at least one gun.

c. The estimate of the population proportion is $\hat{p} = .26$. The estimate of the population percentage is $.26(100\%) = 26\%$.

d. The sample size is large enough if both $n\hat{p} \geq 15$ and $n\hat{q} \geq 15$.

$n\hat{p} = 2,770(.26) = 720.2$ and $n\hat{q} = 2,770(.74) = 2,049.8$. Since both of these numbers are greater than or equal to 15, the sample size is sufficiently large to conclude the normal approximation is reasonable.

For confidence coefficient .99, $\alpha = .01$ and $\alpha/2 = .01/2 = .005$. From Table III, Appendix A, $z_{.005} = 2.58$. The 99% confidence interval is:

$$\hat{p} \pm z_{.005}\sqrt{\frac{pq}{n}} \Rightarrow \hat{p} \pm 2.58\sqrt{\frac{\hat{p}\hat{q}}{n}} \Rightarrow .26 \pm 2.58\sqrt{\frac{.26(.74)}{2,770}} \Rightarrow .26 \pm .02 \Rightarrow (.24, .28)$$

The 99% confidence interval for the true percentage is (24%, 28%).

e. We are 99% confident that the true percentage of adults in the U.S who own at least one gun is between 24% and 28%.

f. "99% confidence" means that in repeated sampling 99% of all confidence intervals constructed in this manner will contain the true percentage.

5.59 a. The sample size is large enough if both $n\hat{p} \geq 15$ and $n\hat{q} \geq 15$.

$n\hat{p} = 1{,}000(.63) = 630$ and $n\hat{q} = 1{,}000(.37) = 370$. Since both of the numbers are greater than or equal to 15, the sample size is sufficiently large to conclude the normal approximation is reasonable.

For confidence coefficient .95, $\alpha = .05$ and $\alpha / 2 = .05 / 2 = .025$. From Table III, Appendix A, $z_{.025} = 1.96$. The 95% confidence interval is:

$$\hat{p} \pm z_{.025}\sqrt{\frac{pq}{n}} \Rightarrow \hat{p} \pm 1.96\sqrt{\frac{\hat{p}\hat{q}}{n}} \Rightarrow .63 \pm 1.96\sqrt{\frac{.63(.37)}{1{,}000}} \Rightarrow .63 \pm .030 \Rightarrow (.600, \ .660)$$

b.   Yes, we would be surprised. Since .70 is not in the 95% confidence interval, it is not a likely value for the true value of the population proportion of adults who would choose to sleep when home sick.

5.61   The estimate of the true proportion of all U. S. teenagers who have used at least one informal element in a school writing assignment is $\hat{p} = \dfrac{448}{700} = .64$.

The sample size is large enough if both $n\hat{p} \geq 15$ and $n\hat{q} \geq 15$. For this problem, $\hat{p} = .64$. $n\hat{p} = 700(.64) = 448$ and $n\hat{q} = 700(.36) = 252$. Since both of these numbers are greater than or equal to 15, the sample size is sufficiently large to conclude the normal approximation is reasonable.

For confidence coefficient .99, $\alpha = .01$ and $\alpha / 2 = .01 / 2 = .005$. From Table III, Appendix A, $z_{.005} = 2.58$. The 99% confidence interval is:

$$\hat{p} \pm z_{.005}\sqrt{\frac{pq}{n}} \approx \hat{p} \pm 2.58\sqrt{\frac{\hat{p}\hat{q}}{n}} \Rightarrow .64 \pm 2.58\sqrt{\frac{.64(.36)}{700}} \Rightarrow .64 \pm .047 \Rightarrow (.593, .687)$$

We are 99% confident that the true proportion of all U. S. teenagers who have used at least one informal element in a school writing assignment is between .593 and .687.

5.63   a.   First, we compute $\hat{p}$. There were 1333 useable responses. Of these, 450 chose "Bible is the actual word of God and is to be taken literally".

$$\hat{p} = \frac{x}{n} = \frac{450}{1333} = .338$$

The sample size is large enough if both $n\hat{p} \geq 15$ and $n\hat{q} \geq 15$.

$n\hat{p} = 1{,}333(.338) = 450.6$ and $n\hat{q} = 1{,}333(.662) = 882.4$. Since both of the numbers are greater than or equal to 15, the sample size is sufficiently large to conclude the normal approximation is reasonable.

For confidence coefficient .95, $\alpha = .05$ and $\alpha / 2 = .05 / 2 = .025$. From Table III, Appendix A, $z_{.025} = 1.96$. The 95% confidence interval is:

$$\hat{p} \pm z_{.025}\sqrt{\frac{pq}{n}} \Rightarrow \hat{p} \pm 1.96\sqrt{\frac{\hat{p}\hat{q}}{n}} \Rightarrow .338 \pm 1.96\sqrt{\frac{.338(.662)}{1,333}}$$

$$\Rightarrow .338 \pm .025 \Rightarrow (.313, \ .363)$$

b.   We are 95% confident that the true proportion of all Americans who believe that the "Bible is the actual word of God and is to be taken literally" is between .313 and .363.

c.   If the responses from the survey are not representative of the population, then the validity of the results will be questionable. If the survey is sent out to a representative sample, but the response rate is low, then again, the results may be questionable.

5.65   a.   The estimate of the true proportion of all mountain casualties that require a femoral shaft splint is $\hat{p} = \dfrac{1}{333} = .003$.

The sample size is large enough if both $n\hat{p} \geq 15$ and $n\hat{q} \geq 15$. For this problem, $\hat{p} = .003$. $n\hat{p} = 333(.003) = .999$ and $n\hat{q} = 333(.997) = 332.001$. Since the first of these numbers is less than to 15, the sample size is not sufficiently large to conclude the normal approximation is reasonable.

b.   The Wilson adjusted sample proportion is

$$\tilde{p} = \frac{x+2}{n+4} = \frac{1+2}{333+4} = \frac{3}{337} = .009$$

For confidence coefficient .95, $\alpha = .05$ and $\alpha / 2 = .05 / 2 = .025$. From Table III, Appendix A, $z_{.025} = 1.96$. The Wilson adjusted 95% confidence interval is:

$$\tilde{p} \pm z_{.025}\sqrt{\frac{\tilde{p}(1-\tilde{p})}{n+4}} \Rightarrow .009 \pm 1.96\sqrt{\frac{.009(.991)}{333+4}} \Rightarrow .009 \pm .010 \Rightarrow (-.001, \ .019)$$

We are 95% confident that the true proportion of all mountain casualties that require a femoral shaft splint is between 0 and .019.

5.67   The point estimate for the proportion of health care workers with latex allergy who suspect that he/she has the allergy is $\hat{p} = \dfrac{x}{n} = \dfrac{36}{83} = .434$.

The sample size is large enough if both $n\hat{p} \geq 15$ and $n\hat{q} \geq 15$.

$n\hat{p} = 83(.434) = 36$ and $n\hat{q} = 83(.566) = 47$. Since both of the numbers are greater than or equal to 15, the sample size is sufficiently large to conclude the normal approximation is reasonable.

Since no confidence limit is given, we will use 95%. For confidence coefficient .95, $\alpha = .05$ and $\alpha / 2 = .05 / 2 = .025$. From Table III, Appendix A, $z_{.025} = 1.96$. The 95% confidence interval is:

$$\hat{p} \pm z_{.025}\sigma_{\hat{p}} \Rightarrow \hat{p} \pm 1.96\sqrt{\frac{\hat{p}\hat{q}}{n}} \Rightarrow .434 \pm 1.96\sqrt{\frac{.434(.566)}{83}} \Rightarrow .434 \pm .107 \Rightarrow (.327, \ .541)$$

We are 95% confident that the proportion of health care workers with latex allergy who suspect that he/she has the allergy is between .327 and .541.

5.69   The statement "For a specified sampling error SE, increasing the confidence level $(1-\alpha)$ will lead to a larger n when determining sample size" is True.

5.71   To compute the necessary sample size, use

$$n = \frac{\left(z_{\alpha/2}\right)^2 \sigma^2}{(SE)^2} \quad \text{where } \alpha = 1 - .95 = .05 \text{ and } \alpha / 2 = .05 / 2 = .025$$

From Table III, Appendix A, $z_{.025} = 1.96$. Thus,

$$n = \frac{(1.96)^2 (5.4)}{.2^2} = 518.616 \approx 519$$

You would need to take 519 samples.

5.73   a.   Range $= 39 - 31 = 8$. $\sigma \approx \dfrac{\text{Range}}{4} = \dfrac{8}{4} = 2$

For confidence coefficient .90, $\alpha = .10$ and $\alpha / 2 = .10 / 2 = .05$. From Table III, Appendix A, $z_{.05} = 1.645$.

The sample size is $n = \dfrac{z_{\alpha/2}^2 \sigma^2}{(SE)^2} = \dfrac{1.645^2 (2^2)}{.15^2} = 481.07 \approx 482$

b. $\sigma \approx \dfrac{\text{Range}}{6} = \dfrac{8}{6} = 1.333$

The sample size is $n = \dfrac{z_{\alpha/2}^2 \sigma^2}{(SE)^2} = \dfrac{1.645^2 (1.333^2)}{.15^2} = 213.7 \approx 214$

5.75   For confidence coefficient .90, $\alpha = .10$ and $\alpha / 2 = .10 / 2 = .05$. From Table III, Appendix A, $z_{.05} = 1.645$.

We know $\hat{p}$ is in the middle of the interval, so $\hat{p} = \dfrac{.54 + .26}{2} = .4$

The confidence interval is $\hat{p} \pm z_{.05}\sqrt{\dfrac{\hat{p}\hat{q}}{n}} \Rightarrow .4 \pm 1.645\sqrt{\dfrac{.4(.6)}{n}}$

We know $.4 - 1.645\sqrt{\dfrac{.4(.6)}{n}} = .26$

$$\Rightarrow .4 - \dfrac{.8059}{\sqrt{n}} = .26$$

$$\Rightarrow .4 - .26 = \dfrac{.8059}{\sqrt{n}} \Rightarrow \sqrt{n} = \dfrac{.8059}{.14} = 5.756$$

$$\Rightarrow n = 5.756^2 = 33.1 \approx 34$$

5.77 a. The width of a confidence interval is $2(SE) = 2z_{\alpha/2}\dfrac{\sigma}{\sqrt{n}}$

For confidence coefficient .95, $\alpha = .05$ and $\alpha/2 = .05/2 = .025$. From Table III, Appendix A, $z_{.025} = 1.96$.

For $n = 16$, $W = 2z_{\alpha/2}\dfrac{\sigma}{\sqrt{n}} = 2(1.96)\dfrac{1}{\sqrt{16}} = 0.98$

For $n = 25$, $W = 2z_{\alpha/2}\dfrac{\sigma}{\sqrt{n}} = 2(1.96)\dfrac{1}{\sqrt{25}} = 0.784$

For $n = 49$, $W = 2z_{\alpha/2}\dfrac{\sigma}{\sqrt{n}} = 2(1.96)\dfrac{1}{\sqrt{49}} = 0.56$

For $n = 100$, $W = 2z_{\alpha/2}\dfrac{\sigma}{\sqrt{n}} = 2(1.96)\dfrac{1}{\sqrt{100}} = 0.392$

For $n = 400$, $W = 2z_{\alpha/2}\dfrac{\sigma}{\sqrt{n}} = 2(1.96)\dfrac{1}{\sqrt{400}} = 0.196$

b.

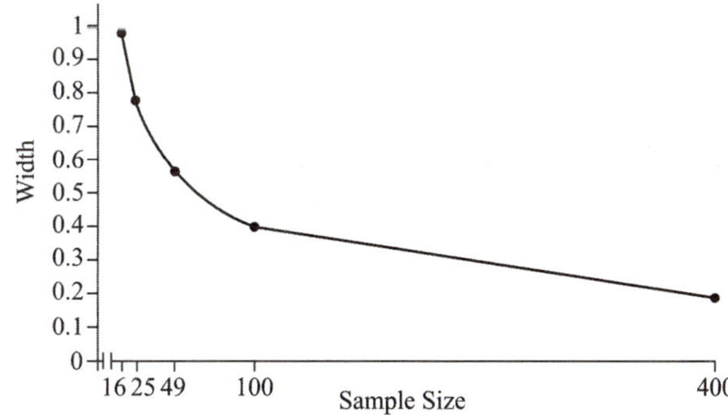

5.79   From Exercise 5.35, the sample standard deviation is $s = 11.63$. We will use this to estimate the population standard deviation. The necessary sample size is:

$$n = \frac{z_{\alpha/2}^2 \sigma^2}{(SE)^2} = \frac{1.96^2 (11.63)^2}{5^2} = 20.78 \approx 21$$

Thus, we will need to sample approximately 21 teams of fishermen.

5.81   a.   The sample size may not be large enough. The sample size is large enough if both $n\hat{p} \geq 15$ and $n\hat{q} \geq 15$.

$$\hat{p} = \frac{x}{n} = \frac{52}{60} = .867$$

$n\hat{p} = 60(.867) = 52$ and $n\hat{q} = 60(.133) = 8$. Since the second number is not greater than or equal to 15, the sample size is not sufficiently large to conclude the normal approximation is reasonable.

b.   For confidence coefficient .90, $\alpha = .10$ and $\alpha/2 = .10/2 = .05$. From Table III, Appendix A, $z_{.05} = 1.645$. We will use the sample proportion to estimate the population proportion:

$$\hat{p} = \frac{x}{n} = \frac{52}{60} = .867$$

$$n = \frac{z_{\alpha/2}^2 pq}{(SE)^2} = \frac{1.645^2 (.867)(.133)}{.05^2} = 124.8 \approx 125$$

5.83   For confidence coefficient .90, $\alpha = .10$ and $\alpha/2 = .10/2 = .05$. From Table III, Appendix A, $z_{.05} = 1.645$.

$$n = \frac{z_{\alpha/2}^2 \sigma^2}{(SE)^2} = \frac{1.645^2 10.9^2}{4^2} = 20.09 \approx 21$$

5.85   From Exercise 5.41, the standard deviation was $s = .19$. We will use this to estimate the population standard deviation. For confidence coefficient .90, $\alpha = .10$ and $\alpha/2 = .10/2 = .05$. From Table III, Appendix A, $z_{.05} = 1.645$. The necessary sample size is:

$$n = \frac{z_{\alpha/2}^2 \sigma^2}{(SE)^2} = \frac{1.645^2 (.19)^2}{.05^2} = 39.075 \approx 40$$

Thus, we will need to sample approximately 40 amusiacs.

5.87   For confidence coefficient .99, $\alpha = .01$ and. From Table III, Appendix A, $z_{.005} = 2.58$. From the previous estimate, we will use $\hat{p} = .333$ to estimate p.

$$n = \frac{z_{\alpha/2}^2 pq}{(SE)^2} = \frac{2.58^2(.333)(.667)}{.01^2} = 14,784.5966 \approx 14,785$$

5.89   From Exercise 5.64, our estimate of p is $\hat{p} = .021$. We will use this to estimate the sample size.

For confidence coefficient .99, $\alpha = .01$ and $\alpha/2 = .01/2 = .005$. From Table III, Appendix A, $z_{.005} = 2.58$.

$$n = \frac{z_{\alpha/2}^2 pq}{(SE)^2} = \frac{2.58^2(.021)(.979)}{.04^2} = 85.5 \approx 86$$

We would need to sample 86 U.S. adults.

Now, the sample from Japan may not be representative because college students were used. To be extremely conservative, we could estimate the needed sample size by estimating p with .5. Now the needed sample size would be:

$$n = \frac{z_{\alpha/2}^2 pq}{(SE)^2} = \frac{2.58^2(.5)(.5)}{.04^2} = 1040.06 \approx 1041$$

This is much larger than the sample size found using the estimate from Exercise 5.64.

5.91   For confidence coefficient .90, $\alpha = .10$ and $\alpha/2 = .10/2 = .05$. From Table III, Appendix A, $z_{.05} = 1.645$.

The sample size is $n = \dfrac{\left(z_{\alpha/2}\right)^2 \sigma^2}{(SE)^2} = \dfrac{(1.645)^2(10^2)}{1^2} = 270.6 \approx 271$

5.93   The sampling distribution used to find interval estimates for σ^2 is the chi-square distribution.

5.95   The degrees of freedom associated with the chi-square sampling distribution for a sample of size n is $n - 1$.

5.97   a.   For confidence level .90, $\alpha = .10$ and $\alpha/2 = .10/2 = .05$. From Table V, Appendix A, with df $= n - 1 = 50\text{-}1 = 49$, $\chi_{.05,49}^2 \approx 67.5048$ and $\chi_{.95,49}^2 \approx 34.7642$. The 90% confidence interval is:

$$\frac{(n-1)s^2}{\chi_{.05}^2} \le \sigma^2 \le \frac{(n-1)s^2}{\chi_{.95}^2} \Rightarrow \frac{(50-1)2.5^2}{67.5048} \le \sigma^2 \le \frac{(50-1)2.5^2}{34.7642} \Rightarrow 4.537 \le \sigma^2 \le 8.809$$

b.   For confidence level .90, $\alpha = .10$ and $\alpha/2 = .10/2 = .05$. From Table V, Appendix A, with df $= n - 1 = $ 15-1 = 14, $\chi^2_{.05,14} = 23.6848$ and $\chi^2_{.95,14} = 6.57063$. The 90% confidence interval is:

$$\frac{(n-1)s^2}{\chi^2_{.05}} \le \sigma^2 \le \frac{(n-1)s^2}{\chi^2_{.95}} \Rightarrow \frac{(15-1).02^2}{23.6848} \le \sigma^2 \le \frac{(15-1).02^2}{6.57063} \Rightarrow .00024 \le \sigma^2 \le .00085$$

c.   For confidence level .90, $\alpha = .10$ and $\alpha/2 = .10/2 = .05$. From Table V, Appendix A, with df $= n - 1 = $ 22-1 = 21, $\chi^2_{.05,21} = 32.6705$ and $\chi^2_{.95,21} = 11.5913$. The 90% confidence interval is:

$$\frac{(n-1)s^2}{\chi^2_{.05}} \le \sigma^2 \le \frac{(n-1)s^2}{\chi^2_{.95}} \Rightarrow \frac{(22-1)31.6^2}{32.6705} \le \sigma^2 \le \frac{(22-1)31.6^2}{11.5913} \Rightarrow 641.86 \le \sigma^2 \le 1{,}809.09$$

d.   For confidence level .90, $\alpha = .10$ and $\alpha/2 = .10/2 = .05$. From Table V, Appendix A, with df $= n - 1 = $ 5-1 = 4, $\chi^2_{.05,4} = 9.48773$ and $\chi^2_{.95,4} = .710721$. The 90% confidence interval is:

$$\frac{(n-1)s^2}{\chi^2_{.05}} \le \sigma^2 \le \frac{(n-1)s^2}{\chi^2_{.95}} \Rightarrow \frac{(5-1)1.5^2}{9.48773} \le \sigma^2 \le \frac{(5-1)1.5^2}{.710721} \Rightarrow .94859 \le \sigma^2 \le 12.6632$$

5.99   Using MINITAB, the descriptive statistics are:

Descriptive Statistics: x

```
Variable  N   Mean   StDev   Minimum    Q1   Median    Q3   Maximum
x         6   6.17   3.31      2.00   2.75    6.50   8.75    11.00
```

For confidence level .95, $\alpha = .05$ and $\alpha/2 = .05/2 = .025$. From Table V, Appendix A, with df $= n - 1 = 6 - 1 = 5$, $\chi^2_{.025,5} = 12.8325$ and $\chi^2_{.975,5} = .831211$. The 95% confidence interval is:

$$\frac{(n-1)s^2}{\chi^2_{.025}} \le \sigma^2 \le \frac{(n-1)s^2}{\chi^2_{.975}} \Rightarrow \frac{(6-1)3.31^2}{12.8325} \le \sigma^2 \le \frac{(6-1)3.31^2}{.831211} \Rightarrow 4.2689 \le \sigma^2 \le 65.9044$$

5.101   a.   For confidence level .95, $\alpha = .05$ and $\alpha/2 = .05/2 = .025$. From Table V, Appendix A, with df $= n - 1 = 106 - 1 = 105$, $\chi^2_{.025,105} \approx 129.561$ and $\chi^2_{.975,105} = 74.2219$. The 95% confidence interval is:

$$\frac{(n-1)s^2}{\chi^2_{.025}} \le \sigma^2 \le \frac{(n-1)s^2}{\chi^2_{.975}} \Rightarrow \frac{(106-1)19.5^2}{129.561} \le \sigma^2 \le \frac{(106-1)19.5^2}{74.2219} \Rightarrow 308.166 \le \sigma^2 \le 537.931$$

We are 95% confident that the true variance of the times per day that laptops are used for taking notes for all middle school students is between 308.166 and 537.931.

b.   Yes. One of the assumptions necessary for this confidence interval to be valid is that the population being sampled from is approximately normal. Since the population of times is not normal, the validity of this confidence interval is questionable.

5.103 a.   From the printout, the 95% confidence interval for the variance is: (8.6, 45.7). We are 95% confident that the true variance of rebound lengths is between 8.6 and 45.7.

b.   From the printout, the 95% confidence interval for the standard deviation is: (2.94, 6.76). We are 95% confident that the true standard deviation of rebound lengths is between 2.94 and 6.76.

c.   We must assume that the sample was a random sample from the target population and that the target population is approximately normal.

5.105   Using MINITAB, the descriptive statistics are:

Descriptive Statistics: Depth

```
Variable    N     Mean   StDev   Minimum       Q1   Median       Q3   Maximum
Depth      18   16.499   1.970    13.250   15.285   16.160   18.123    19.700
```

For confidence level .95, $\alpha = .05$ and $\alpha / 2 = .05 / 2 = .025$. From Table V, Appendix A, with $df = n - 1 = 18 - 1 = 17$, $\chi^2_{.025,17} = 30.1910$ and $\chi^2_{.975,17} = 7.56418$. The 95% confidence interval for the variance is:

$$\frac{(n-1)s^2}{\chi^2_{.025}} \le \sigma^2 \le \frac{(n-1)s^2}{\chi^2_{.975}} \Rightarrow \frac{(18-1)1.97^2}{30.1910} \le \sigma^2 \le \frac{(18-1)1.97^2}{7.56418} \Rightarrow 2.185 \le \sigma^2 \le 8.722$$

The 95% confidence interval for the standard deviation is:

$$\sqrt{2.185} \le \sigma \le \sqrt{8.722} \Rightarrow 1.478 \le \sigma \le 2.953$$

We are 95% confident that the true standard deviation of the molar depths for the population of cheek teeth in extinct primates is between 1.478 and 2.953.

We must assume that a random sample of observations was selected from a normal distribution.

Using MINITAB, a histogram of the data is:

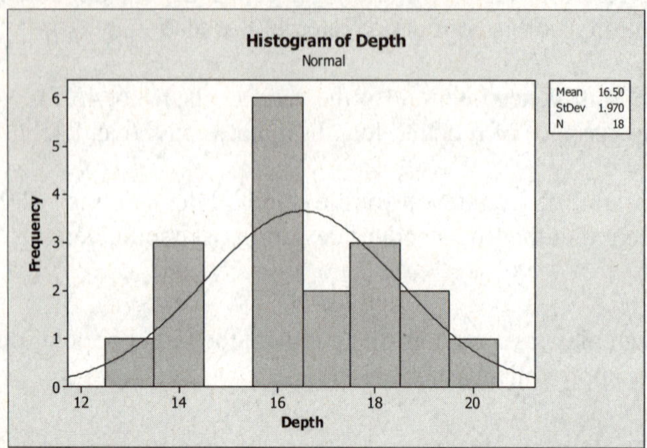

The data do not look particularly normal. The confidence interval for the standard deviation may not be valid.

5.107   From Exercise 5.35, $s = 11.63$.

For confidence level .99, $\alpha = .01$ and $\alpha / 2 = .01 / 2 = .005$. From Table V, Appendix A, with df $= n - 1 = 7 - 1 = 6$, $\chi^2_{.005,6} = 18.5476$ and $\chi^2_{.995,6} = .675727$. The 99% confidence interval for the variance is:

$$\frac{(n-1)s^2}{\chi^2_{.005}} \leq \sigma^2 \leq \frac{(n-1)s^2}{\chi^2_{.995}} \Rightarrow \frac{(7-1)11.63^2}{18.5476} \leq \sigma^2 \leq \frac{(7-1)11.63^2}{.675727} \Rightarrow 43.755 \leq \sigma^2 \leq 1,200.990$$

We are 99% confident that the true variance of the trap spacing measurements of red spiny lobster fishermen fishing in Baja California Sur, Mexico is between 43.755 and 1,200.990.

5.109   a.   The target parameter for the average score on the SAT is μ.

b.   The target parameter for the mean time waiting at a supermarket checkout lane is μ.

c.   The target parameter for the proportion of voters in favor of legalizing marijuana is p.

b.   The target parameter for the percentage of NFL players who have ever made the Pro Bowl is p.

e.   The target parameter for the dropout rate of American college students is p.

f.   The target parameter for the variation in IQ scores of sociopaths is σ^2.

5.111   a.   For a small sample from a normal distribution with unknown standard deviation, we use the t statistic. For confidence coefficient .95, $\alpha = .05$ and $\alpha / 2 = .05 / 2 = .025$. From Table IV, Appendix A, with df $= n - 1 = 21 - 1 = 20$, $t_{.025} = 2.086$.

b. For a large sample from a distribution with an unknown standard deviation, we can estimate the population standard deviation with s and use the z statistic. For confidence coefficient .95, $\alpha = .05$ and $\alpha/2 = .05/2 = .025$. From Table III, Appendix A, $z_{.025} = 1.96$.

c. For a small sample from a normal distribution with known standard deviation, we use the z statistic. For confidence coefficient .95, $\alpha = .05$ and $\alpha/2 = .05/2 = .025$. From Table III, Appendix A, $z_{.025} = 1.96$.

d. For a large sample from a distribution about which nothing is known, we can estimate the population standard deviation with s and use the z statistic. For confidence coefficient .95, $\alpha = .05$ and $\alpha/2 = .05/2 = .025$. From Table III, Appendix A, $z_{.025} = 1.96$.

e. For a small sample from a distribution about which nothing is known, we can use neither z nor t.

5.113 a. The point estimate for the proportion of measurements in the population with characteristic A is $\hat{p} = \dfrac{x}{n} = \dfrac{227}{400} = .5675$.

The sample size is large enough if both $n\hat{p} \geq 15$ and $n\hat{q} \geq 15$.

$n\hat{p} = 400(.5675) = 227$ and $n\hat{q} = 400(.4325) = 173$. Since both of the numbers are greater than or equal to 15, the sample size is sufficiently large to conclude the normal approximation is reasonable.

For confidence coefficient .95, $\alpha = .05$ and $\alpha/2 = .05/2 = .025$. From Table III, Appendix A, $z_{.025} = 1.96$. The 95% confidence interval is:

$$\hat{p} \pm z_{.025}\sigma_{\hat{p}} \Rightarrow \hat{p} \pm 1.96\sqrt{\frac{\hat{p}\hat{q}}{n}} \Rightarrow .5675 \pm 1.96\sqrt{\frac{.5675(.4325)}{400}}$$

$$\Rightarrow .5675 \pm .0486 \Rightarrow (.5189, \ .6161)$$

We are 95% confident that the proportion of measurements in the population with characteristic A is between .5189 and .6161.

b. We will use $\hat{p} = .5675$ to estimate p.

For confidence coefficient .95, $\alpha = .05$ and $\alpha/2 = .05/2 = .025$. From Table III, Appendix A, $z_{.025} = 1.96$.

The sample size is $n = \dfrac{(z_{.025})^2 \, pq}{(SE)^2} = \dfrac{1.96^2(.5675)(.4325)}{.02^2} = 2{,}357.2 \approx 2{,}358$.

5.115   The parameters of interest for the problems are:

(1)   The question requires a categorical response. One parameter of interest might be the proportion, p, of all Americans over 18 years of age who think their health is generally very good or excellent.

(2)   A parameter of interest might be the mean number of days, μ, in the previous 30 days that all Americans over 18 years of age felt that their physical health was not good because of injury or illness.

(3)   A parameter of interest might be the mean number of days, μ, in the previous 30 days that all Americans over 18 years of age felt that their mental health was not good because of stress, depression, or problems with emotions.

(4)   A parameter of interest might be the mean number of days, μ, in the previous 30 days that all Americans over 18 years of age felt that their physical or mental health prevented them from performing their usual activities.

5.117   First, we compute $\hat{p}$: $\hat{p} = \dfrac{x}{n} = \dfrac{183}{837} = .219$

The sample size is large enough if both $n\hat{p} \geq 15$ and $n\hat{q} \geq 15$.

$n\hat{p} = 837(.219) = 183.303$ and $n\hat{q} = 837(.781) = 653.697$. Since both of the numbers are greater than or equal to 15, the sample size is sufficiently large to conclude the normal approximation is reasonable.

For confidence coefficient .90, $\alpha = .10$ and $\alpha / 2 = .10 / 2 = .05$. From Table III, Appendix A, $z_{.05} = 1.645$. The confidence interval is:

$$\hat{p} \pm z_{.05}\sqrt{\frac{pq}{n}} \approx \hat{p} \pm 1.645\sqrt{\frac{\hat{p}\hat{q}}{n}} \Rightarrow .219 \pm 1.645\sqrt{\frac{.219(.781)}{837}} \Rightarrow .219 \pm .024 \Rightarrow (.195, \ .243)$$

We are 95% confident that the population proportion of all pottery artifacts at Phylakopi that are painted is between .195 and .243.

5.119   a.   The point estimate of p, the true driver phone cell use rate, is $\hat{p} = \dfrac{x}{n} = \dfrac{35}{1,165} = .030$.

b.   The sample size is large enough if both $n\hat{p} \geq 15$ and $n\hat{q} \geq 15$.

$n\hat{p} = 1,165(.030) = 35$ and $n\hat{q} = 1,165(.970) = 1130$. Since both of the numbers are greater than or equal to 15, the sample size is sufficiently large to conclude the normal approximation is reasonable.

For confidence coefficient .95, $\alpha = .05$ and $\alpha / 2 = .05 / 2 = .025$. From Table III, Appendix A, $z_{.025} = 1.96$. The 95% confidence interval is:

$$\hat{p} \pm z_{.025}\sigma_{\hat{p}} \Rightarrow \hat{p} \pm 1.96\sqrt{\frac{\hat{p}\hat{q}}{n}} \Rightarrow .030 \pm 1.96\sqrt{\frac{.030(.970)}{1,165}} \Rightarrow .030 \pm .010 \Rightarrow (.020, \ .040)$$

c.   We are 95% confident that the true driver cell phone use rate is between .020 and .040.

d.   For confidence coefficient .95, $\alpha = .05$ and $\alpha / 2 = .05 / 2 = .025$. From Table III, Appendix A, $z_{.025} = 1.96$. We will use $\hat{p} = .030$ from part **a** to estimate p.

$$n = \frac{z_{\alpha/2}^2 pq}{(SE)^2} = \frac{1.96^2 (.03)(.97)}{.005^2} = 4,471.6 \approx 4,472$$

We would need to sample 4,472 drivers.

5.121   a.   The confidence level desired by the researchers is .90.

b.   The sampling error desired by the researchers is $SE = .05$.

c.   Since no value was given for p, we will use $\hat{p} = \dfrac{x}{n} = \dfrac{64}{106} = .604$ (from Exercise 5.120).

For confidence coefficient .90, $\alpha = .10$ and $\alpha / 2 = .10 / 2 = .05$. From Table III, Appendix A, $z_{.05} = 1.645$.

The sample size is $n = \dfrac{(z_{.05})^2 pq}{(SE)^2} = \dfrac{1.645^2 (.604)(.396)}{.05^2} = 258.9 \approx 259$.

5.123   a. $\bar{x} = \dfrac{\sum x}{n} = \dfrac{39}{21} = 1.857$ $s^2 = \dfrac{\sum x^2 - \dfrac{(\sum x)^2}{n}}{n-1} = \dfrac{101 - \dfrac{39^2}{21}}{21-1} = \dfrac{28.57143}{20} = 1.4286$

$s = \sqrt{1.4286} = 1.1952$

b.   Using MINITAB, a histogram of the data is:

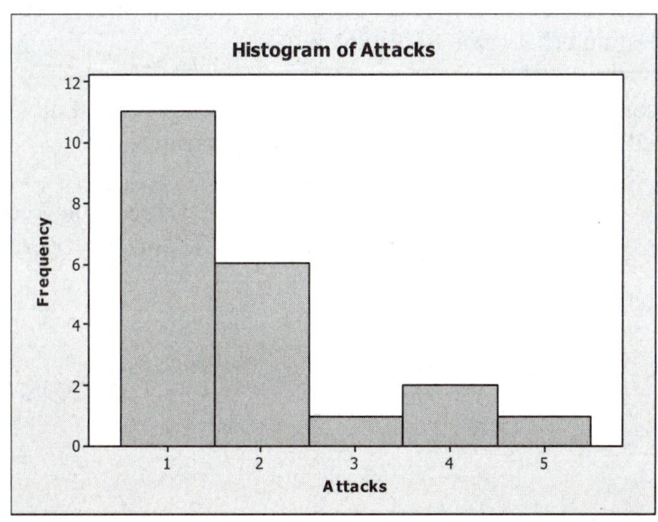

Although this is a histogram of the sample, it should reflect the distribution of the population fairly well. The population distribution appears to be skewed to the right. However, the distribution is somewhat mound-shaped. Since the sample size of $n = 21$ is somewhat close to 30, the sampling distribution of $\bar{x}$ may be approximately normal.

c.   For confidence coefficient .90, $\alpha = .10$ and $\alpha / 2 = .10 / 2 = .05$. From Table IV, Appendix A, with df $= n - 1 = 21 - 1 = 20$, $t_{.05} = 1.725$. The 90% confidence interval is:

$$\bar{x} \pm t_{.05,20} \frac{s}{\sqrt{n}} \Rightarrow 1.857 \pm 1.725 \frac{1.1952}{\sqrt{21}} \Rightarrow 1.857 \pm .450 \Rightarrow (1.407,\ 2.307)$$

d.   We are 90% confident that the true mean number of individual suicide bombings and/or attacks per incident is between 1.407 and 2.307.

e.   With repeated sampling, 90% of all confidence intervals constructed in this manner will contain the true mean and 10% will not.

5.125   Some preliminary calculations:

$$\bar{x} = \frac{\sum x}{n} = \frac{6.44}{6} = 1.073 \qquad s^2 = \frac{\sum x^2 - \frac{(\sum x)^2}{n}}{n-1} = \frac{7.1804 - \frac{6.44^2}{6}}{6-1} = .0536$$

$$s = \sqrt{.0536} = .2316$$

a.   For confidence coefficient .95, $\alpha = .05$ and $\alpha / 2 = .05 / 2 = .025$. From Table IV, Appendix A, with df $= n - 1 = 6 - 1 = 5$, $t_{.025} = 2.571$. The confidence interval is:

$$\bar{x} \pm t_{\alpha/2} \frac{s}{\sqrt{n}} \Rightarrow 1.073 \pm 2.571 \frac{.2316}{\sqrt{6}} \Rightarrow 1.073 \pm .243 \Rightarrow (.830,\ 1.316)$$

We are 95% confident that the true average decay rate of fine particles produced from oven cooking or toasting is between .830 and 1.316.

b.   The phrase "95% confident" means that in repeated sampling, 95% of all confidence intervals constructed in this manner will contain the true mean.

c.   For confidence level .95, $\alpha = .05$ and $\alpha / 2 = .05 / 2 = .025$. From Table V, Appendix A, with df $= n - 1 = 6 - 1 = 5$, $\chi^2_{.025,5} = 12.8325$ and $\chi^2_{.975,5} = .831211$. The 95% confidence interval for the variance is:

$$\frac{(n-1)s^2}{\chi^2_{.025}} \le \sigma^2 \le \frac{(n-1)s^2}{\chi^2_{.975}} \Rightarrow \frac{(6-1).2316^2}{12.8325} \le \sigma^2 \le \frac{(6-1).2316^2}{.831211} \Rightarrow .0209 \le \sigma^2 \le .3227$$

The 95% confidence interval for the standard deviation is:

$$\sqrt{.0209} \le \sigma \le \sqrt{.3227} \Rightarrow .1446 \le \sigma \le .5681$$

We are 95% confident that the true standard deviation of the decay rate is between .1446 and .5681.

d. In order for the inferences above to be valid, the distribution of decay rates must be normally distributed.

e. For confidence coefficient .95, $\alpha = .05$ and $\alpha/2 = .05/2 = .025$. From Table III, Appendix A, $z_{.025} = 1.96$.

$$n = \frac{z_{\alpha/2}^2 \alpha^2}{(SE)^2} = \frac{1.96^2(.2316)^2}{.04^2} = 128.786 \approx 129$$

We would need to sample 129 decay rates.

5.127 a. First, we compute $\hat{p}$: $\hat{p} = \frac{x}{n} = \frac{665}{1007} = .660$

The sample size is large enough if both $n\hat{p} \ge 15$ and $n\hat{q} \ge 15$.

$n\hat{p} = 1,007(.660) = 665$ and $n\hat{q} = 1,007(.340) = 342$. Since both of the numbers are greater than or equal to 15, the sample size is sufficiently large to conclude the normal approximation is reasonable.

For confidence coefficient .95, $\alpha = .05$ and $\alpha/2 = .05/2 = .025$. From Table III, Appendix A, $z_{.025} = 1.96$. The confidence interval is:

$$\hat{p} \pm z_{.025}\sqrt{\frac{pq}{n}} \approx \hat{p} \pm 1.96\sqrt{\frac{\hat{p}\hat{q}}{n}} \Rightarrow .660 \pm 1.96\sqrt{\frac{.660(.340)}{1007}} \Rightarrow .660 \pm .029 \Rightarrow (.631, .689)$$

We are 95% confident that the proportion of U.S. workers who take their lunch to work is between .631 and .689.

b. First, we compute $\hat{p}$: $\hat{p} = \frac{x}{n} = \frac{200}{665} = .301$

The sample size is large enough if both $n\hat{p} \ge 15$ and $n\hat{q} \ge 15$.

$n\hat{p} = 665(.301) = 200$ and $n\hat{q} = 665(.699) = 465$. Since both of the numbers are greater than or equal to 15, the sample size is sufficiently large to conclude the normal approximation is reasonable.

For confidence coefficient .95, $\alpha = .05$ and $\alpha / 2 = .05 / 2 = .025$. From Table III, Appendix A, $z_{.025} = 1.96$. The confidence interval is:

$$\hat{p} \pm z_{.025} \sqrt{\frac{pq}{n}} \approx \hat{p} \pm 1.96 \sqrt{\frac{\hat{p}\hat{q}}{n}} \Rightarrow .301 \pm 1.96 \sqrt{\frac{.301(.699)}{665}} \Rightarrow .301 \pm .035 \Rightarrow (.266, .336)$$

We are 95% confident that the proportion of U.S. workers who take their lunch to work who brown-bag their lunch is between .266 and .336.

5.129   a.   For confidence coefficient .99, $\alpha = .01$ and $\alpha / 2 = .01 / 2 = .005$. From Table III, Appendix A, $z_{.005} = 2.58$. The confidence interval is:

$$\bar{x} \pm z_{.005} \frac{s}{\sqrt{n}} \Rightarrow .044 \pm 2.58 \frac{.884}{\sqrt{197}} \Rightarrow .044 \pm .162 \Rightarrow (-.118, .206)$$

We are 99% confident that the mean inbreeding coefficient for this species of wasps is between $-.118$ and .206.

b.   A coefficient of 0 indicates that the wasp has no tendency to inbreed. Since 0 is in the 99% confidence interval, it is a likely value for the true mean inbreeding coefficient. Thus, there is no evidence to indicate that this species of wasps has a tendency to inbreed.

5.131   a.   Using MINITAB, the descriptive statistics are:

Descriptive Statistics: CPU Time

```
Variable   N   Mean   StDev  Minimum     Q1  Median     Q3  Maximum
CPU Time  52  0.812  1.505   0.0360  0.136   0.275  0.595    8.788
```

For confidence coefficient .95, $\alpha = .05$ and $\alpha / 2 = .05 / 2 = .025$. From Table III, Appendix A, $z_{.025} = 1.96$. The 95% confidence interval is:

$$\bar{x} \pm z_{.025} \sigma_{\bar{x}} \Rightarrow \bar{x} \pm 1.96 \frac{\sigma}{\sqrt{n}} \Rightarrow .812 \pm 1.96 \frac{1.505}{\sqrt{52}} \Rightarrow .812 \pm .409 \Rightarrow (.403, \ 1.221)$$

We are 95% confident that the mean solution time for the hybrid algorithm is between .403 and 1.221.

b.   The sample size needed would be $n = \dfrac{(z_{.025})^2 \sigma^2}{(SE)^2} = \dfrac{1.96^2 (1.505)^2}{.25^2} = 139.2 \approx 140$.

c. For confidence level .95, $\alpha = .05$ and $\alpha / 2 = .05 / 2 = .025$. From Table V, Appendix A, with df $= n - 1 = 52 - 1 = 51$, $\chi^2_{.025,51} \approx 71.4202$ and $\chi^2_{.975,51} \approx 32.3574$. The 95% confidence interval for the variance is:

$$\frac{(n-1)s^2}{\chi^2_{.025}} \leq \sigma^2 \leq \frac{(n-1)s^2}{\chi^2_{.975}} \Rightarrow \frac{(52-1)1.505^2}{71.4202} \leq \sigma^2 \leq \frac{(52-1)1.505^2}{32.3574} \Rightarrow 1.6174 \leq \sigma^2 \leq 3.5700$$

The 95% confidence interval for the standard deviation is:

$$\sqrt{1.6174} \leq \sigma \leq \sqrt{3.5700} \Rightarrow 1.272 \leq \sigma \leq 1.889$$

We are 95% confident that the true standard deviation of the solution time for the hybrid algorithm is between 1.272 and 1.889.

5.133   For all parts to this problem, we will use a 95% confidence interval. For confidence coefficient .95, $\alpha = .05$ and $\alpha / 2 = .05 / 2 = .025$. From Table III, Appendix A, $z_{.025} = 1.96$.

a. Some preliminary calculations:

$$\hat{p} = \frac{x}{n} = \frac{14}{37} = .378$$

The sample size is large enough if both $n\hat{p} \geq 15$ and $n\hat{q} \geq 15$.

$n\hat{p} = 37(.378) = 14$ and $n\hat{q} = 37(.622) = 23$. Since the first number is not greater than or equal to 15, the sample size is not sufficiently large to conclude the normal approximation is reasonable.

Since the normal approximation is not reasonable, we will use the Wilson adjusted confidence interval.

$$\tilde{p} = \frac{x+2}{n+4} = \frac{14+2}{37+4} = \frac{16}{41} = .390$$

The Wilson adjusted 95% confidence interval is:

$$\tilde{p} \pm z_{.025} \sqrt{\frac{\tilde{p}(1-\tilde{p})}{n+4}} \Rightarrow .390 \pm 1.96 \sqrt{\frac{.390(.610)}{37+4}} \Rightarrow .390 \pm .149 \Rightarrow (.241, \ .539)$$

We are 95% confident that the true proportion of suicides at the jail that are committed by inmates charged with murder/manslaughter is between .241 and .539.

b. Some preliminary calculations:

$$\hat{p} = \frac{x}{n} = \frac{26}{37} = .703$$

The sample size is large enough if both $n\hat{p} \geq 15$ and $n\hat{q} \geq 15$.

$n\hat{p} = 37(.703) = 26$ and $n\hat{q} = 37(.297) = 11$. Since the second number is not greater than or equal to 15, the sample size is not sufficiently large to conclude the normal approximation is reasonable.

Since the normal approximation is not reasonable, we will use the Wilson adjusted confidence interval.

$$\tilde{p} = \frac{x+2}{n+4} = \frac{26+2}{37+4} = \frac{28}{41} = .683$$

The Wilson adjusted 95% confidence interval is:

$$\tilde{p} \pm z_{.025}\sqrt{\frac{\tilde{p}(1-\tilde{p})}{n+4}} \Rightarrow .683 \pm 1.96\sqrt{\frac{.683(.317)}{37+4}} \Rightarrow .683 \pm .142 \Rightarrow (.541,\ .825)$$

We are 95% confident that the true proportion of suicides at the jail that are committed at night is between .541 and .825.

c. Some preliminary calculations are:

$$\bar{x} = \frac{\sum x}{n} = \frac{1532}{37} = 41.405$$

$$s^2 = \frac{\sum x^2 - \frac{\left(\sum x\right)^2}{n}}{n-1} = \frac{223,606 - \frac{(1532)^2}{37}}{37-1} = 4,449.2477$$

$$s = \sqrt{s^2} = \sqrt{4,449.2477} = 66.703$$

The confidence interval is:

$$\bar{x} \pm z_{.05}\frac{s}{\sqrt{n}} \Rightarrow 41.405 \pm 1.96\frac{66.703}{\sqrt{37}} \Rightarrow 41.405 \pm 21.493 \Rightarrow (19.912,\ 62.898)$$

We are 95% confident that the true average length of time an inmate is in jail before committing suicide is between 19.912 and 62.898 days.

d. Some preliminary calculations:

$$\hat{p} = \frac{x}{n} = \frac{14}{37} = .378$$

The sample size is large enough if both $n\hat{p} \geq 15$ and $n\hat{q} \geq 15$.

$n\hat{p} = 37(.378) = 14$ and $n\hat{q} = 37(.622) = 23$. Since the first number is not greater than or equal to 15, the sample size is not sufficiently large to conclude the normal approximation is reasonable.

Since the normal approximation is not reasonable, we will use the Wilson adjusted confidence interval.

$$\tilde{p} = \frac{x+2}{n+4} = \frac{14+2}{37+4} = \frac{16}{41} = .390$$

The Wilson adjusted 95% confidence interval is:

$$\tilde{p} \pm z_{.025}\sqrt{\frac{\tilde{p}(1-\tilde{p})}{n+4}} \Rightarrow .390 \pm 1.96\sqrt{\frac{.390(.610)}{37+4}} \Rightarrow .390 \pm .149 \Rightarrow (.241,\ .539)$$

We are 95% confident that the true percentage of suicides at the jail that are committed by white inmates is between 24.1% and 53.9%.

5.135　a.　The point estimate for the fraction of the entire market that refuses to purchase bars is:

$$\hat{p} = \frac{x}{n} = \frac{23}{244} = .094$$

　　　b.　The sample size is large enough if both $n\hat{p} \geq 15$ and $n\hat{q} \geq 15$.

$n\hat{p} = 244(.094) = 23$ and $n\hat{q} = 244(.906) = 221$. Since both of the numbers are greater than or equal to 15, the sample size is sufficiently large to conclude the normal approximation is reasonable.

　　　c.　For confidence coefficient .95, $\alpha = .05$ and $\alpha/2 = .05/2 = .025$. From Table III, Appendix A, $z_{.025} = 1.96$. The confidence interval is:

$$\hat{p} \pm z_{.025}\sqrt{\frac{\hat{p}\hat{q}}{n}} \Rightarrow .094 \pm 1.96\sqrt{\frac{.094(.906)}{224}} \Rightarrow .094 \pm .037 \Rightarrow (.057,\ 131)$$

　　　d.　The best estimate of the true fraction of the entire market that refuses to purchase bars six months after the poisoning is .094. We are 95% confident the true fraction of the entire market that refuses to purchase bars six months after the poisoning is between .057 and .131.

5.137　a.　For confidence coefficient .99, $\alpha = .01$ and $\alpha/2 = .01/2 = .005$. From Table IV, Appendix A, with df $= n - 1 = 3 - 1 = 2$, $t_{.005} = 9.925$. The confidence interval is:

$$\bar{x} \pm t_{.005}\frac{s}{\sqrt{n}} \Rightarrow 49.3 \pm 9.925\frac{1.5}{\sqrt{3}} \Rightarrow 49.3 \pm 8.60 \Rightarrow (40.70,\ 57.90)$$

　　　b.　We are 99% confident that the mean percentage of B(a)p removed from all soil specimens using the poison is between 40.70% and 57.90%.

c. We must assume that the distribution of the percentages of B(a)p removed from all soil specimens using the poison is normal.

d. For confidence coefficient .99, $\alpha = .01$ and $\alpha / 2 = .01 / 2 = .005$. From Table III, Appendix A, $z_{.005} = 2.575$.

$$n = \frac{z_{\alpha/2}^2 \sigma^2}{(SE)^2} = \frac{2.575^2 (1.5)^2}{.5^2} = 59.68 \approx 60$$

e. For confidence level .90, $\alpha = .10$ and $\alpha / 2 = .10 / 2 = .05$. From Table V, Appendix A, with df $= n - 1 = 3 - 1 = 2$, $\chi_{.05,2}^2 = 5.99147$ and $\chi_{.95,2}^2 = .102587$. The 90% confidence interval for the variance is:

$$\frac{(n-1)s^2}{\chi_{.05}^2} \le \sigma^2 \le \frac{(n-1)s^2}{\chi_{.95}^2} \Rightarrow \frac{(3-1)1.5^2}{5.99147} \le \sigma^2 \le \frac{(3-1)1.5^2}{.102587} \Rightarrow .7511 \le \sigma^2 \le 43.8652$$

We are 90% confidence that the true variance of the percentage of B(a)p removed is between .7511 and 43.8652.

5.139 a. From Chebyshev's Rule, we know that at least $1 - \frac{1}{k^2}$ of the observations will fall within k standard deviations of the mean. We need to find k so that $1 - \frac{1}{k^2} = .60$.

$$1 - \frac{1}{k^2} = .60 \Rightarrow .40 = \frac{1}{k^2} \Rightarrow k^2 = \frac{1}{.40} = 2.5 \Rightarrow k = \sqrt{2.5} = 1.58$$

$$s = \frac{80^{th} \text{ percentile} - 60^{th} \text{ percentile}}{2k} = \frac{73,000 - 35,100}{2(1.58)} = 11,993.67$$

For confidence coefficient .98, $\alpha = .02$ and $\alpha / 2 = .02 / 2 = .01$. From Table III, Appendix A, $z_{.01} = 2.33$. The sample size required is:

$$n = \frac{(z_{.01})^2 \sigma^2}{(SE)^2} = \frac{2.33^2 (11,993.67)^2}{2,000^2} = 195.23 \approx 196$$

b. See part **a**.

c. We must assume that any sample selected will be random.

5.141 a. As long as the sample is random (and thus representative), a reliable estimate of the mean weight of all the scallops can be obtained.

b. The government is using only the sample mean to make a decision. Rather than using a point estimate, they should probably use a confidence interval to estimate the true mean weight of the scallops.

c.   We will form a 95% confidence interval for the mean weight of the scallops. Using
MINITAB, the descriptive statistics are:

Descriptive Statistics: Weight

```
Variable    N    Mean    StDev  Minimum       Q1  Median     Q3    Maximum
Weight     18  0.9317   0.0753   0.8400   0.8800  0.9100   9800     1.1400
```

For confidence coefficient .95, $\alpha = .05$ and $\alpha / 2 = .05 / 2 = .025$. From Table IV,
Appendix A, with df $= n - 1 = 18 - 1 = 17$, $t_{.025} = 2.110$.

The 95% confidence interval is:

$$\bar{x} \pm t_{.025} \frac{s}{\sqrt{n}} \Rightarrow .9317 \pm 2.110 \frac{.0753}{\sqrt{18}} \Rightarrow .9317 \pm .0374 \Rightarrow (.8943, \ .9691)$$

We are 95% confident that the true mean weight of the scallops is between .8943 and
.9691. Recall that the weights have been scaled so that a mean weight of 1 corresponds
to 1/36 of a pound. Since the above confidence interval does not include 1, we have
sufficient evidence to indicate that the minimum weight restriction was violated.

Inferences Based on a Single Sample
Tests of Hypothesis

6.1   The null hypothesis is the "status quo" hypothesis, while the alternative hypothesis is the research hypothesis.

6.3   The "level of significance" of a test is α. This is the probability that the test statistic will fall in the rejection region when the null hypothesis is true.

6.5   The four possible results are:

1.   Rejecting the null hypothesis when it is true. This would be a Type I error.
2.   Accepting the null hypothesis when it is true. This would be a correct decision.
3.   Rejecting the null hypothesis when it is false. This would be a correct decision.
4.   Accepting the null hypothesis when it is false. This would be a Type II error.

6.7   When you reject the null hypothesis in favor of the alternative hypothesis, this does not prove the alternative hypothesis is correct. We are $100(1-\alpha)\%$ confident that there is sufficient evidence to conclude that the alternative hypothesis is correct.

We know the hypothesis-testing process will lead to this conclusion (reject H_0) incorrectly only $100\alpha\%$ of the time when H_0 is true.

6.9   a.   Probability of Type I error
        $= P(z > 1.96) = .5 - .4750 = .0250$
        (From Table III, Appendix A)

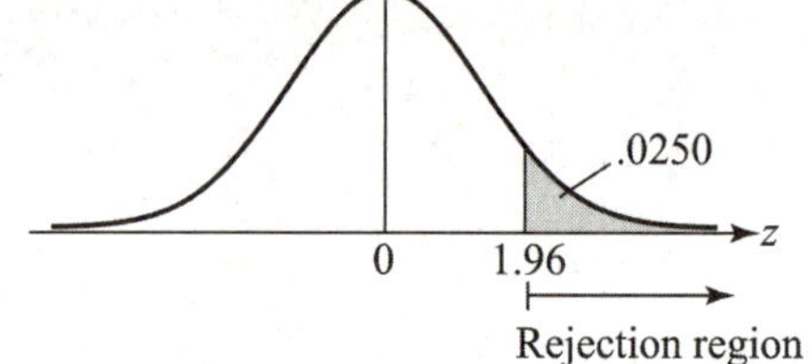

b.   Probability of Type I error
        $= P(z > 1.645) = .5 - .4500 = .05$
        (From Table III, Appendix A)

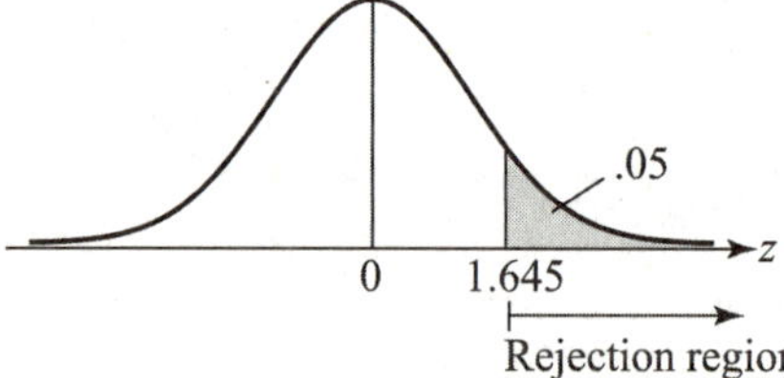

c.   Probability of Type I error
        $= P(z > 2.575) = .5 - .4950 = .0050$
        (From Table III, Appendix A)

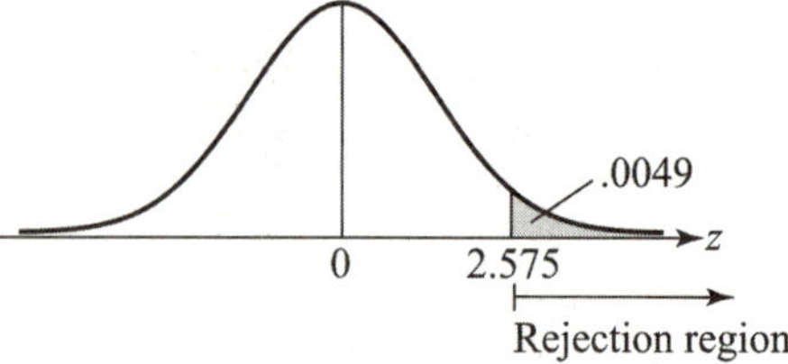

d. Probability of Type I error
$$= P(z < -1.28) = .5 - .3997 = .1003$$
(From Table IV, Appendix A)

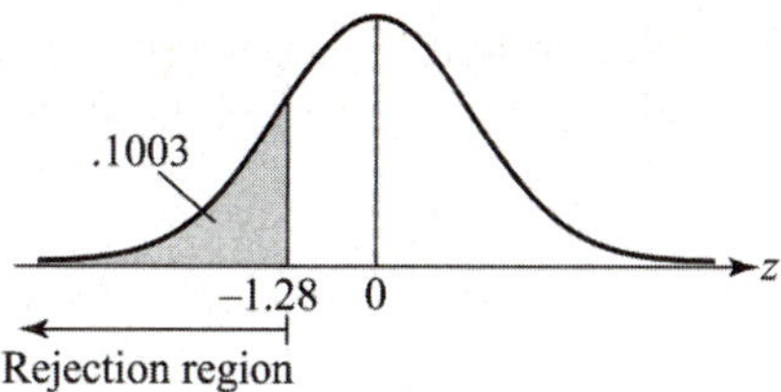

e. Probability of Type I error
$$= P(z < -1.645) + P(z > 1.645)$$
$$= .5 - .4500 + .5 - .4500 = .05 + .05 = .10$$
(From Table IV, Appendix A)

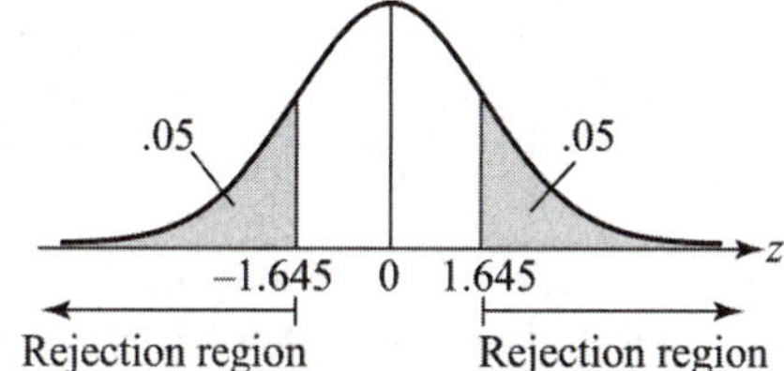

f. Probability of Type I error
$$= P(z < -2.575) + P(z > 2.575)$$
$$= .5 - .4950 + .5 - .4950 = .0050 + .0050$$
$$= .0100 \quad \text{(From Table IV, Appendix A)}$$

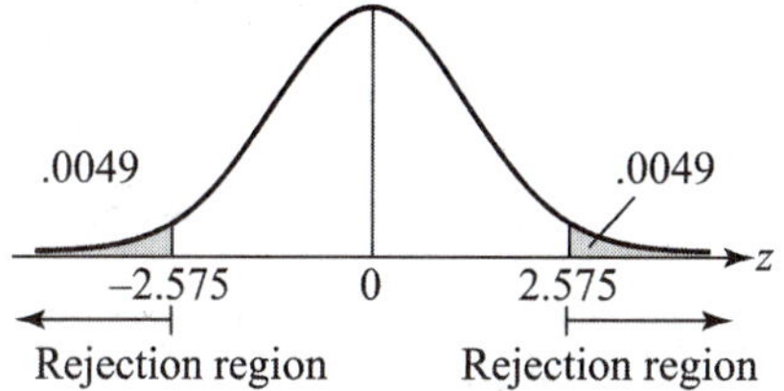

6.11 Let μ = average gain in green fees, lessons, or equipment expenditures for participating facilities.

a. To determine if the average gain exceeds $2,400, we test:

H_0: $\mu = 2400$
H_a: $\mu > 2400$

b. For this problem, α = the probability of concluding the average gain in green fees, lessons, or equipment expenditures for participating facilities is greater than $2,400 when, in fact, it is not.

c. The rejection region requires $\alpha = .05$ in the upper tail of the z distribution. From Table III, Appendix A, $z_{.05} = 1.645$. The rejection region is $z > 1.645$.

6.13 Let p = proportion of liars correctly detected by the new thermal imaging camera. To determine if the camera can correctly detect liars 75% of the time, we test:

H_0: $p = .75$

6.15 Let p = error rate of the DNA-reading device. To determine if the error rate of the DNA-reading device is less than 5%, we test:

H_0: $p = .05$
H_a: $p < .05$

6.17   a.   To determine if the average level of mercury uptake in wading birds in the Everglades today is less than 15 parts per million, we test:

$$H_0: \mu = 15$$
$$H_a: \mu < 15$$

b.   A Type I error is rejecting H_0 when H_0 is true. In terms of this problem, we would be concluding that the average level of mercury uptake in wading birds in the Everglades today is less than 15 parts per million, when in fact, the average level of mercury uptake in wading birds in the Everglades today is equal to 15 parts per million.

c.   A Type II error is accepting H_0 when H_0 is false. In terms of this problem, we would be concluding that the average level of mercury uptake in wading birds in the Everglades today is equal to 15 parts per million, when in fact, the average level of mercury uptake in wading birds in the Everglades today is less than 15 parts per million.

6.19   a.   The null hypothesis is: H_0: No intrusion occurs

b.   The alternative hypothesis is: H_a: Intrusion occurs

c. α = Probability of a Type I error = Probability of rejecting H_o when it is true =
P(system
       provides warning when no intrusion occurs) = 1 / 1000 = .001.

β = Probability of a Type II error = Probability of accepting H_o when it is false =
P(system does not provide warning when an intrusion occurs) = 500 / 1000 = .5.

6.21   There are 2 conditions required for a valid large-sample hypothesis test for μ. They are:

1.   A random sample is selected from a target population.
2.   The sample size n is large, i.e., $n \geq 30$. (Due to the Central Limit Theorem, this condition guarantees that the test statistic will be approximately normal regardless of the shape of the underlying probability distribution of the population.)

6.23   a. $H_0: \mu = 100$
$H_a: \mu > 100$

The test statistic is $z = \dfrac{\bar{x} - \mu_0}{\sigma_{\bar{x}}} = \dfrac{\bar{x} - \mu_0}{\sigma / \sqrt{n}} = \dfrac{110 - 100}{60 / \sqrt{100}} = 1.67$

The rejection region requires $\alpha = .05$ in the upper tail of the z distribution. From Table III, Appendix A, $z_{.05} = 1.645$. The rejection region is $z > 1.645$.

Since the observed value of the test statistic falls in the rejection region, ($z = 1.67 > 1.645$), H_0 is rejected. There is sufficient evidence to indicate the true population mean is greater than 100 at $\alpha = .05$.

b. H_0: $\mu = 100$
H_a: $\mu \neq 100$

The test statistic is $z = \dfrac{\overline{x} - \mu_0}{\sigma_{\overline{x}}} = \dfrac{110 - 100}{60 / \sqrt{100}} = 1.67$

The rejection region requires $\alpha / 2 = .05 / 2 = .025$ in each tail of the z distribution. From Table III, Appendix A, $z_{.025} = 1.96$. The rejection region is $z < -1.96$ or $z > 1.96$.

Since the observed value of the test statistic does not fall in the rejection region, $(z = 1.67 \ngtr 1.96)$, H_0 is not rejected. There is insufficient evidence to indicate μ differs from 100 at $\alpha = .05$.

c.   In part **a**, we rejected H_0 and concluded the mean was greater than 100. In part **b**, we did not reject H_0. There was insufficient evidence to conclude the mean was different from 100. Because the alternative hypothesis in part **a** is more specific than the one in **b**, it is easier to reject H_0.

6.25   a. $z = \dfrac{\overline{x} - \mu}{\sigma_{\overline{x}}} = \dfrac{215 - 200}{80 / \sqrt{100}} = 1.88$

The decision rule is: Reject H_0 if $z > 1.88$.

b. $\alpha = P(z > 1.88) = .5 - .4699 = .0301$ (using Table III, Appendix A)

6.27   a.   A Type I error would be to conclude that the mean response for the population of all New York City public school children is not 3 when, in fact, the mean response is equal to 3.

A Type II error would be to conclude that the mean response for the population of all New York City public school children is equal to 3 when, in fact, the mean response is not equal to 3.

b.   To determine if the mean response for the population of all New York City public school children is different from 3, we test:

H_0: $\mu = 3$
H_a: $\mu \neq 3$

The test statistic is $z = \dfrac{\overline{x} - \mu_o}{\sigma_{\overline{x}}} = \dfrac{2.15 - 3}{\dfrac{1.05}{\sqrt{11,160}}} = -85.52$

The rejection region requires $\alpha / 2 = .05 / 2 = .025$ in each tail of the z distribution. From Table III, Appendix A, $z_{.025} = 1.96$. The rejection region is $z < -1.96$ or $z > 1.96$.

Since the observed value of the test statistic falls in the rejection region ($z = -85.52 < -1.96$), H_0 is rejected. There is sufficient evidence to indicate the mean response for the population of all New York City public school children different from 3 at $\alpha = .05$.

c.   To determine if the mean response for the population of all New York City public school children is different from 3, we test:

H_0: $\mu = 3$

H_a: $\mu \neq 3$

The test statistic is $z = \dfrac{\bar{x} - \mu_o}{\sigma_{\bar{x}}} = \dfrac{2.15 - 3}{\dfrac{1.05}{\sqrt{11,160}}} = -85.52$

The rejection region requires $\alpha / 2 = .10 / 2 = .05$ in each tail of the z distribution. From Table III, Appendix A, $z_{.05} = 1.645$. The rejection region is $z < -1.645$ or $z > 1.645$.

Since the observed value of the test statistic falls in the rejection region ($z = -85.52 < -1.645$), H_0 is rejected. There is sufficient evidence to indicate the mean response for the population of all New York City public school children different from 3 at $\alpha = .10$.

6.29   Let $\mu =$ mean Mach rating score of all purchasing managers.

a.   To determine if the true mean Mach rating score of all purchasing managers is different from 85, we test,

H_0: $\mu = 85$

H_a: $\mu \neq 85$

b.   A Type I error would be concluding that the true mean Mach rating score is different from 85 when, in fact, it is equal to 85.

c.   By definition, $\alpha =$ probability of committing a Type I error. With a value of $\alpha = .10$, this indicates that in repeated sampling, we will reject H_0 when it is true about 10% of the time.

d.   The rejection region requires $\alpha / 2 = .10 / 2 = .05$ in each tail of the z distribution. From Table III, Appendix A, $z_{.05} = 1.645$. The rejection region is $z > 1.645$ or $z < -1.645$.

e.   The test statistic is $z = \dfrac{\bar{x} - \mu}{\sigma_{\bar{x}}} \approx \dfrac{99.6 - 85}{12.6 / \sqrt{122}} = 12.80.$

f.   Since the observed value of the test statistic falls in the rejection region ($z = 12.80 > 1.645$), H_0 is rejected. There is sufficient evidence to indicate the true mean Mach rating score is different from 85 at $\alpha = .10$.

g.   No. Because the sample size is sufficiently large ($n = 122$), the Central Limit Theorem says that the distribution of $\bar{x}$ will be approximately normal.

6.31   Let μ = mean disclosure score. To determine if the true mean disclosure score exceeds 3, we test:

$$H_0:\ \mu = 3$$
$$H_a:\ \mu > 3$$

The test statistic is $z = \dfrac{\bar{x} - \mu_o}{\sigma_{\bar{x}}} \approx \dfrac{3.26 - 3}{\dfrac{.93}{\sqrt{222}}} = 4.17$

The rejection region requires $\alpha = .01$ in the upper tail of the z distribution. From Table III, Appendix A, $z_{.01} = 2.33$. The rejection region is $z > 2.33$.

Since the observed value of the test statistic falls in the rejection region ($z = 4.17 > 2.33$), H_0 is rejected. There is sufficient evidence to indicate the true mean disclosure score exceeds 3 at $\alpha = .01$.

6.33   a.   Using MINITAB, the descriptive statistics are:

Descriptive Statistics: Bones

Variable	N	Mean	StDev	Minimum	Q1	Median	Q3	Maximum
Bones	41	9.258 8	1.204	6.230	8.615	9.200	9.935	12.000

To determine if the population mean ratio of all bones of this particular species differs from 8.5, we test:

$$H_0:\ \mu = 8.5$$
$$H_a:\ \mu \neq 8.5$$

The test statistic is $z = \dfrac{\bar{x} - \mu_0}{\sigma_{\bar{x}}} = \dfrac{9.258 - 8.5}{1.204 / \sqrt{41}} = 4.03$

The rejection region requires $\alpha / 2 = .01 / 2 = .005$ in each tail of the z distribution. From Table III, Appendix A, $z_{.005} = 2.58$. The rejection region is $z < -2.58$ or $z > 2.58$.

Since the observed value of the test statistic falls in the rejection region ($z = 4.03 > 2.58$), H_0 is rejected. There is sufficient evidence to indicate that the true population mean ratio of all bones of this particular species differs from 8.5 at $\alpha = .01$.

b.   The practical implications of the test in part **a** is that the species from which these particular bones came from is probably not species A.

6.35   a.   Since the hypothesized value of μ_M ($60,000) falls in the 95% confidence interval, it is not an unusual value. There is no evidence to reject it. There is no evidence to indicate that the mean salary of all males with post-graduate degrees differs from $60,000.

b. To determine if the mean salary of all males with post-graduate degrees differs from $60,000, we test:

H_0: $\mu_M = 60,000$

H_a: $\mu_M \neq 60,000$

The test statistic is $z = \dfrac{\bar{x} - \mu_0}{\sigma_{\bar{x}}} = \dfrac{61,340 - 60,000}{2,185} = 0.61$

The rejection region requires $\alpha / 2 = .05 / 2 = .025$ in each tail of the z distribution. From Table III, Appendix A, $z_{.025} = 1.96$. The rejection region is $z < -1.96$ or $z > 1.96$.

Since the observed value of the test statistic does not fall in the rejection region ($z = 0.61 \not> 1.96$), H_0 is not rejected. There is insufficient evidence to indicate the mean salary of all males with post-graduate degrees differs from $60,000 at $\alpha = .05$.

c. The inferences in parts **a** and **b** agree because the same values are used for both. The z value used in the 95% confidence interval is the same z used for the rejection region. The value of $\bar{x}$ and $s_{\bar{x}}$ are the same in both the confidence interval and the test statistic.

d. Since the hypothesized value of μ_F ($33,000) falls in the 95% confidence interval, it is not an unusual value. There is no evidence to reject it. There is no evidence to indicate that the mean salary of all females with post-graduate degrees differs from $33,000.

e. To determine if the mean salary of all females with post-graduate degrees differs from $33,000, we test:

H_0: $\mu_M = 33,000$

H_a: $\mu_M \neq 33,000$

The test statistic is $z = \dfrac{\bar{x} - \mu_0}{\sigma_{\bar{x}}} = \dfrac{32,227 - 33,000}{932} = -0.83$

The rejection region requires $\alpha / 2 = .05 / 2 = .025$ in each tail of the z distribution. From Table III, Appendix A, $z_{.025} = 1.96$. The rejection region is $z < -1.96$ or $z > 1.96$.

Since the observed value of the test statistic does not fall in the rejection region ($z = -0.83 \not< -1.96$), H_0 is not rejected. There is insufficient evidence to indicate the mean salary of all females with post-graduate degrees differs from $33,000 at $\alpha = .05$.

f. The inferences in parts **d** and **e** agree because the same values are used for both. The z value used in the 95% confidence interval is the same z used for the rejection region. The value of $\bar{x}$ and $s_{\bar{x}}$ are the same in both the confidence interval and the test statistic.

6.37   a.   To determine if the mean social interaction score of all Connecticut mental health patients differs from 3, we test:

$$H_0: \ \mu = 3$$
$$H_a: \ \mu \neq 3$$

The test statistic is $z = \dfrac{\bar{x} - \mu_0}{\sigma_{\bar{x}}} = \dfrac{2.95 - 3}{1.10 / \sqrt{6,681}} = -3.72$

The rejection region requires $\alpha / 2 = .01 / 2 = .005$ in each tail of the z distribution. From Table III, Appendix A, $z_{.005} = 2.58$. The rejection region is $z < -2.58$ or $z > 2.58$.

Since the observed value of the test statistic falls in the rejection region ($z = -3.72 < -2.58$), H_0 is rejected. There is sufficient evidence to indicate that the mean social interaction score of all Connecticut mental health patients differs from 3 at $\alpha = .01$.

   b.   From the test in part a, we found that the mean social interaction score was statistically different from 3. However, the sample mean score was 2.95. Practically speaking, 2.95 is very similar to 3.0. The very large sample size, $n = 6,681$, makes it very easy to find statistical significance, even when no practical significance exists.

   c.   Because the variable of interest is measured on a 5-point scale, it is very unlikely that the population of the ratings will be normal. However, because the sample size was extremely large, ($n = 6,681$), the Central Limit Theorem will apply. Thus, the distribution of $\bar{x}$ will be normal, regardless of the distribution of x. Thus, the analysis used above is appropriate.

6.39   In general, small p-values support the alternative hypothesis, H_a. The p-value is the probability of observing your test statistic or anything more unusual, given H_0 is true. If the p-value is small, it would be very unusual to observe your test statistic, given H_0 is true. This would indicate that H_0 is probably false and H_a is probably true.

6.41   a.   Since the p-value $= .10$ is greater than $\alpha = .05$, H_0 is not rejected.

   b.   Since the p-value $= .05$ is less than $\alpha = .10$, H_0 is rejected.

   c.   Since the p-value $= .001$ is less than $\alpha = .01$, H_0 is rejected.

   d.   Since the p-value $= .05$ is greater than $\alpha = .025$, H_0 is not rejected.

   e.   Since the p-value $= .45$ is greater than $\alpha = .10$, H_0 is not rejected.

6.43 p-value $= P(z \geq 2.17) = .5 - P(0 < z < 2.17) = .5 - .4850 = .0150$

(using Table III, Appendix A)

6.45   The test statistic is $z = \dfrac{\bar{x} - \mu_0}{\sigma_{\bar{x}}} = \dfrac{49.4 - 50}{4.1 / \sqrt{100}} = -1.46$

The p-value $= p = P(z \geq -1.46) = .5 + .4279 = .9279$

Since the p-value is so large, there is no evidence to reject H_0 for any reasonable value of α.

6.47   a.   The p-value reported by SPSS is for a two-tailed test. Thus, $P(z \leq -1.63) + P(z \geq 1.63)$
= .1032. For this one-tailed test, the p-value $= P(z \leq -1.63) = .1032/2 = .0516$.

Since the p-value $= .0516 > \alpha = .05$, H_0 is not rejected. There is insufficient evidence to indicate μ is less than 75 at $\alpha = .05$.

b.   For this one-tailed test, the p-value $= P(z \leq 1.63)$. Since $P(z \leq -1.63) = .1032/2 = .0516$, $P(z \leq 1.63) = 1 - .0516 = .9484$.

Since the p-value $= .9484 > \alpha = .10$, H_0 is not rejected. There is insufficient evidence to indicate μ is less than 75 at $\alpha = .10$.

c.   For this one-tailed test, the p-value $= P(z \geq 1.63) = .1032/2 = .0516$.

Since the p-value $= p = .0516 < \alpha = .10$, H_0 is rejected. There is sufficient evidence to indicate μ is greater than 75 at $\alpha = .10$.

d.   For this two-tailed test, the p-value $= .1032$.

Since the p-value $= .1032 > \alpha = .01$, H_0 is not rejected. There is insufficient evidence to indicate μ differs from 75 at $\alpha = .01$.

6.49   a.   From Exercise 6.27, $z = -85.52$. The p-value is $p = P(z \leq -85.52) + P(z \geq 85.52)$
= (.5 − .5) + (.5 − .5) = 0 (using Table III, Appendix A).

b.   The p-value is $p = 0$. Since the p-value is less than $\alpha = .10$, H_0 is rejected. There is sufficient evidence to indicate the mean response for the population of all New York City public school children different from 3 at $\alpha = .10$.

6.51   From the printout, the p-value $< .0001$. Since the p-value $< .0001 < \alpha = .01$, H_0 is rejected. There is sufficient evidence to indicate that the true population mean ratio of all bones of this particular species differs from 8.5 at $\alpha = .01$.

6.53   a.   Let $\mu =$ mean forecast error for buy-side analysts. To determine if the mean forecast error for buy-side analysts is positive, we test:

H_0: $\mu = 0$
H_a: $\mu > 0$

The test statistic is $z = \dfrac{\bar{x} - \mu_o}{\sigma_{\bar{x}}} \approx \dfrac{.85 - 0}{\dfrac{1.93}{\sqrt{3,526}}} = 26.15$

The p-value is $p = P(z \geq 26.51) \approx .5 - .5 = 0$.

Since the p-value is less than $\alpha = .01$ ($p = 0 < .01$), H_0 is rejected. There is sufficient evidence to indicate the mean forecast error for buy-side analysts is positive at $\alpha = .01$.

b. Let μ = mean forecast error for sell-side analysts. To determine if the mean forecast error for sell-side analysts is negative, we test:

$H_0 : \mu = 0$
$H_a : \mu < 0$

The test statistic is $z = \dfrac{\bar{x} - \mu_o}{\sigma_{\bar{x}}} \approx \dfrac{-.05 - 0}{\dfrac{.85}{\sqrt{58,562}}} = -14.24$

The p-value is $p = P(z \leq -14.24) \approx .5 - .5 = 0$.

Since the p-value is less than $\alpha = .01$ ($p = 0 < .01$), H_0 is rejected. There is sufficient evidence to indicate the mean forecast error for sell-side analysts is negative at $\alpha = .01$.

6.55 a. The test statistic is $z = \dfrac{\bar{x} - \mu_0}{\sigma_{\bar{x}}} = \dfrac{10.2 - 0}{31.3 / \sqrt{50}} = 2.3$

b. To determine if the mean level of feminization differs from 0%, we test:

$H_0: \mu = 0$
$H_a: \mu \neq 0$

Since the alternative hypothesis contains $\neq$, this is a two-tailed test. The p-value is
$p = P(z \leq -2.3) + P(z \geq 2.3) = .5 - .4893 + .5 - .4893 = .0214$.
(Using Table III, Appendix A)

Since the p-value is so small, there is evidence to reject H_0. There is sufficient evidence to indicate the mean level of feminization is different from 0% for any value of $\alpha < .0214$.

c.   The test statistic is $z = \dfrac{\bar{x} - \mu_0}{\sigma_{\bar{x}}} = \dfrac{15 - 0}{25.1/\sqrt{50}} = 4.23$.

Since the alternative hypothesis contains $\neq$, this is a two-tailed test. The p-value is
$p = P(z \leq -4.23) + P(z \geq 4.23) \approx .5 - .5 + .5 - .5 = 0$.

Since the p-value is so small, there is evidence to reject H_0. There is sufficient evidence
to indicate the mean level of feminization is different from 0% for any value of α.

6.57   The z and t distributions are alike in that they are both symmetric, bell-shaped distributions
centered at 0. They differ in variability, though. The z distribution always has $\sigma^2 = 1$, while
the variance of the t distribution is always larger and depends on the degrees of freedom.

6.59   a. $P(t > 1.440) = .10$
(Using Table VI, Appendix A, with df = 6)

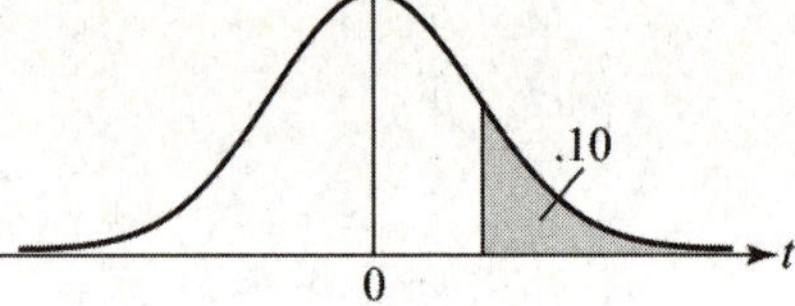

b. $P(t < -1.782) = .05$
(Using Table VI, Appendix A, with df = 12)

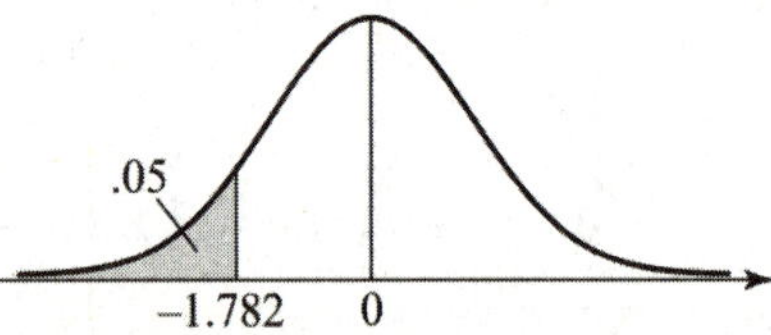

c. $P(t < -2.060) = P(t > 2.060) = .025$
(Using Table VI, Appendix A, with df =
25)

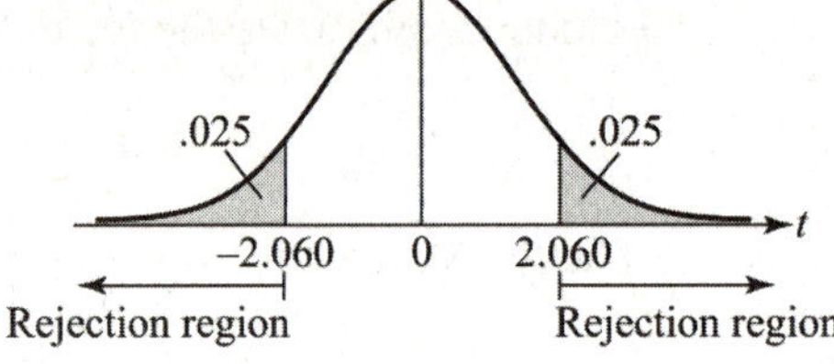

6.61   a.   The rejection region requires $\alpha / 2 = .05 / 2 = .025$ in each tail of the t distribution with
df $= n - 1 = 14 - 1 = 13$. From Table IV, Appendix A, $t_{.025} = 2.160$. The rejection
region is $t < -2.160$ or $t > 2.160$.

b.   The rejection region requires $\alpha = .01$ in the upper tail of the t distribution with df $=$
$n - 1 = 24 - 1 = 23$. From Table IV, Appendix A, $t_{.01} = 2.500$. The rejection region is
$t > 2.500$.

c.   The rejection region requires $\alpha = .10$ in the upper tail of the t distribution with df $=$
$n - 1 = 9 - 1 = 8$. From Table IV, Appendix A, $t_{.10} = 1.397$. The rejection region is
$t > 1.397$.

d.   The rejection region requires $\alpha = .01$ in the lower tail of the t distribution with df $=$
$n - 1 = 12 - 1 = 11$. From Table IV, Appendix A, $t_{.01} = 2.718$. The rejection region is
$t < -2.718$.

e.   The rejection region requires $\alpha/2 = .10/2 = .05$ in each tail of the t distribution with df $= n - 1 = 20 - 1 = 19$. From Table IV, Appendix A, $t_{.05} = 1.729$. The rejection region is $t < -1.729$ or $t > 1.729$.

f.   The rejection region requires $\alpha = .05$ in the lower tail of the t distribution with df $= n - 1 = 4 - 1 = 3$. From Table IV, Appendix A, $t_{.05} = 2.353$. The rejection region is $t < -2.353$.

6.63   a. H_0: $\mu = 6$

          H_a: $\mu < 6$

The test statistic is $t = \dfrac{\bar{x} - \mu_0}{s/\sqrt{n}} = \dfrac{4.8 - 6}{1.3/\sqrt{5}} = -2.064$.

The rejection region requires $\alpha = .05$ in the lower tail of the t distribution with df $= n - 1 = 5 - 1 = 4$. From Table IV, Appendix A, $t_{.05} = 2.132$. The rejection region is $t < -2.132$.

Since the observed value of the test statistic does not fall in the rejection region $(t = -2.064 \not< -2.132)$, H_0 is not rejected. There is insufficient evidence to indicate the mean is less than 6 at $\alpha = .05$.

b. H_0: $\mu = 6$

        H_a: $\mu = 6$

The test statistic is $t = -2.064$ (from **a**).

The rejection region requires $\alpha/2 = .05/2 = .025$ in each tail of the t distribution with df $= n - 1 = 5 - 1 = 4$. From Table IV, Appendix A, $t_{.025} = 2.776$. The rejection region is $t < -2.776$ or $t > 2.776$.

Since the observed value of the test statistic does not fall in the rejection region ($t = -2.064 \not< -2.776$), H_0 is not rejected. There is insufficient evidence to indicate the mean is different from 6 at $\alpha = .05$.

c.   For part **a**, the p-value $= P(t \le -2.064)$.

From Table IV, with df $= 4$, $.05 < P(t \le -2.064) < .10$ or $.05 < p\text{-value} < .10$. Using MINITAB, the p-value is $p = .0539809$.

For part **b**, the p-value $= P(t \le -2.064) + P(t \ge 2.064)$.

From Table IV, with df $= 4$, $2(.05) < p\text{-value} < 2(.10)$ or $.10 < p\text{-value} < .20$. Using MINITAB, the p-value is $p = 2(.0539809) = .1079618$.

6.65   a.   Let $\mu =$ average trap spacing measurement. To determine if the true average trap spacing measurement is different from 95, we test:

$$H_0 : \mu = 95$$

$$H_a : \mu \neq 95$$

b. For this example, the sample mean is 89.9. If another sample of size 7 was selected, chances are, the sample mean would not be 89.9. The next sample could possibly have a sample mean greater than 95. We need to determine how unusual a value 89.9 is for $\bar{x}$ if the true mean is 95. Also, we need to take into account the variability in the data.

c. The test statistic is $t = \dfrac{\bar{x} - \mu_0}{\dfrac{s}{\sqrt{n}}} = \dfrac{89.9 - 95}{\dfrac{11.6}{\sqrt{7}}} = -1.163$

d. To find the p-value, we need to find $p = P(t \leq -1.163) + P(t \geq 1.163)$. From Table IV, Appendix A, with df $= n - 1 = 7 - 1 = 6$, $P(t \geq 1.440) = .10$. Thus, $P(t \geq 1.163) > .10$. The *p*-value is

$$p = P(t \leq -1.163) + P(t \geq 1.163) > .10 + .10 > .20.$$

Using MINITAB, the exact *p*-value is .289.

e. Let $\alpha = .10$. Thus, the probability of concluding that the average strap spacing measurement is different from 95 when, in fact, it is equal to 95 is .10.

f. Since the *p*-value is greater than $\alpha = .10$ $(p = .289 > .10)$, H_0 is not rejected. There is insufficient evidence to indicate the average trap spacing measurement is different from 95 at $\alpha = .10$.

g. We must assume that a random sample was drawn from a normal population of trap spacing measurements.

h. From Exercise 5.35, the 95% confidence interval is (79.095, 100.625). Since 95 falls in the interval, it is a likely value for μ. We would not reject it. This agrees with the results in part f.

6.67   To determine if the mean Dental Anxiety Scale score differs from 11, we test:

H_0: $\mu = 11$
H_a: $\mu \neq 11$

The test statistic is $t = \dfrac{\bar{x} - \mu_0}{s/\sqrt{n}} = \dfrac{10.7 - 11}{3.6/\sqrt{15}} = -.32$

The rejection region requires $\alpha / 2 = .05 / 2 = .025$ in each tail of the t distribution with df $= n - 1 = 15 - 1 = 14$. From Table IV, Appendix A, $t_{.025} = 2.145$. The rejection region is $t < -2.145$ or $t > 2.145$.

Since the observed value of the test statistic does not fall in the rejection region
($t = -.32 \not< -2.145$), H_0 is not rejected. There is insufficient evidence to indicate
the mean Dental Anxiety Scale score differs from 11 at $\alpha = .05$.

6.69   Let μ = mean dentary depth of molars. To determine if the sample of 18 cheek teeth come
from some other extinct primate species, we test:

$$H_0: \ \mu = 15$$
$$H_a: \ \mu \neq 15$$

The test statistic is $t = \dfrac{\overline{x} - \mu_0}{\dfrac{s}{\sqrt{n}}} = 3.229 \qquad$ (From the printout.)

From the printout, the p-value is $p = .005$. Since the p-value is so small, we would reject H_0
for any reasonable value of α.   There is sufficient evidence to indicate that the sample of 18
cheek teeth come from some other extinct primate species other than species A.

6.71   To determine if the mean amount of cesium in lichen specimens differs from .003, we test:

$$H_0: \ \mu = .003$$
$$H_a: \ \mu \neq .003$$

From the printout, the test statistic is $t = 3.725$ and the p-value is $p = .0058$.

Since the p-value is less than α ($p = .0058 < \alpha = .10$), H_0 is rejected. There is sufficient
evidence to indicate the mean amount of cesium in lichen specimens differs from .003 at
$\alpha = .10$.

6.73   From Exercise 5.44, $\overline{x} = 358.45$ and s = 117.8172. To determine if the mean skidding
distance is less than 425 meters, we test:

$$H_0: \ \mu = 425$$
$$H_a: \ \mu < 425$$

The test statistic is $t = \dfrac{\overline{x} - \mu_o}{s / \sqrt{n}} = \dfrac{358.45 - 425}{\dfrac{117.8172}{\sqrt{20}}} = -2.53$

The rejection region requires $\alpha = .10$ in the lower tail of the t-distribution with df $= n - 1$
$= 20 - 1 = 19$. From Table IV, Appendix A, $t_{.10} = 1.328$. The rejection region is $t < -1.328$.

Since the observed value of the test statistic falls in the rejection region ($t = -2.53 < -1.328$),
H_0 is rejected. There is sufficient evidence to indicate the mean skidding distance is less than
425 meters at $\alpha = .10$. Thus, there is enough evidence to refute the claim.

6.75   a.   To determine if the true mean bias differs from 0, we test:

$$H_0\text{: } \mu = 0$$
$$H_a\text{: } \mu \neq 0$$

From the printout, the test statistics is $t = .21$ and the p-value is $p = .840$.

Since the p-value is larger than $\alpha = .10$, $(p = .840 > .10)$, H_0 is not rejected. There is insufficient evidence to indicate that the true mean bias differs from 0.

b.   The scores given are the average scores of different people. There are several people with averages scores that are positive and several with average scores which are negative. These scores will tend to cancel each other out. Also, each individual could travel in circles to the right and circles to the left. Again, these would tend to cancel each other out.

6.77   Typically, qualitative data are associated with making inferences about a population proportion.

6.79   The sample size is large enough if both np_o and nq_o are greater than or equal to 15.

a. $np_o = 500(.05) = 25$ and $nq_o = 500(1 - .05) = 500(.95) = 475$. Since both of these values are greater than 15, the sample size is large enough to use the normal approximation.

b. $np_o = 100(.99) = 99$ and $nq_o = 100(1 - .99) = 100(.01) = 1$. Since the second number is less than 15, the sample size is not large enough to use the normal approximation.

c. $np_o = 50(.2) = 10$ and $nq_o = 50(1 - .2) = 50(.8) = 40$. Since the first number is less than 15, the sample size is not large enough to use the normal approximation.

d. $np_o = 20(.2) = 4$ and $nq_o = 20(1 - .2) = 20(.8) = 16$. Since the first number is less than 15, the sample size is not large enough to use the normal approximation.

e. $np_o = 10(.4) = 4$ and $nq_o = 10(1 - .4) = 10(.6) = 6$. Since both numbers are less than 15, the sample size is not large enough to use the normal approximation.

6.81   a. $$z = \frac{\hat{p} - p_0}{\sqrt{\dfrac{p_0 q_0}{n}}} = \frac{.84 - .9}{\sqrt{\dfrac{.9(.1)}{100}}} = -2.00$$

b.   The denominator in Exercise 6.80 is $\sqrt{\dfrac{.75(.25)}{100}} = .0433$ as compared to $\sqrt{\dfrac{.9(.1)}{100}} = .03$ in part **a**. Since the denominator in this problem is smaller, the absolute value of z is larger.

c.   The rejection region requires $\alpha = .05$ in the lower tail of the z distribution. From Table III, Appendix A, $z_{.05} = 1.645$. The rejection region is $z < -1.645$.

Since the observed value of the test statistic falls in the rejection region $(z = -2.00 < -1.645)$, H_0 is rejected. There is sufficient evidence to indicate the population proportion is less than .9 at $\alpha = .05$.

d.   The p-value $= P(z \le -2.00) = .5 - .4772 = .0228$ (from Table III, Appendix A).

6.83   From Exercise 5.54, $n = 50$ and since p is the proportion of consumers who do not like the snack food, $\hat{p}$ will be:

$$\hat{p} = \frac{\text{Number of 0's in sample}}{n} = \frac{29}{50} = .58$$

In order for the inference to be valid, the sample size must be large enough. The sample size is large enough if both np_o and nq_o are greater than or equal to 15.

$np_o = 50(.5) = 25$ and $nq_o = 50(1 - .5) = 50(.5) = 25$. Since both of these values are greater than 15, the sample size is large enough to use the normal approximation.

a. H_0: $p = .5$
       H_a: $p > .5$

The test statistic is $z = \dfrac{\hat{p} - p_0}{\sigma_{\hat{p}}} = \dfrac{\hat{p} - p_0}{\sqrt{\dfrac{p_0 q_0}{n}}} = \dfrac{.58 - .5}{\sqrt{\dfrac{.5(1 - .5)}{50}}} = 1.13$

The rejection region requires $\alpha = .10$ in the upper tail of the z distribution. From Table III, Appendix A, $z_{.10} = 1.28$. The rejection region is $z > 1.28$.

Since the observed value of the test statistic does not fall in the rejection region $(z = 1.13 \ngtr 1.28)$, H_0 is not rejected. There is insufficient evidence to indicate the proportion of customers who do not like the snack food is greater than .5 at $\alpha = .10$.

b. p-value $= P(z \ge 1.13) = .5 - .3708 = .1292$ (Using Table III, Appendix A.)

6.85   a.   The population parameter of interest is p, the proportion of items that had the wrong price when scanned through the electronic checkout scanner at California Wal-Mart stores.

b.   To determine if the true proportion of items incorrectly scanned at California Wal-Mart stores exceeds the 2% NIST standard, we test:

H_0: $p = .02$
H_a: $p > .02$

c.   The test statistic is $z = \dfrac{\hat{p} - p_o}{\sqrt{\dfrac{p_o q_o}{n}}} = \dfrac{.083 - .02}{\sqrt{\dfrac{.02(.98)}{1000}}} = 14.23$

The rejection region requires $\alpha = .05$ in the upper tail of the z distribution. From Table III, Appendix A, $z_{.05} = 1.645$. The rejection region is $z > 1.645$.

d.   Since the observed value of the test statistic falls in the rejection region ($z = 14.23 > 1.645$), H_o is rejected. There is sufficient evidence to indicate the true proportion of items incorrectly scanned at California Wal-Mart stores exceeds the 2% NIST standard at $\alpha = .05$.

e.   In order for the inference to be valid, the sample size must be large enough. The sample size is large enough if both np_o and nq_o are greater than or equal to 15.

$np_o = 1000(.02) = 20$ and $nq_o = 1000(1 - .02) = 1000(.98) = 980$. Since both of these values are greater than 15, the sample size is large enough to use the normal approximation.

6.87   Some preliminary calculations:

$$\hat{p} = \frac{x}{n} = \frac{401}{835} = .48$$

In order for the inference to be valid, the sample size must be large enough. The sample size is large enough if both np_o and nq_o are greater than or equal to 15.

$np_o = 835(.45) = 375.75$ and $nq_o = 835(1 - .45) = 835(.55) = 459.25$. Since both of these values are greater than 15, the sample size is large enough to use the normal approximation.

To determine if more than 45% of male youths are raised in a single-parent family, we test:

H_0: $p = .45$
H_a: $p > .45$

The test statistic is $z = \dfrac{\hat{p} - p_0}{\sqrt{\dfrac{p_0 q_0}{n}}} = \dfrac{.48 - .45}{\sqrt{\dfrac{.45(.55)}{835}}} = 1.74$

The rejection region requires $\alpha = .05$ in the upper tail of the z distribution. From Table III, Appendix A, $z_{.05} = 1.645$. The rejection region is $z > 1.645$.

Since the observed value of the test statistic falls in the rejection region ($z = 1.74 > 1.645$), H_0 is rejected. There is sufficient evidence that more than 45% of male youths are raised in a single-parent family at $\alpha = .05$.

6.89   Let p = true proportion of students who prefer the new, computerized method of identifying common conifers.

The sample proportion is $\hat{p} = \dfrac{138}{171} = .81$.

In order for the inference to be valid, the sample size must be large enough. The sample size is large enough if both np_o and nq_o are greater than or equal to 15.

$np_o = 171(.7) = 119.7$ and $nq_o = 171(1 - .7) = 171(.3) = 51.3$. Since both of these values are greater than 15, the sample size is large enough to use the normal approximation.

To determine if more than 70% of all students prefer the new, computerized method of identifying common conifers, we test:

H_0: $p = .70$
H_a: $p > .70$

The test statistic is $z = \dfrac{\hat{p} - p_0}{\sigma_{\hat{p}}} = \dfrac{\hat{p} - p_0}{\sqrt{\dfrac{p_0 q_0}{n}}} = \dfrac{.81 - .70}{\sqrt{\dfrac{.70(1 - .70)}{171}}} = 3.14$

The p-value for the test is $p = P(z \geq 3.14) \approx .5 - .5 = 0$. (From Table III, Appendix A.) Since the p-value is so small, we would reject H_0 for any reasonable value of α. There is sufficient evidence to indicate that more than 70% of all students prefer the new, computerized method of identifying common conifers. Thus, Confir ID should be added to the curriculum at SRU.

6.91   First, we calculate the estimate for p: $\hat{p} = \dfrac{x}{n} = \dfrac{37}{148} = .25$

In order for the inference to be valid, the sample size must be large enough. The sample size is large enough if both np_o and nq_o are greater than or equal to 15.

$np_o = 148(.20) = 29.6$ and $nq_o = 148(1 - .20) = 148(.80) = 118.4$. Since both of these values are greater than 15, the sample size is large enough to use the normal approximation.

To determine if more than 20% of all freshman college students believe in the Big Bang Theory, we test:

H_0: $p = .20$
H_a: $p > .20$

The test statistic is $z = \dfrac{\hat{p} - p_o}{\sqrt{\dfrac{p_o q_o}{n}}} = \dfrac{.25 - .20}{\sqrt{\dfrac{.20(.80)}{148}}} = 1.52$

Since no value of α was given, we will choose $\alpha = .05$. The rejection region requires $\alpha = .05$ in the upper tail of the z-distribution. From Table III, Appendix A, $z_{.05} = 1.645$. The rejection region is $z > 1.645$.

Since the observed value of the test statistic does not fall in the rejection region
($z = 1.52 \not> 1.645$), H_0 is not rejected. There is insufficient evidence to indicate that more
than 20% of all freshman college students believe in the Big Bang Theory at $\alpha = .05$.

Thus, we would not be willing to state that more than 20% of all freshman college students
believe in the Big Bang Theory. We are 95% confident of this decision.

6.93   a.   The point estimate for p is $\hat{p} = \dfrac{x}{n} = \dfrac{24}{33} = .727$

In order for the inference to be valid, the sample size must be large enough. The sample
size is large enough if both np_0 and nq_0 are greater than or equal to 15.

$np_0 = 33(.60) = 19.8$ and $nq_0 = 33(1 - .60) = 33(.40) = 13.2$. Since the second value is
not greater than 15, the sample size is may not be large enough to use the normal
approximation.

To determine if the cream will improve the skin of more than 60% of middle-aged
women, we test:

H_0: $p = .60$
H_a: $p > .60$

The test statistic is $z = \dfrac{\hat{p} - p_0}{\sqrt{\dfrac{p_0 q_0}{n}}} = \dfrac{.727 - .60}{\sqrt{\dfrac{.60(.40)}{33}}} = 1.49$

The rejection region requires $\alpha = .05$ in the upper tail of the z distribution. From Table
III, Appendix A, $z_{.05} = 1.645$. The rejection region is $z > 1.645$.

Since the observed value of the test statistic does not fall in the rejection region
($z = 1.49 \not> 1.645$), H_0 is not rejected. There is insufficient evidence to indicate the
cream will improve the skin of more than 60% of middle-aged women at $\alpha = .05$.

b.   The p-value of the test is $p = P(z \geq 1.49) = .5 - .4319 = .0681$.

Since the p-value is not less than α ($p = .0681 \not< .05$), H_0 is not rejected. There is
insufficient evidence to indicate the cream will improve the skin of more than 60% of
middle-aged women at $\alpha = .05$.

6.95   We know that the miscarriage rate for pregnant women expecting a boy in the United States
is 3 in 1,000 or .003. Let $p =$ miscarriage rate for pregnant women expecting a boy in the
United States during September 2001.

To determine if the male fetal death rate in September, 2001 was higher than expected, we would test:

$$H_0: \quad p = .003$$
$$H_a: \quad p > .003$$

The rejection region requires $\alpha = .05$ in the upper tail of the z distribution. From Table III, Appendix A, $z_{.05} = 1.645$. The rejection region is $z > 1.645$.

The test statistic would be $z = \dfrac{\hat{p} - p_o}{\sqrt{\dfrac{p_o q_o}{n}}} = \dfrac{\hat{p} - .003}{\sqrt{\dfrac{.003(.997)}{2000}}} = 1.645$.

We now need to solve this for $\hat{p}$.

$$\frac{\hat{p} - .003}{\sqrt{\dfrac{.003(.997)}{2000}}} = 1.645 \Rightarrow \frac{\hat{p} - .003}{.001223} = 1.645 \Rightarrow \hat{p} - .003 = .002 \Rightarrow \hat{p} = .005$$

Thus, there would need to be $2000(.005) = 10$ miscarriages in the 2,000 pregnant women to support the claim.

6.97   The conditions required for a valid test for σ^2 are:

1. A random sample is selected from the target population.
2. The population from which the sample is selected has a distribution that is approximately normal.

6.99   The statement "When the sample size n is large, no assumptions about the population are necessary to test the population variance σ^2" is false. The population from which the sample is drawn must be approximately normal.

6.101   a.   df $= n - 1 = 16 - 1 = 15$; reject H_0 if $\chi^2 < 6.26214$ or $\chi^2 > 27.4884$

       b.   df $= n - 1 = 23 - 1 = 22$; reject H_0 if $\chi^2 > 40.2894$

       c.   df $= n - 1 = 15 - 1 = 14$; reject H_0 if $\chi^2 > 21.0642$

       d.   df $= n - 1 = 13 - 1 = 12$; reject H_0 if $\chi^2 < 3.57056$

       e.   df $= n - 1 = 7 - 1 = 6$; reject H_0 if $\chi^2 < 1.63539$ or $\chi^2 > 12.5916$

       f.   df $= n - 1 = 25 - 1 = 24$; reject H_0 if $\chi^2 < 13.8484$

6.103　a.

$$H_0: \sigma^2 = 1$$
$$H_a: \sigma^2 > 1$$

The test statistic is $\chi^2 = \dfrac{(n-1)s^2}{\sigma_0^2} = \dfrac{(100-1)(1.84)}{1} = 182.16$

The rejection region requires $\alpha = .05$ in the upper tail of the χ^2 distribution with df = $n - 1 = 100 - 1 = 99$. From Table V, Appendix A, $\chi^2_{.05} \approx 124.324$. The rejection region is $\chi^2 > 124.324$.

Since the observed value of the test statistic falls in the rejection region ($\chi^2 = 182.16 > 124.324$), H_0 is rejected. There is sufficient evidence to indicate the variance is larger than 1 at $\alpha = .05$.

　b.　In part **b** of Exercise 6.102, the test statistic was $\chi^2 = 11.04$. In Exercise 6.102, we did not reject H_0, but we did in this Exercise. As the sample size increases, it becomes easier to reject H_0.

6.105　a.　The rejection region requires $\alpha / 2 = .01 / 2 = .005$ in each tail of the χ^2 distribution with df $= n - 1 = 46 - 1 = 45$. From Table V, Appendix A, $\chi^2_{.005} \approx 66.7659$ and $\chi^2_{.995} \approx 20.7065$. The rejection region is $\chi^2 < 20.7065$ or $\chi^2 > 66.7659$.

　b.　The test statistic is $\chi^2 = \dfrac{(n-1)s^2}{\sigma_o^2} = \dfrac{(46-1)11.9^2}{100} = 63.7245$

　c.　Since the observed value of the test statistic does not fall in the rejection region ($\chi^2 = 63.7245 \ngtr 66.7659$ and $\chi^2 = 63.7245 \nless 20.7065$), H_0 is not rejected. There is insufficient evidence to indicate the variance is different from 100 at $\alpha = .01$.

6.107　a.　The parameter of interest is σ^2, the variance of the population of rock bounces.

　b.　To determine if the variance differs from 10 m^2, we test

$$H_o: \sigma^2 = 10$$
$$H_a: \sigma^2 \neq 10$$

　c.　The test statistic is $\chi^2 = \dfrac{(n-1)s^2}{\sigma_0^2} = \dfrac{(13-1)(4.0947291)^2}{10} = 20.12$

　d.　The rejection region requires $\alpha = .10 / 2 = .05$ in each tail of the χ^2 distribution with df = $n - 1 = 13 - 1 = 12$. From Table V, Appendix A, $\chi^2_{.05} = 21.0261$ and $\chi^2_{.95} = 5.22603$. The rejection region is $\chi^2 > 21.0261$ or $\chi^2 < 5.22603$.

e.   Since the observed value of the test statistic does not fall in the rejection region ($\chi^2 = 20.12 \not> 21.0261$ and $\chi^2 = 20.12 \not< 5.22603$), H_0 is not rejected. There is insufficient evidence to indicate the variance differs from 10 m^2 at $\alpha = .10$.

f.   We must assume that the population of rock bounces is normally distributed.

6.109   To determine if the true standard deviation of the point-spread errors exceed 15 (variance exceeds 225), we test:

$$H_0:\ \sigma^2 = 225$$
$$H_a:\ \sigma^2 > 225$$

The test statistic is $\chi^2 = \dfrac{(n-1)s^2}{\sigma_0^2} = \dfrac{(240-1)13.3^2}{225} = 187.896$

The rejection region requires $\alpha = .05$ in the upper tail of the χ^2 distribution with df $= n - 1 = 240 - 1 = 239$. The maximum value of df in Table V is 100. Thus, we cannot find the rejection region using Table V. Using a statistical package, $\chi^2_{.05} = 276.062$. The rejection region is $\chi^2 > 276.062$.

Since the observed value of the test statistics does not fall in the rejection region ($\chi^2 = 187.896 \not> 276.062$), H_0 is not rejected. There is insufficient evidence to indicate that the true standard deviation of the point-spread errors exceed 15 at $\alpha = .05$.

(Since the observed variance (or standard deviation) is less than the hypothesized value of the variance (or standard deviation) under H_0, there is no way H_0 will be rejected for any reasonable value of α.)

6.111   To determine if the variance of birthweights of babies delivered by cocaine-dependent women is less than 200,000, we test:

$$H_0:\ \sigma^2 = 200,000$$
$$H_a:\ \sigma^2 < 200,000$$

The test statistic is $\chi^2 = \dfrac{(n-1)s^2}{\sigma_0^2} = \dfrac{(16-1)410^2}{200,000} = 12.6075$

The rejection region requires $\alpha = .01$ in the lower tail of the χ^2 distribution with df $= n - 1 = 16 - 1 = 15$. From Table V, Appendix A, $\chi^2_{.99} = 5.22935$. The rejection region is $\chi^2 < 5.22935$.

Since the observed value of the test statistic does not fall in the rejection region ($\chi^2 = 12.6075 \not< 5.22935$), H_0 is not rejected. There is insufficient evidence to indicate that the variance of birthweights of babies delivered by cocaine-dependent women is less than 200,000 at $\alpha = .01$.

6.113   To determine if the true conduction time standard deviation is less than 7 nanoseconds, we test:

$$H_0:\ \sigma^2 = 7^2 = 49$$
$$H_a:\ \sigma^2 < 49$$

The test statistic is $\chi^2 = \dfrac{(n-1)s^2}{\sigma_o^2} = \dfrac{(18-1)6.3^2}{7^2} = 13.77$

The rejection region requires $\alpha = .01$ in the lower tail of the χ^2 distribution with $df = n - 1 = 18 - 1 = 17$. From Table V, Appendix A, $\chi^2_{.99} = 6.40776$. The rejection region is $\chi^2 < 6.40776$.

Since the observed value of the test statistic does not fall in the rejection region ($\chi^2 = 13.77 \not< 6.40776$), H_0 is not rejected. There is insufficient evidence to indicate the true conduction time standard deviation is less than 7 nanoseconds at $\alpha = .01$.

6.115   The sign test is preferred to the t-test when the population from which the sample is selected is not normal.

6.117   a. $P(x \geq 6) = 1 \leq P(x - 5) = 1 - .937 = .063$

   b. $P(x \geq 5) = 1 \leq P(x - 4) = 1 - .500 = .500$

   c. $P(x \geq 8) = 1 \leq P(x - 7) = 1 - .996 = .004$

   d. $P(x \geq 10) = 1 \leq P(x - 9) = 1 - .849 = .151$

   $\mu = np = 15(.5) = 7.5$ and $\sigma = \sqrt{npq} = \sqrt{15(.5)(.5)} = 1.9365$

   $P(x \geq 10) \approx P\left(z \geq \dfrac{(10 - .5) - 7.5}{1.9365}\right) = P(z \geq 1.03) = .5 - .3485 = .1515$

   e. $P(x \geq 15) = 1 \leq P(x \leq 14) = 1 - .788 = .212$

   $\mu = np = 25(.5) = 12.5$ and $\sigma = \sqrt{npq} = \sqrt{25(.5)(.5)} = 2.5$

   $P(x \geq 15) \approx P\left(z \geq \dfrac{(15 - .5) - 12.5}{2.5}\right) = P(z \geq .80) = .5 - .2881 = .2119$

6.119   To determine if the median is greater than 80, we test:

$$H_0:\ \eta = 80$$
$$H_a:\ \eta > 80$$

The test statistic is $S =$ number of measurements greater than $80 = 16$.

The p-value $= P(x \geq 16)$ where x is a binomial random variable with $n = 25$ and $p = .5$. From Table II,

$$p\text{-value} = P(x \geq 16) = 1 - P(x \leq 15) = 1 - .885 = .115$$

Since the p-value $= .115 > \alpha = .10$, H_0 is not rejected. There is insufficient evidence to indicate the median is greater than 80 at $\alpha = .10$.

We must assume the sample was randomly selected from a continuous probability distribution.

Note: Since $n \geq 10$, we could use the large-sample approximation.

6.121   a.   To determine if the median molar depth of all cheek teeth differs from 15, we test:

$$H_0: \ \eta = 15$$
$$H_a: \ \eta \neq 15$$

b.   The sign test is appropriate because the data do not appear to be normally distributed and the sample size is only $n = 18$.

c.   The test statistic is $S = 14$.

d.   From the printout, the p-value is $p = .0309$. Since the p-value is less than α ($p = .0309 < .05$), H_0 is rejected. There is sufficient evidence to indicate the median molar depth of all cheek teeth differs from 15 at $\alpha = .05$.

6.123   a.   To determine if female college students have a median emotional scale score higher than 2.8, we test:

$$H_0: \ \eta = 2.8$$
$$H_a: \ \eta > 2.8$$

b.   The test statistics is $S = \{$number of observations greater than 2.8$\} = 21$. Because $n = 30$, we will use the large sample approximation:

$$z = \frac{(S - .5) - .5n}{.5\sqrt{n}} = \frac{(21 - .5) - .5(30)}{.5\sqrt{30}} = 2.01$$

c.   The p-value $= P(z \geq 2.01) = .5 - .4778 = .0222$ using Table II, Appendix A.

d.   Since the p-value is not less than α ($p = .0222 \not< .01$), H_0 is not rejected. There is insufficient evidence to indicate that female college students have a median emotional scale score higher than 2.8 at $\alpha = .01$.

6.125   a.   If the data do not come from a normal distribution, then the test performed in Exercise 6.35 is not valid.

b.   If the data are not normal, then we could use the sign test for the median with

$$H_0: \ \eta = 95$$
$$H_a: \ \eta \neq 95$$

c. S_1 = {Number of observations less than 95} = 5 and
S_2 = {Number of observations greater than 95} = 2.

The test statistic is S = larger of S_1 and S_2 = 5.

d.   The p-value = $2P(x \geq 5)$ where x is a binomial random variable with $n = 7$ and $p = .5$. From Table II,

$$p\text{-value} = 2P(x \geq 5) = 2(1 - P(x \leq 4)) = 2(1 - .773) = .454$$

e.   In Exercise 6.65, we used $\alpha = .10$. Since the p-value is not less than α ($p = .454 \not< .10$), H_0 is not rejected. There is insufficient evidence to indicate the median trap spacing measurement differs from 95 at $\alpha = .10$.

6.127   To determine if 50% of the ingots have a freckle index of 10 or higher, we test:

$$H_o: \ \eta = 10$$
$$H_a: \ \eta > 10$$

S = number of measurements greater than 10 = 10.

The p-value = $P(x \geq 10)$ where x is a binomial random variable with $n = 17$ and $p = .5$. (One observation is eliminated because it is equal to 10.) Using MINITAB, the p-value = $P(x \geq 10) = 1 - P(x \leq 9) = 1 - .6855 = .3145$.

Since the p-value is greater than $\alpha = .01$ ($p = .3145 > .01$), H_0 is not rejected. There is insufficient evidence to indicate that 50% of the ingots have a freckle index of 10 or higher at $\alpha = .01$.

6.129   To determine if the population median skidding distance is more than 400 meters, we test:

$$H_0: \ \eta = 400$$
$$H_a: \ \eta > 400$$

The test statistic is S = number of measurements greater than 400 = 8.

The p-value = $P(x \geq 8)$ where x is a binomial random variable with $n = 19$ and $p = .5$. (One observation is eliminated because it is equal to 400.) Using MINITAB,

$$P(x \geq 8) = 1 - P(x \leq 7) = 1 - .180 = .820$$

Since the p-value is greater than $\alpha = .10$ ($p = .820 > .10$), H_0 is not rejected. There is insufficient evidence to indicate the population median skidding distance is more than 400 meters at $\alpha = .10$.

6.131   The smaller the p-value associated with a test of hypothesis, the stronger the support for the **alternative** hypothesis. The p-value is the probability of observing your test statistic or anything more unusual, given the null hypothesis is true. If this value is small, it would be very unusual to observe this test statistic if the null hypothesis were true. Thus, it would indicate the alternative hypothesis is true.

6.133   The larger the p-value associated with a test of hypothesis, the stronger the support for the **null** hypothesis. The p-value is the probability of observing your test statistic or anything more unusual, given the null hypothesis is true. If this value is large, it would not be very unusual to observe this test statistic if the null hypothesis were true. Thus, it would lend support that the null hypothesis is true.

6.135   a. H_a: $p = .35$
H_a: $p < .35$

The test statistic is $z = \dfrac{\hat{p} - p_0}{\sqrt{\dfrac{p_0 q_0}{n}}} = \dfrac{.29 - .35}{\sqrt{\dfrac{.35(.65)}{200}}} = -1.78$

The rejection region requires $\alpha = .05$ in the lower tail of the z distribution. From Table III, Appendix A, $z_{.05} = 1.645$. The rejection region is $z < -1.645$.

Since the observed value of the test statistic falls in the rejection region ($z = -1.78 < -1.645$), H_0 is rejected. There is sufficient evidence to indicate $p < .35$ at $\alpha = .05$.

   b. H_0: $p = .35$
H_a: $p \neq .35$

The test statistic is $z = -1.78$ (from **a**).

The rejection region requires $\alpha / 2 = .05 / 2 = .025$ in each tail of the z distribution. From Table III, Appendix A, $t_{.025} = 1.96$. The rejection region is $z < -1.96$ or $z > 1.96$.

Since the observed value of the test statistic does not fall in the rejection region ($z = -1.78 \not< -1.96$), H_0 is not rejected. There is insufficient evidence to indicate p is different from .35 at $\alpha = .05$.

6.137   a. H_0: $\sigma^2 = 30$
H_a: $\sigma^2 > 30$

The test statistic is $\chi^2 = \dfrac{(n-1)s^2}{\sigma_0^2} = \dfrac{(41-1)(6.9)^2}{30} = 63.48$

The rejection region requires $\alpha = .05$ in the upper tail of the χ^2 distribution with $df = n - 1 = 40$. From Table V, Appendix A, $\chi^2_{.05} = 55.7585$. The rejection region is $\chi^2 > 55.7585$.

Since the observed value of the test statistic falls in the rejection region ($\chi^2 = 63.48 >$ 55.7585), H_0 is rejected. There is sufficient evidence to indicate the variance is larger than 30 at $\alpha = .05$.

b. H_0: $\sigma^2 = 30$
H_a: $\sigma^2 \neq 30$

The test statistic is $\chi^2 = 63.48$ (from part **a**).

The rejection region requires $\alpha / 2 = .05 / 2 = .025$ in each tail of the χ^2 distribution with df $= n - 1 = 40$. From Table V, Appendix A, $\chi^2_{.025} = 59.3417$ and $\chi^2_{.975} = 24.4331$. The rejection region is $\chi^2 < 24.4331$ or $\chi^2 > 59.3417$.

Since the observed value of the test statistic falls in the rejection region $\left(\chi^2 = 63.48 > 59.3417 \right)$, H_0 is rejected. There is sufficient evidence to indicate the variance is not 30 at $\alpha = .05$.

6.139   a.   The null hypothesis would be: H_0: $p = .45$

b.   The null hypothesis would be: H_0: $\mu = 2.5$

6.141   a.   To determine if the true mean score of all sleep-deprived subjects is less than 80, we test:

H_0: $\mu = 80$
H_a: $\mu < 80$

The test statistic is $t = \dfrac{\bar{x} - \mu_0}{s / \sqrt{n}} = \dfrac{63 - 80}{17 / \sqrt{12}} = -3.46$

The rejection region requires $\alpha = .05$ in the lower tail of the t distribution with df $= n - 1$ $= 12 - 1 = 11$. From Table IV, Appendix A, $t_{.05} = 1.796$. The rejection region is $t < -1.796$.

Since the observed value of the test statistic falls in the rejection region ($t = -3.46 <$ -1.796), H_0 is rejected. There is sufficient evidence to indicate that the true mean score of all sleep-deprived subjects is less than 80 at $\alpha = .05$.

b.   We must assume that the overall test scores of sleep-deprived students are normally distributed.

6.143   To determine if the median cesium amount of lichen differs from .003, we test:

H_0: $\eta = .003$
H_a: $\eta \neq .003$

The test statistic is S = {number of observations greater than .003} = 8   (from printout). The p-value is $p = .0391$. Since the p-value is less than $\alpha = .10$, H_0 is rejected. There is sufficient evidence to indicate the median cesium amount of lichen differs from .003 at $\alpha = .10$.

This result agrees with the t-test in Exercise 6.71.

6.145   a.   Let p = proportion of the subjects who felt that the masculinization of face shape decreased attractiveness of the male face. If there is no preference for the unaltered or morphed male face, then $p = .5$.

For this problem, $\hat{p} = x / n = 58 / 67 = .866$.

In order for the inference to be valid, the sample size must be large enough. The sample size is large enough if both np_o and nq_o are greater than or equal to 15.

$np_o = 67(.5) = 33.5$ and $nq_o = 67(1 - .5) = 67(.5) = 33.5$. Since both of these values are greater than 15, the sample size is large enough to use the normal approximation.

To determine if the subjects showed a preference for either the unaltered or morphed face, we test:

H_0: $p = .5$
H_a: $p \neq .5$

b.   The test statistic is $z = \dfrac{\hat{p} - p_0}{\sqrt{\dfrac{p_0 q_0}{n}}} = \dfrac{.866 - .50}{\sqrt{\dfrac{.50(.50)}{67}}} = 5.99$

c.   By definition, the p-value is the probability of observing our test statistic or anything more unusual, given H_0 is true. For this problem, $p = P(z \leq -5.99) + P(z \geq 5.99) \approx .5 - .5 + .5 - .5 = 0$. This corresponds to what is listed in the Exercise.

d.   The rejection region requires $\alpha / 2 = .01 / 2 = .005$ in each tail of the z distribution. From Table III, Appendix A, $z_{.005} = 2.58$. The rejection region is $z < -2.58$ or $z > 2.58$.

Since the observed value of the test statistic falls in the rejection region ($z = 5.99 > 2.58$), H_0 is rejected. There is sufficient evidence to indicate that the subjects showed a preference for either the unaltered or morphed face at $\alpha = .01$.

6.147   To determine if the mean alkalinity level of water in the tributary exceeds 50 mpl, we test:

H_0: $\mu = 50$
H_a: $\mu > 50$

The test statistic is $z = \dfrac{\bar{x} - \mu_0}{\sigma_{\bar{x}}} = \dfrac{67.8 - 50}{14.4 / \sqrt{100}} = 12.36$

The rejection region requires $\alpha = .01$ in the upper tail of the z distribution. From Table III, Appendix A, $z_{.01} = 2.33$. The rejection region is $z > 2.33$.

Since the observed value of the test statistic falls in the rejection region ($z = 12.36 > 2.33$), H_0 is rejected. There is sufficient evidence to indicate that the mean alkalinity level of water in the tributary exceeds 50 mpl at $\alpha = .01$.

6.149   To determine if the true mean PTSD score of all World War II aviator POWs is less than 16, we test:

H_0: $\mu = 16$
H_a: $\mu < 16$

The test statistic is $z = \dfrac{\bar{x} - \mu_o}{\sigma_{\bar{x}}} = \dfrac{9.00 - 16}{9.32/\sqrt{33}} = -4.31$

The rejection region requires $\alpha = .10$ in the lower tail of the z distribution. From Table III, Appendix A, $z_{.10} = 1.28$. The rejection region is $z < -1.28$.

Since the observed value of the test statistic falls in the rejection region ($z = -4.31 < -1.28$), H_0 is rejected. There is sufficient evidence to indicate that the true mean PTSD score of all World War II aviator POWs is less than 16 at $\alpha = .10$.

6.151   a.   To determine if the true mean inbreeding coefficient μ for this species of wasp exceeds 0, we test:

H_0: $\mu = 0$
H_a: $\mu > 0$

The test statistic is $z = \dfrac{\bar{x} - \mu_0}{\sigma_{\bar{x}}} = \dfrac{.044 - 0}{.884 / \sqrt{197}} = 0.70$

The rejection region requires $\alpha = .05$ in the upper tail of the z distribution. From Table III, Appendix A, $z_{.05} = 1.645$. The rejection region is $z > 1.645$.

Since the observed value of the test statistic does not fall in the rejection region ($z = 0.70 \not> 1.645$), H_0 is not rejected. There is insufficient evidence to indicate that the true mean inbreeding coefficient μ for this species of wasp exceeds 0 at $\alpha = .05$.

b.   This result agrees with that of Exercise 5.129. The confidence interval in Exercise 5.129 was $(-.118, .206)$. Since this interval contains 0, there is no evidence to indicate that the mean is different from 0. This is the same conclusion that was reached in part **a.**

6.153   a.   To determine if the production process should be halted, we test:

H_0: $\mu = 3$
H_a: $\mu > 3$

where μ = mean amount of PCB in the effluent.

The test statistic is $z = \dfrac{\bar{x} - \mu_0}{\sigma_{\bar{x}}} = \dfrac{3.1 - 3}{.5 / \sqrt{50}} = 1.41$

The rejection region requires $\alpha = .01$ in the upper tail of the z distribution. From Table III, Appendix A, $z_{.01} = 2.33$. The rejection region is $z > 2.33$.

Since the observed value of the test statistic does not fall in the rejection region, ($z = 1.41 \not> 2.33$), H_0 is not rejected. There is insufficient evidence to indicate the mean amount of PCB in the effluent is more than 3 parts per million at $\alpha = .01$. Do not halt the manufacturing process.

b.   As plant manager, I do not want to shut down the plant unnecessarily. Therefore, I want $\alpha = P(\text{shut down plant when } \mu = 3)$ to be small.

6.155   The point estimate for p is $\hat{p} = \dfrac{x}{n} = \dfrac{85}{124} = .685$

In order for the inference to be valid, the sample size must be large enough. The sample size is large enough if both np_o and nq_o are greater than or equal to 15.

$np_o = 124(.167) = 20.708$ and $nq_o = 124 \, (1 - .167) = 124 \, (.833) = 103.292$. Since both of these values are greater than 15, the sample size is large enough to use the normal approximation.

To determine if the students will tend to choose the three-grill display so that Grill #2 is a compromise between a more and a less desirable grill ($p > .167$), we test:

H_0: $p = .167$
H_a: $p > .167$

The test statistic is $z = \dfrac{\hat{p} - p_0}{\sqrt{\dfrac{p_0 q_0}{n}}} = \dfrac{.685 - .167}{\sqrt{\dfrac{.167(.822)}{124}}} = 15.46$

The rejection region requires $\alpha = .05$ in the upper tail of the z distribution. From Table III, Appendix A, $z_{.05} = 1.645$. The rejection region is $z > 1.645$.

Since the observed value of the test statistic falls in the rejection region ($z = 15.46 > 1.645$), H_0 is rejected. There is sufficient evidence to indicate the students will tend to choose the three-grill display so that Grill #2 is a compromise between a more and a less desirable grill ($p > .167$) at $\alpha = .05$.

6.157   a.   A Type I error is rejecting the null hypothesis when it is true. In this problem, we would be concluding that the individual is a liar when, in fact, the individual is telling the truth.

A Type II error is accepting the null hypothesis when it is false. In this problem, we would be concluding that the individual is telling the truth when, in fact, the individual is a liar.

b.   The probability of a Type I error would be the probability of concluding the individual is a liar when he/she is telling the truth. From the problem, it stated that the polygraph would indicate that of 500 individuals who were telling the truth, 185 would be liars. Thus, an estimate of the probability of a Type I error would be $185/500 = .37$.

The probability of a Type II error would be the probability of concluding the individual is telling the truth when he/she is a liar. From the problem, it stated that the polygraph would indicate that of 500 individuals who were liars, 120 would be telling the truth. Thus, an estimate of the probability of a Type II error would be $120/500 = .24$.

6.159   a.   Using MINITAB, the descriptive statistics are:

Descriptive Statistics: PCB

```
Variable   N  Mean   StDev      Minimum     Q1   Median      Q3  Maximum
PCB       48 9.292   2.103   0.000000000  9.000   9.000  10.000   13.000
```

To determine if the data can refute the manufacturer's claim, we test:

$$H_0:\ \mu = 10$$
$$H_a:\ \mu < 10$$

The test statistic is $z = \dfrac{\bar{x} - \mu_0}{\sigma_{\bar{x}}} = \dfrac{9.292 - 10}{2.103/\sqrt{48}} = -2.33$

The rejection region requires $\alpha = .05$ in the lower tail of the z distribution. From Table III, Appendix A, $z_{.05} = 1.645$. The rejection region is $z < -1.645$.

Since the observed value of the test statistic falls in the rejection region ($z = -2.33 < -1.645$), H_0 is rejected. There is sufficient evidence to indicate that the true average number of inspections is less than 10 at $\alpha = .05$.

We agree with the potential buyer who doubts the manufacturer's claim.

b.   First, we must compute the value of $\bar{x}$ corresponding to the border of the rejection region. The rejection region is $z < -1.645$ with $\alpha = .05$. The border value for the one-sided test is:

$$\bar{x}_0 = \mu_0 - z_\alpha \sigma_{\bar{x}} = \mu_0 - z_\alpha \frac{\sigma}{\sqrt{n}} = 10 - 1.645\frac{1.2}{\sqrt{48}} = 10 - .285 = 9.715$$

Next, we convert this border value to a z-score using the alternative distribution with mean $\mu_a = 9.5$. The z-score is:

$$z = \frac{\bar{x}_0 - \mu_a}{\sigma_{\bar{x}}} = \frac{9.715 - 9.5}{\dfrac{1.2}{\sqrt{48}}} = 1.24$$

The probability of rejecting H_0 is:

$$Power = P(\bar{x} \leq 9.715 \mid \mu_a = 9.5) = P(z \leq 1.24) = .5 + .3925 = .8925$$

6.161   a.   We want to show that more than 17,000 of the 18,200 signatures are valid. Thus, we want to show that the proportion of valid signatures is greater than $17{,}000/18{,}200 = .934$.

From the Exercise, $\hat{p} = 98/100 = .98$.

In order for the inference to be valid, the sample size must be large enough. The sample size is large enough if both np_o and nq_o are greater than or equal to 15.

$np_o = 100(.934) = 93.4$ and $nq_o = 100\,(1 - .934) = 100\,(.066) = 6.6$. Since the second value is not greater than 15, the sample size may not be large enough to use the normal approximation.

To determine if more than .934 of the signatures are valid, we test:

H_0: $p = .934$
H_a: $p > .934$

The test statistic is $z = \dfrac{\hat{p} - p_0}{\sqrt{\dfrac{p_0 q_0}{n}}} = \dfrac{.98 - .934}{\sqrt{\dfrac{.934(.066)}{100}}} = 1.85$

Since no α is given in the problem, we will choose $\alpha = .05$. The rejection region requires $\alpha = .05$ in the upper tail of the z distribution. From Table III, Appendix A, $z_{.05} = 1.645$. The rejection region is $z > 1.645$.

Since the observed value of the test statistic falls in the rejection region ($z = 1.85 > 1.645$), H_0 is rejected. There is sufficient evidence to indicate that the true proportion of valid signatures is greater than .934 at $\alpha = .05$. This indicates that more than 17,000 of the 18,200 signatures are valid.

b.   We want to show that more than 16,000 of the 18,200 signatures are valid. Thus, we want to show that the proportion of valid signatures is greater than $16{,}000/18{,}200 = .879$.

Again, from the Exercise, $\hat{p} = 98/100 = .98$.

In order for the inference to be valid, the sample size must be large enough. The sample size is large enough if both np_o and nq_o are greater than or equal to 15.

$np_o = 100(.879) = 87.9$ and $nq_o = 100\,(1 - .879) = 100\,(.121) = 12.1$. Since the second value is not greater than 15, the sample size may not be large enough to use the normal approximation.

To determine if more than .879 of the signatures are valid, we test:

$$H_0: \ p = .879$$
$$H_a: \ p > .879$$

The test statistic is $z = \dfrac{\hat{p} - p_0}{\sqrt{\dfrac{p_0 q_0}{n}}} = \dfrac{.98 - .879}{\sqrt{\dfrac{.879(.121)}{100}}} = 3.10$

Since no α is given in the problem, we will choose $\alpha = .05$. The rejection region requires $\alpha = .05$ in the upper tail of the z distribution. From Table III, Appendix A, $z_{.05} = 1.645$. The rejection region is $z > 1.645$.

Since the observed value of the test statistic falls in the rejection region ($z = 3.10 > 1.645$), H_0 is rejected. There is sufficient evidence to indicate that the true proportion of valid signatures is greater than .879 at $\alpha = .05$. This indicates that more than 16,000 of the 18,200 signatures are valid.

Chapter

7

Comparing Population Means

7.1 When the samples are large, the sampling distribution of $(\bar{x}_1 - \bar{x}_2)$ is approximately normal by the Central Limit Theorem. The mean of the sampling distribution is $(\mu_1 - \mu_2)$ and the standard deviation is $\sqrt{\dfrac{\sigma_1^2}{n_1} + \dfrac{\sigma_2^2}{n_2}}$.

7.3 a. No. Both populations must be normal.

  b. No. Both population variances must be equal.

  c. No. Both populations must be normal.

  d. Yes.

  e. No. Both populations must be normal.

7.5 The confidence interval for $(\mu_1 - \mu_2)$ is $(-10, -4)$. The correct inference is **b**: $\mu_1 < \mu_2$.
Since all the values in the interval are negative, there is evidence that $\mu_1 < \mu_2$.

7.7 a. $\mu_{\bar{x}_1} = \mu_1 = 14$

$$\sigma_{\bar{x}} = \frac{\sigma_1}{\sqrt{n_1}} = \frac{4}{\sqrt{100}} = .4$$

  b. $\mu_{\bar{x}_2} = \mu_2 = 10$

$$\sigma_{\bar{x}_2} = \frac{\sigma_2}{\sqrt{n_2}} = \frac{3}{\sqrt{100}} = .3$$

  c. $\mu_{\bar{x}_1 - \bar{x}_2} = \mu_1 - \mu_2 = 14 - 10 = 4$

$$\sigma_{\bar{x}_1 - \bar{x}_2} = \sqrt{\frac{\sigma_1^2}{n_1} + \frac{\sigma_2^2}{n_2}} = \sqrt{\frac{4^2}{100} + \frac{3^2}{100}} = \sqrt{\frac{25}{100}} = .5$$

  d. Since $n_1 \geq 30$ and $n_2 \geq 30$, the sampling distribution of $\bar{x}_1 - \bar{x}_2$ is approximately normal by the Central Limit Theorem.

7.9   Some preliminary calculations are:

$$\bar{x}_1 = \frac{\sum x_1}{n_1} = \frac{11.8}{5} = 2.36 \qquad s_1^2 = \frac{\sum x_1^2 - \dfrac{\left(\sum x_1\right)^2}{n_1}}{n_1 - 1} = \frac{30.78 - \dfrac{(11.8)^2}{5}}{5-1} = .733$$

$$\bar{x}_2 = \frac{\sum x_2}{n_2} = \frac{14.4}{4} = 3.6 \qquad s_2^2 = \frac{\sum x_2^2 - \dfrac{\left(\sum x_2\right)^2}{n_2}}{n_2 - 1} = \frac{53.1 - \dfrac{(14.4)^2}{4}}{4-1} = .42$$

a. $$s_p^2 = \frac{(n_1 - 1)s_1^2 + (n_2 - 1)s_2^2}{n_1 + n_2 - 2} = \frac{(5-1).773 + (4-1).42}{5+4-2} = \frac{4.192}{7} = .5989$$

b. H_0: $\mu_1 - \mu_2 = 0$

     H_a: $\mu_1 - \mu_2 < 0$

The test statistic is $t = \dfrac{(\bar{x}_1 - \bar{x}_2) - D_0}{\sqrt{s_p^2\left(\dfrac{1}{n_1} + \dfrac{1}{n_2}\right)}} = \dfrac{(2.36 - 3.6) - 0}{\sqrt{.5989\left(\dfrac{1}{5} + \dfrac{1}{4}\right)}} = \dfrac{-1.24}{.5191} = -2.39$

The rejection region requires $\alpha = .10$ in the lower tail of the t distribution with df = $n_1 + n_2 - 2 = 5 + 4 - 2 = 7$. From Table IV, Appendix A, $t_{.10} = 1.415$. The rejection region is $t < -1.415$.

Since the test statistic falls in the rejection region ($t = -2.39 < -1.415$), H_0 is rejected. There is sufficient evidence to indicate that $\mu_2 > \mu_1$ at $\alpha = .10$.

c.   A small sample confidence interval is needed because $n_1 = 5 < 30$ and $n_2 = 4 < 30$.

For confidence coefficient .90, $\alpha = .10$ and $\alpha / 2 = .05$. From Table IV, Appendix A, with df = $n_1 + n_2 - 2 = 5 + 4 - 2 = 7$, $t_{.05} = 1.895$. The 90% confidence interval for $(\mu_1 - \mu_2)$ is:

$$(\bar{x}_1 - \bar{x}_2) \pm t_{.05}\sqrt{s_p^2\left(\frac{1}{n_1} + \frac{1}{n_2}\right)} \Rightarrow (2.36 - 3.6) \pm 1.895\sqrt{.5989\left(\frac{1}{5} + \frac{1}{4}\right)}$$

$$\Rightarrow -1.24 \pm .98 \Rightarrow (-2.22, -0.26)$$

d.   The confidence interval in part **c** provides more information about $(\mu_1 - \mu_2)$ than the test of hypothesis in part **b**. The test in part **b** only tells us that μ_2 is greater than μ_1. However, the confidence interval estimates what the difference is between μ_1 and μ_2.

7.11   a.   Let μ_1 = mean test score of students in classrooms using software products and μ_2 = mean test score of students in classrooms not using software products. The parameter of interest is $\mu_1 - \mu_2$.

      b.   To determine if the mean test score of students in classrooms using software products is higher than the mean test score of students in classrooms not using software products, we test:

$$H_0: \ \mu_1 - \mu_2 = 0$$
$$H_a: \ \mu_1 - \mu_2 > 0$$

      c.   Since the p-value is so large ($p = .62$), H_0 is not rejected for any reasonable value of α. There is insufficient evidence to indicate the mean test score of students in classrooms using software products is higher than the mean test score of students in classrooms not using software products. This agrees with the conclusion of the DOE.

7.13   a.   Let μ_1 = mean recall of those who received only visual aspects of the ad and μ_2 = mean recall of those who received an audiovisual presentation. To determine if the mean recall of ad information by those who receive an audiovisual presentation is different from those who receive only the visual aspects of the ad, we test:

$$H_0: \ \mu_1 - \mu_2 = 0$$
$$H_a: \ \mu_1 - \mu_2 \neq 0$$

      b.   Some preliminary calculations are:

$$s_p^2 = \frac{(n_1 - 1)s_1^2 + (n_2 - 1)s_2^2}{n_1 + n_2 - 2} = \frac{(20-1)1.98^2 + (20-1)2.13^2}{20 + 20 - 2} = 4.22865$$

The test statistic is
$$t = \frac{\overline{x}_1 - \overline{x}_2}{\sqrt{s_p^2\left(\dfrac{1}{n_1} + \dfrac{1}{n_2}\right)}} = \frac{3.70 - 3.30}{\sqrt{4.22865\left(\dfrac{1}{20} + \dfrac{1}{20}\right)}} = .62$$

      c.   The rejection region requires $\alpha/2 = .10/2 = .05$ in each tail of the t distribution with df $= n_1 + n_2 - 2 = 20 + 20 - 2 = 38$. From Table IV, Appendix A, $t_{.05} \approx 1.684$. The rejection region is $t > 1.684$ or $t < -1.684$.

      d.   Since the observed value of the test statistic does not fall in the rejection region ($t = .62 \not> 1.684$), H_0 is not rejected. There is insufficient evidence to indicate the mean recall of ad information by those who receive an audiovisual presentation is different from those who receive only the visual aspects of the ad at $\alpha = .10$.

         Although we cannot prove the researchers' theory is correct (Accept H_o), there is evidence that supports the theory.

e.   A *p*-value of .62 is the probability of observing our test statistic or anything more unusual. For this problem, the *p*-value = $P(t > .62) + P(t < -.62) = .62$. Since the *p*-value is so large, H_o will not be rejected for any reasonable value of α.

f.   In order for the inference to be valid, the following conditions must be met:

    1. Both populations being sampled from are normal
    2. Both population variances must be the same
    3. The samples must be random and independent.

7.15   a.   Let μ_1 = mean number of books read by students who earned an "A" grade and μ_2 = mean number of books read by students who earned a "B" or "C" grade. The parameter of interest for this problem is $\mu_1 - \mu_2$, the difference in the mean number of books read between students who received an "A" grade and those who received a "B or C" grade.

b.   Using MINITAB, the descriptive statistics are:

Descriptive Statistics: A, B-C

```
Variable   N    Mean   StDev  Variance  Minimum  Median  Maximum
A          8   37.00   8.70     75.71    24.00    36.50    53.00
B-C        6   24.50   8.53     72.70    16.00    21.50    40.00
```

Some preliminary calculations are:

$$s_p^2 = \frac{(n_1 - 1)s_1^2 + (n_2 - 1)s_2^2}{n_1 + n_2 - 2} = \frac{(8-1)75.71 + (6-1)72.70}{8 + 6 - 2} = 74.456$$

For confidence coefficient .95, $\alpha = .05$ and $\alpha / 2 = .05 / 2 = .025$. From Table IV, Appendix A, with $df = n_1 + n_2 - 2 = 8 + 6 - 2 = 12$, $t_{.025} = 2.179$. The confidence interval is:

$$(\bar{x}_1 - \bar{x}_2) \pm t_{.025} \sqrt{s_p^2 \left(\frac{1}{n_1} + \frac{1}{n_2} \right)} \Rightarrow (37 - 24.5) \pm 2.179 \sqrt{74.456 \left(\frac{1}{8} + \frac{1}{6} \right)}$$

$$\Rightarrow 12.5 \pm 10.154 \Rightarrow (2.346, 22.654)$$

c.   We are 95% confident that the true difference in the mean number of books read between the students who received an "A" grade and those who received a "B or C" grade is between 2.346 and 22.654.

d.   In Exercise 2.33, the stem-and-leaf display indicated that the students who earned A's tended to read more books than those who earned B's and C's. This agrees with the confidence interval in part b. Because all the values in the confidence interval are positive, the mean number of books read by the students earning A's is larger than the mean number read by students earning B's or C's.

7.17   a.   From the printout, the 95% confidence interval is (-0.60, 7.95). We are 95% confident that the difference in mean FNE scores for bulimic and normal female students is between -0.60 and 7.95.

b.   We must assume that the distribution of FNE scores for bulimic female students and the distribution of the FNE scores for the normal female students are normally distributed. We must also assume that the variances of the two populations are equal and the samples are random and independent.

Both sample distributions look somewhat mound-shaped and the sample variances are fairly close in value. Thus, both assumptions appear to be reasonably satisfied.

7.19   Let μ_1 = mean drug concentration for site 1 and μ_2 = mean drug concentration for site 2. To determine if there is a difference in the mean drug concentration of the 2 sites, we test:

H_0: $\mu_1 - \mu_2 = 0$
H_a: $\mu_1 - \mu_2 \neq 0$

From the printout, the test statistic is $t = .57$ and the p-value is $p = .573$. Since the p-value is so large, H_0 is not rejected. There is insufficient evidence to indicate a difference in the mean drug concentration of the 2 sites for any reasonable value of α.

7.21   Let μ_1 = mean total rating for group 1 (strengthen favored position) and μ_2 = mean total rating for group 2 (weaken opposing position).

Some preliminary calculations are:

$$s_p^2 = \frac{(n_1 - 1)s_1^2 + (n_2 - 1)s_2^2}{n_1 + n_2 - 2} = \frac{(26 - 1)12.5^2 + (26 - 1)12.2^2}{26 + 26 - 2} = \frac{7,627.25}{50} = 152.545$$

To determine if the difference in the mean total ratings for the two groups differ, we test:

H_0: $\mu_1 - \mu_2 = 0$
H_a: $\mu_1 - \mu_2 \neq 0$

The test statistic is $t = \dfrac{(\bar{x}_1 - \bar{x}_2) - 0}{\sqrt{s_p^2\left(\dfrac{1}{n_1} + \dfrac{1}{n_2}\right)}} = \dfrac{(28.6 - 24.9) - 0}{\sqrt{152.545\left(\dfrac{1}{26} + \dfrac{1}{26}\right)}} = 1.08$

The rejection region requires $\alpha / 2 = .05 / 2 = .025$ in each tail of the t distribution. From Table IV, Appendix A, with df $= n_1 + n_2 - 2 = 26 + 26 - 2 = 50$, $t_{.025} \approx 2.010$. The rejection region is $t < -2.010$ or $t > 2.010$.

Since the observed value of the test statistic does not fall in the rejection region ($t = 1.08 \not> 2.010$), H_0 is not rejected. There is insufficient evidence to indicate there is a difference in the mean total ratings for the two groups at $\alpha = .05$.

7.23   Using MINITAB, the descriptive statistics are:

Descriptive Statistics: Species

```
Variable  Region        N    Mean  StDev Minimum     Q1  Median      Q3  Maximum
Species   Dry Steppe    5   14.00  21.31    3.00   3.00    5.00   29.50    52.00
          Gobi Desert   6   11.83  18.21    4.00   4.00    4.50   16.00    49.00

Variable  Region             Q3  Maximum
Species   Dry Steppe      29.50    52.00
          Gobi Desert     16.00    49.00
```

Let μ_1 = average number of ant species found in the Dry Steppe and μ_2 = average number of ant species found in the Gobi Desert.

$$s_p^2 = \frac{(n_1-1)s_1^2 + (n_2-1)s_2^2}{n_1 + n_2 - 2} = \frac{(5-1)21.31^2 + (6-1)18.21^2}{5+6-2} = 386.0539$$

To determine if there is a difference in the average number of ant species found at sites in the two regions, we test:

H_0: $\mu_1 - \mu_2 = 0$
H_a: $\mu_1 - \mu_2 \neq 0$

The test statistic is $t = \dfrac{(\bar{x}_1 - \bar{x}_2) - D_o}{\sqrt{s_p^2\left(\dfrac{1}{n_1} + \dfrac{1}{n_2}\right)}} = \dfrac{(14 - 11.83) - 0}{\sqrt{386.0539\left(\dfrac{1}{5} + \dfrac{1}{6}\right)}} = 0.18$

The rejection region requires $\alpha/2 = .05/2 = .025$ in each tail of the t-distribution with df = $n_1 + n_2 - 2 = 5 + 6 - 2 = 9$. From Table IV, Appendix B, $t_{.025} = 2.262$. The rejection region is $t < -2.262$ or $t > 2.262$.

Since the observed value of the test statistic does not fall in the rejection region ($t = 0.18 \not> 2.262$), H_0 is not rejected. There is insufficient evidence to indicate a difference in the average number of ant species found at sites in the two regions at $\alpha = .05$.

7.25   a.   Let μ_1 = mean MFS score for the violent event group and μ_2 = mean MFS score for the avoided-violent event group. The target parameter of interest is $\mu_1 - \mu_2$ = difference in the mean MFS score between the 2 groups.

b.   We are given the sample mean scores for the two groups. However, in order to determine if there is a difference in the 2 population means, we also need to know the population standard deviations for the 2 groups or the variances for the 2 groups. This information was not provided.

c.   The p-value of the test is

$$p = P(z \leq -1.21) + P(z \geq 1.21) = (.5 - .3869) + (.5 - .3869) = .1131 + .1131 = .2262$$

d.   Since the observed p-value is greater than α ($p = .2262 > \alpha = .10$), H_0 is not rejected. There is insufficient evidence to indicate a difference in the mean MFS score between the violent event group and the avoided-violent event group at $\alpha = .10$.

7.27   a.   From the information given, we have no idea what the standard deviations are. Whether the population means are different or not depends on how variable the data are.

b.   Let μ_1 = mean pain intensity rating for blacks and μ_2 = mean pain intensity rating for whites.

To determine if blacks, on average, have a higher pain intensity rating than whites, we test:

$$H_0: \ \mu_1 - \mu_2 = 0$$
$$H_a: \ \mu_1 - \mu_2 > 0$$

The rejection region requires $\alpha = .05$ in the upper tail of the z distribution. From Table III, Appendix A, $z_{.05} = 1.645$. Thus, in order to reject H_0, the test statistic would have to be greater than 1.645.

Substituting known values into the test statistic, we can solve for s_p.

$$z = \frac{\bar{x}_1 - \bar{x}_2}{\sqrt{s_p^2\left(\frac{1}{n_1} + \frac{1}{n_2}\right)}} > 1.645 \qquad \Rightarrow \frac{8.2 - 6.9}{\sqrt{s_p^2\left(\frac{1}{55} + \frac{1}{159}\right)}} > 1.645$$

$$\Rightarrow 1.3 > 1.645\left(s_p\right)\sqrt{\left(\frac{1}{55} + \frac{1}{159}\right)} \qquad \Rightarrow \frac{1.3}{1.645(.1564)} > s_p \qquad \Rightarrow 5.053 > s_p$$

Thus, if both s_1 and s_2 were less than 5.053, then we would conclude that blacks, on average would have a higher pain intensity than whites. (It is possible that either s_1 or s_2 could be greater than 5.053 and we would still reject H_0. As long as $s_p < 5.053$, we would reject H_0.)

c.   From part **b**, we know we would reject H_0 if $s_p < 5.053$. Thus, we would not reject H_0 if $s_p \geq 5.053$. This could happen if both s_1 and s_2 are greater than 5.053. (It is possible that either s_1 or s_2 could be less than 5.053 and we would still reject H_0. As long as $s_p \geq 5.053$, we would reject H_0.)

7.29   In a paired difference experiment, the observations should be paired before collecting the data.

7.31   a.   The rejection region requires $\alpha = .05$ in the upper tail of the t distribution with df = $n_d - 1 = 10 - 1 = 9$. From Table IV, Appendix A, $t_{.05} = 1.833$. The rejection region is $t > 1.833$.

b.   From Table IV, with df = $n_d - 1 = 20 - 1 = 19$, $t_{.10} = 1.328$. The rejection region is $t > 1.328$.

c.   From Table IV, with df = $n_d - 1 = 5 - 1 = 4$, $t_{.025} = 2.776$. The rejection region is $t > 2.776$.

d. From Table IV, with df $= n_d - 1 = 9 - 1 = 8$, $t_{.01} = 2.896$. The rejection region is $t > 2.896$.

7.33   a.

Pair	Difference
1	3
2	2
3	2
4	4
5	0
6	1

$$\bar{x}_d = \frac{\sum x_d}{n_d} = \frac{12}{6} = 2 \qquad s_d^2 = \frac{\sum x_d^2 - \dfrac{\left(\sum x_d\right)^2}{n_d}}{n_d - 1} = \frac{\left(34 - \dfrac{(12)^2}{6}\right)}{5} = 2$$

b. $\mu_d = \mu_1 - \mu_2$

c. For confidence coefficient .95, $\alpha = .05$ and $\alpha/2 = .025$. From Table IV, Appendix A, with df $= n_d - 1 = 6 - 1 = 5$, $t_{.025} = 2.571$. The confidence interval is:

$$\bar{x}_d \pm t_{\alpha/2}\frac{s_d}{\sqrt{n_d}} \Rightarrow 2 \pm 2.571\frac{\sqrt{2}}{\sqrt{6}} \Rightarrow 2 \pm 1.484 \Rightarrow (.516, 3.484)$$

d. H_0: $\mu_d = 0$

H_a: $\mu_d \neq 0$

The test statistic is $t = \dfrac{\bar{x}_d - 0}{s_d / \sqrt{n_d}} = \dfrac{2}{\sqrt{2}/\sqrt{6}} = 3.46$

The rejection region requires $\alpha/2 = .05/2 = .025$ in each tail of the t distribution with df $= n_d - 1 = 6 - 1 = 5$. From Table IV, Appendix A, $t_{.025} = 2.571$. The rejection region is $t < -2.571$ or $t > 2.571$.

Since the observed value of the test statistic falls in the rejection region ($t = 3.46 > 2.571$), H_0 is rejected. There is sufficient evidence to indicate that the mean difference is different from 0 at $\alpha = .05$.

7.35   a. H_0: $\mu_d = 10$

H_a: $\mu_d \neq 10$

The test statistic is $z = \dfrac{\bar{x}_d - 10}{s_d / \sqrt{n_d}} = \dfrac{11.7 - 10}{6 / \sqrt{40}} = 1.79$

The rejection region requires $\alpha / 2 = .05 / 2 = .025$ in each tail of the z distribution. From Table III, Appendix A, $z_{.025} = 1.96$. The rejection region is $z < -1.96$ or $z > 1.96$.

Since the observed value of the test statistic does not fall in the rejection region ($z = 1.79 \ngtr 1.96$), H_0 is not rejected. There is insufficient evidence to indicate the difference in the population means is different from 10 at $\alpha = .05$.

b. The p-value is

$$p = P(z \le -1.79) + P(z \ge 1.79) = (.5 - .4633) + (.5 - .4633) = .0367 + .0367 = .0734$$

7.37 a. The health status is measured on all patients two times, both before and after handling museum objects. Thus, the two sets of measurements are not independent. The data need to be analyzed as paired differences.

b. The differences between the "before" and "after" measurements are:

Session	Difference	Session	Difference
1	-7	17	0
2	-12	18	-11
3	-9	19	-11
4	-9	20	-10
5	1	21	-12
6	-13	22	-7
7	-16	23	-4
8	-3	24	-2
9	-7	25	-8
10	-2	26	-1
11	-6	27	4
12	-10	28	-17
13	-9	29	-11
14	-9	30	-12
15	1	31	-5
16	-14	32	-13

c. Using MINITAB, the descriptive statistics are:

Descriptive Statistics: Before, After, Diff

```
Variable    N     Mean   StDev   Minimum        Q1   Median        Q3   Maximum
Before     32    51.66   10.09     30.00     43.00    51.00     59.00     73.00
After      32    59.28    9.60     42.00     52.25    59.00     64.75     83.00
Diff       32   -7.625   5.272   -17.000   -11.750   -9.000    -3.250     4.000
```

The mean of the differences is $\mu_d = -7.625$ and the standard deviation of the differences is $s_d = 5.272$.

d.   For confidence coefficient .90, $\alpha = .10$ and $\alpha/2 = .05$. From Table III, Appendix A, $z_{.05} = 1.645$. The confidence interval is:

$$\bar{x}_d \pm z_{.05}\frac{s_d}{\sqrt{n_d}} \Rightarrow -7.625 \pm 1.645\frac{5.272}{\sqrt{32}} \Rightarrow -7.625 \pm 1.533 \Rightarrow (-9.158, -6.092)$$

e.   We are 90% confident that the true difference in mean health status before and after handling museum objects is between -9.158 and -6.092. Since the interval is entirely negative, there is evidence to indicate the mean after handling the museum objects is higher than the mean before handling the museum objects. Thus, handling museum objects has a positive impact on a sick patient's well-being.

7.39   a.   Let μ_1 = mean BOLD score for the first session and μ_2 = mean BOLD score when the placebo was applied. The target parameter is $\mu_d = \mu_1 - \mu_2$, the difference in the mean BOLD scores between the first session and the session involving the placebo.

b.   A paired difference design was used to collect the data. Each experimental person had two observations measured on it.

c.   To determine if there was a placebo effect, we test:

H_o: $\mu_d = 0$
H_a: $\mu_d > 0$

d.   The test statistic is $t = \dfrac{\bar{x}_d - 0}{s_d/\sqrt{n_d}} = \dfrac{0.21 - 0}{.47/\sqrt{24}} = 2.19$

Since no α value was given, we will use $\alpha = .05$. The rejection region requires $\alpha = .05$ in the upper tail of the t distribution with df $= n_d - 1 = 24 - 1 = 23$. From Table IV, Appendix A, $t_{.05} = 1.714$. The rejection region is $t > 1.714$.

Since the observed value of the test statistic falls in the rejection region ($t = 2.19 > 1.714$), H_0 is rejected. There is sufficient evidence to indicate that there is a placebo effect at $\alpha = .05$.

e.   Since the p-value is less than α ($p = .02 < \alpha = .05$), H_0 is rejected. There is sufficient evidence to indicate that there is a placebo effect at $\alpha = .05$.

7.41   Using MINITAB, the descriptive statistics are:

Descriptive Statistics: Initial, Final, Diff

```
Variable  N    Mean    StDev  Minimum      Q1  Median      Q3  Maximum
Initial   3   5.640    1.075    4.560   4.560   5.650   6.710    6.710
Final     3   5.453    1.125    4.270   4.270   5.580   6.510    6.510
Diff      3  0.1867   0.1106   0.0700  0.0700  0.2000  0.2900   0.2900
```

Let μ_1 = mean initial pH and μ_2 = mean final pH. The parameter of interest is $\mu_d = \mu_1 - \mu_2$. To determine if the mean initial pH differs significantly from the mean pH after 30 days, we test:

$$H_0: \ \mu_d = 0$$
$$H_a: \ \mu_d \neq 0$$

The test statistics is $t = \dfrac{\bar{x}_d - 0}{\dfrac{s_d}{\sqrt{n_d}}} = \dfrac{.1867}{\dfrac{.1106}{\sqrt{3}}} = 2.92$

The rejection region requires $\alpha / 2 = .05 / 2 = .025$ in each tail of the t distribution. From Table IV, Appendix A, with $df = n_d - 1 = 3 - 1 = 2$, $t_{.025} = 4.303$. The rejection region is $t < -4.303$ or $t > 4.303$.

Since the observed value of the test statistic does not fall in the rejection region ($t = 2.92 \not> 4.303$), H_0 is not rejected. There is insufficient evidence to indicate the mean initial pH differs significantly from the mean pH after 30 days at $\alpha = .05$.

7.43   a.   Since all the data were collected from twins, the observations from each pair are not independent of each other. Thus, the data were collected as paired data. If the data are collected as paired data, they must be analyzed as paired data.

       b.   Let μ_1 = mean level of overall leisure activity for the control group and μ_2 = mean level of overall leisure activity for the demented group. The target parameter is $\mu_d = \mu_1 - \mu_2$, the difference in the mean level of overall leisure activity for the control group and the demented group.

Using MINITAB, the descriptive statistics are:

Descriptive Statistics: Diff

Variable	N	Mean	StDev	Minimum	Q1	Median	Q3	Maximum
Diff	107	1.953	10.082	-28.000	-4.000	2.000	9.000	29.000

We will use a 95% confidence interval to estimate the difference in the mean level of leisure activity between the control group and the demented group. For confidence coefficient .95, $\alpha = .05$ and $\alpha / 2 = .05 / 2 = .025$. From Table III, appendix A, $z_{.025} = 1.96$. The 95% confidence interval is:

$$\bar{x}_d \pm z_{.025} \frac{s_d}{\sqrt{n_d}} \Rightarrow 1.953 \pm 1.96 \frac{10.082}{\sqrt{107}} \Rightarrow 1.953 \pm 1.910 \Rightarrow (0.043, \ 3.863)$$

We are 95% confident that the difference in the mean level of leisure activity between the control group and the demented group is between 0.043 and 3.863. Since 0 is not contained in this interval, there is evidence of a difference in the mean level of leisure activity between the control group and the demented group at $\alpha = .05$. Since all the values are positive, there is evidence that the mean level of leisure activity for the control group is larger than that for the demented group.

7.45   Using MINITAB, the descriptive statistics are:

Descriptive Statistics: Before, After, Diff

```
Variable   N   Mean  StDev      Minimum      Q1  Median      Q3  Maximum
Before     13  2.513 1.976        0.270   0.805   2.400   3.405    7.350
After      13  1.506 1.448  0.000000000   0.260   1.360   2.380    4.920
Diff       13  1.007 1.209       -0.850   0.265   0.560   2.335    2.780
```

Let μ_1 = mean number of crashes before the camera and μ_2 = mean number of crashes after the camera. The parameter of interest is $\mu_d = \mu_1 - \mu_2$. To determine if the mean number of crashes after the camera is smaller than before the camera, we test:

$$H_0:\ \mu_d = 0$$
$$H_a:\ \mu_d > 0$$

The test statistics is $t = \dfrac{\bar{x}_d - 0}{\dfrac{s_d}{\sqrt{n_d}}} = \dfrac{1.007}{\dfrac{1.209}{\sqrt{13}}} = 3.00$

Since no α is given, we will use $\alpha = .05$. The rejection region requires $\alpha = .05$ in the upper tail of the t distribution. From Table IV, Appendix A, with $df = n_d - 1 = 13 - 1 = 12$, $t_{.05} = 1.782$. The rejection region is $t > 1.782$.

Since the observed value of the test statistic falls in the rejection region ($t = 3.00 > 1.782$), H_0 is rejected. There is sufficient evidence to indicate the mean number of crashes after the camera is smaller than before the camera at $\alpha = .05$.

7.47   Some preliminary calculations are:

Patient	Time 1	Time 2	Difference		Patient	Time 1	Time 2	Difference
1	5	5	0		11	7	10	-3
2	1	3	-2		12	0	3	-3
3	0	0	0		13	3	9	-6
4	1	1	0		14	5	8	-3
5	0	1	-1		15	7	12	-5
6	2	1	1		16	10	16	-6
7	5	6	-1		17	5	5	0
8	1	2	-1		18	6	3	3
9	0	9	-9		19	9	6	3
10	5	8	-3		20	11	8	3

$$\bar{x}_d = \frac{\sum x_d}{n_d} = \frac{-33}{20} = -1.65$$

$$s_d^2 = \frac{\sum x_d^2 - \dfrac{\left(\sum x_d\right)^2}{n_d}}{n_d - 1} = \frac{249 - \dfrac{(-33)^2}{20}}{20 - 1} = \frac{194.55}{19} = 10.23947 \qquad s_d = \sqrt{10.23947} = 3.1999$$

Let μ_1 = mean homophone confusion errors for time 1 and μ_2 = mean homophone confusion errors for time 2. Then $\mu_d = \mu_1 - \mu_2$.

To determine if Alzheimer's patients show a significant increase in mean homophone confusion errors over time, we test:

$$H_0: \ \mu_d = 0$$
$$H_a: \ \mu_d < 0$$

The test statistic is $t = \dfrac{\bar{x}_d - 0}{s_d \Big/ \sqrt{n_d}} = \dfrac{-1.65 - 0}{3.1999 \Big/ \sqrt{20}} = -2.306$

Since no α was given we will use $\alpha = .05$. The rejection region requires $\alpha = .05$ in the lower tail of the t distribution with df $= n_d = 20 - 1 = 19$. From Table IV, Appendix A, $t_{.05} = 1.729$. The rejection region is $t < -1.729$.

Since the observed value of the test statistic falls in the rejection region $(t = -2.306 < -1.729)$, H_0 is rejected. There is sufficient evidence to indicate Alzheimer's patients show a significant increase in mean homophone confusion errors over time at $\alpha = .05$.

We must assume that the population of differences is normally distributed and that the sample was randomly selected. A stem-and-leaf display of the data indicates that the data are mound-shaped. It appears that these assumptions are valid.

7.49   We can obtain estimates of σ_1^2 and σ_2^2 by using the sample variances s_1^2 and s_2^2, or from an educated guess based on the range – s $\approx$ Range / 4.

7.51   We can obtain an estimate of σ_d^2 by using the sample variance s_d^2, or from an educated guess based on the range – $s \approx$ Range (of differences) / 4.

7.53 $n_1 = n_2 = \dfrac{\left(z_{\alpha/2}\right)^2 \left(\sigma_1^2 + \sigma_2^2\right)}{(SE)^2}$

For confidence coefficient .95, $\alpha = 1 - .95 = .05$ and $\alpha / 2 = .05 / 2 = .025$. From Table III, Appendix A, $z_{.025} = 1.96$.

$$n_1 = n_2 = \frac{1.96^2 (15 + 15)}{2.2^2} = 23.8 \approx 24$$

7.55   For confidence coefficient .95, $\alpha = .05$ and $\alpha / 2 = .05 / 2 = .025$. From Table III, Appendix A, $z_{.025} = 1.96$.

$$n_1 = n_2 = \frac{\left(z_{\alpha/2}\right)^2 \left(\sigma_B^2 + \sigma_N^2\right)}{(SE)} = \frac{1.96^2 \left(25 + 25\right)}{2^2} = 48.02 \approx 49$$

7.57   If the range of daily incomes for both groups is \$200, we would estimate σ by:

$$\sigma \approx 200/4 = 50$$

For confidence coefficient .9, $\alpha = 1 - .9 = .1$ and $\alpha/2 = .1/2 = .05$. From Table III, Appendix A, $z_{.05} = 1.645$.

$$n_1 = n_2 = \frac{\left(z_{\alpha/2}\right)^2 \left(\sigma_1^2 + \sigma_2^2\right)}{(SE)^2} = \frac{(1.645)^2 (50^2 + 50^2)}{10^2} = 135.5 \approx 136$$

7.59   For confidence coefficient .95, $\alpha = 1 - .95 = .05$ and $\alpha / 2 = .05 / 2 = .025$. From Table III, Appendix A, $z_{.025} = 1.96$. The standard error is 5.

$$n_1 = n_2 = \frac{\left(z_{\alpha/2}\right)^2 \left(\sigma_1^2 + \sigma_2^2\right)}{(SE)^2} = \frac{(1.96)^2 (9^2 + 9^2)}{5^2} = 24.9 \approx 25$$

We would need to sample 25 subjects from each group.

7.61   The statement "If the rank sum for sample 1 is much larger than the rank sum for sample 2 when $n_1 = n_2$, then the distribution of population 1 is likely to be shifted to the right of the distribution of population 2." is true.

7.63   a.   The test statistic is T_1, the rank sum of population A (because $n_1 < n_2$).

The rejection region is $T_1 \leq 41$ or $T_1 \geq 71$, from Table VI, Appendix A, with $n_1 = 7$, $n_2 = 8$, and $\alpha = .10$.

b.   The test statistic is T_1, the rank sum of population A (because $n_1 = n_2$).

The rejection region is $T_1 \geq 50$, from Table VI, Appendix A, with $n_1 = 6$, $n_2 = 6$, and $\alpha = .05$.

c.   The test statistic is T_1, the rank sum of population A (because $n_1 < n_2$).

The rejection region is $T_1 \leq 43$, from Table VI, Appendix A, with $n_1 = 7$, $n_2 = 10$, and $\alpha = .025$.

d.   Since $n_1 = n_2 = 20$, the test statistic is $z = \dfrac{T_1 - \dfrac{n_1(n_1 + n_2 + 1)}{2}}{\sqrt{\dfrac{n_1 n_2 (n_1 + n_2 + 1)}{12}}}$

The rejection region is $z < -z_{\alpha/2}$ or $z > z_{\alpha/2}$. For $\alpha = .05$ and $\alpha / 2 = .05 / 2 = .025$, $z_{.025} = 1.96$ from Table III, Appendix A. The rejection region is $z < -1.96$ or $z > 1.96$.

7.65   To determine if distribution A is shifted to the left of distribution B, we test:

H_0: The probability distributions of treatments A and B are identical
H_a: The probability distribution of treatment A lies to the left of that for treatment B

The test statistic is

$$z = \frac{T_1 - \dfrac{n_2(n_1 + n_2 + 1)}{2}}{\sqrt{\dfrac{n_1 n_2 (n_1 + n_2 + 1)}{12}}} = \frac{173 - \dfrac{15(15 + 15 + 1)}{2}}{\sqrt{\dfrac{15(15)(15 + 15 + 1)}{12}}} = \frac{-59.5}{24.1091} = -2.47$$

The rejection region requires $\alpha = .05$ in the lower tail of the z distribution. From Table III, Appendix A, $z_{.05} = 1.645$. The rejection region is $z < -1.645$.

Since the observed value of the test statistic falls in the rejection region ($-2.47 < -1.645$), H_0 is rejected. There is sufficient evidence to conclude that distribution A is shifted to the left of distribution B for $\alpha = .05$.

7.67

Sample from Population 1 (A)	Rank	Sample from Population 2 (B)	Rank
15	13	5	2.5
10	8.5	12	10.5
12	10.5	9	6.5
16	14	9	6.5
13	12	8	4.5
8	4.5	4	1
	$T_1 = 62.5$	5	2.5
		10	8.5
			$T_2 = 42.5$

a.   To determine if there is a shift in location for the two distributions, we test:

H_0: The two sampled populations have identical probability distributions
H_a: The probability distribution for population A is shifted to the left or to the right
of that for B

The test statistic is $T_1 = 62.5$ since sample A has the smallest number of measurements.

The null hypothesis will be rejected if $T_1 \leq T_L$ or $T_1 \geq T_U$ where T_L and T_U correspond to $\alpha = .05$ (two-tailed), $n_1 = 6$ and $n_2 = 8$. From Table VI, Appendix A, $T_L = 29$ and $T_U = 61$.

Reject H_0 if $T_1 \leq 29$ or $T_1 \geq 61$.

Since the observed value of the test statistic falls in the rejection region ($T_1 = 62.5 \geq 61$), H_0 is rejected. There is sufficient evidence to indicate population A is shifted to the left or right of population B at $\alpha = .05$.

b. H_0: The two sampled populations have identical probability distributions
H_a: The probability distribution for population A is shifted to the right of population B

The test statistic remains $T_1 = 62.5$.

The null hypothesis will be rejected if $T_1 \geq T_U$ where T_U corresponds to $\alpha = .05$ (one-tailed), $n_1 = 6$ and $n_2 = 8$. From Table VI, Appendix A, $T_U = 58$.

Reject H_0 if $T_1 \geq 58$.

Since the observed value of the test statistic falls in the rejection region ($T_1 = 62.5 \geq 58$), H_0 is rejected. There is sufficient evidence to indicate population A is shifted to the right of population B at $\alpha = .05$.

7.69   a.   The appropriate test would be the Wilcoxon Rank Sum Test.

b.   To determine if the low-handicapped golfers tend to have higher X-factors than high-handicapped golfers, we test:

H_0: The distributions of the X-factors for the two groups of golfers are identical

H_a: The distribution of the X-factors for the low-handicapped golfers is shifted to the right of that for the high-handicapped golfers

c.   Let T_1 = rank sum of low-handicapped golfers and T_2 = rank sum of high-handicapped golfers. Since $n_1 = 8$ and $n_2 = 7$, the test statistic is T_2. The rejection region is $T_2 \leq T_L$ where $\alpha = .05$ (one-tailed). From Table VI, Appendix A, $T_L = 41$. The rejection region is $T_2 \leq 41$.

d.   Since the p-value is not less than α ($p = .487 \not< .05$), H_0 is not rejected. There is insufficient evidence to indicate the low-handicapped golfers tend to have higher X-factors than high-handicapped golfers at $\alpha = .05$.

7.71   a.   To determine if the distribution of FNE scores for bulimic students is shifted above the corresponding distribution for female students with normal eating habits, we test:

H_0: The distribution of FNE scores for bulimic students is the same as the corresponding distribution for female students with normal eating habits

H_a: The distribution of FNE scores for bulimic students is shifted above the corresponding distribution for female students with normal eating habits

b.   The data ranked are:

Bulimic Students	Rank	Normal Students	Rank
21	21.5	13	8.5
13	8.5	6	1
10	4.5	16	13.5
20	19.5	13	8.5
25	25	8	3
19	17	19	17
16	13.5	23	23
21	21.5	18	15
24	24	11	6
13	8.5	19	17
14	11	7	2
		10	4.5
	$T_1 = 174.5$	15	12
		20	19.5
			$T_2 = 150.5$

c.   The sum of the ranks of the 11 FNE scores for bulimic students is $T_1 = 174.5$.

d.   The sum of the ranks of the 14 FNE scores for normal students is $T_2 = 150.5$.

e.   The test statistic is $z = \dfrac{T_1 - \dfrac{n_1(n_1 + n_2 + 1)}{2}}{\sqrt{\dfrac{n_1 n_2 (n_1 + n_2 + 1)}{12}}} = \dfrac{174.5_1 - \dfrac{11(11 + 14 + 1)}{2}}{\sqrt{\dfrac{11(14)(11 + 14 + 1)}{12}}} = 1.72$

The rejection region requires $\alpha = .10$ in the upper tail of the z distribution. From Table III, Appendix A, $z_{.10} = 1.28$. The rejection region is $z > 1.28$.

f.   Since the observed value of the test statistic falls in the rejection region ($z = 1.72 > 1.28$), H_0 is rejected. There is sufficient evidence to indicate that the distribution of FNE scores for bulimic students is shifted above the corresponding distribution for female students with normal eating habits at $\alpha = .10$.

7.73   First, we rank the data:

Honey Dosage				DM Dosage			
Honey	Rank	Honey	Rank	DM	Rank	DM	Rank
12	52	12	52	4	5	6	11.5
11	44	8	21	6	11.5	8	21
15	65	12	52	9	28	12	52
11	44	9	28	4	5	12	52
10	37	11	44	7	16	4	5
13	59.5	15	65	7	16	12	52
10	37	10	37	7	16	13	59.5
4	5	15	65	9	28	7	16
15	65	9	28	12	52	10	37
16	68	13	59.5	10	37	13	59.5
9	28	8	21	11	44	9	28
14	62	12	52	6	11.5	4	5
10	37	10	37	3	1	4	5
6	11.5	8	21	4	5	10	37
10	37	9	28	9	28	15	65
8	21	5	9	12	52	9	28
11	44	12	52	7	16		
12	52						
		$T_1 = 1440.5$				$T_2 = 905.5$	

To determine if honey is a preferable treatment for the cough and sleep difficulty, we test:

H_0: The distributions of the cough improvement scores for the two groups of children
are identical

H_a: The distribution of the cough improvement scores for the honey dosage children is
shifted to the right of that for the DM dosage children

The test statistic is

$$z = \frac{T_1 - \frac{n_1(n_1 + n_2 + 1)}{2}}{\sqrt{\frac{n_1 n_2 (n_1 + n_2 + 1)}{12}}} = \frac{1440.5 - \frac{35(35 + 33 + 1)}{2}}{\sqrt{\frac{35(33)(35 + 33 + 1)}{12}}} = \frac{233}{81.4939} = 2.86$$

The rejection region requires $\alpha = .05$ in the upper tail of the z distribution. From Table III,
Appendix A, $z_{.05} = 1.645$. The rejection region is $z > 1.645$.

Since the observed value of the test statistic falls in the rejection region ($z = 2.86 > 1.645$), H_0
is rejected. There is sufficient evidence to indicate that honey is a preferable treatment for the
cough and sleep difficulty over DM dosage at $\alpha = .05$.

7.75 a. Since the data are probably not from a normal distribution the Wilcoxon rank sum test should be used.

b. To determine if the relational intimacy scores for the participants in the CMC group are lower than the relational intimacy scores for the participants in the FTF group, we test:

H_0: The probability distributions of those in the CMC group and those in the FTF group are identical

H_a: The probability distribution of the CMC group is shifted to the left of that for the FTF group

c. The rejection region requires $\alpha = .10$ in the lower tail of the z distribution. From Table III, Appendix A, $z_{.10} = 1.28$. The rejection region is $z < -1.28$.

d. First, we rank the data:

CMC Group	Rank	FTF Group	Rank
4	34.5	5	
3	13.5	4	34.5
3	13.5	4	34.5
4	34.5	4	34.5
3	13.5	3	13.5
3	13.5	3	13.5
3	13.5	3	13.5
3	13.5	4	34.5
4	34.5	3	13.5
4	34.5	3	13.5
3	13.5	3	13.5
4	34.5	3	13.5
3	13.5	4	34.5
3	13.5	4	34.5
2	2	4	34.5
4	34.5	4	34.5
2	2	4	34.5
4	34.5	3	13.5
5	47	3	13.5
4	34.5	3	13.5
4	34.5	4	34.5
4	34.5	4	34.5
5	47	2	2
3	13.5	4	34.5
	$T_1 = 578$		$T_2 = 598$

The test statistic is

$$z = \frac{T_1 - \dfrac{n_2(n_1 + n_2 + 1)}{2}}{\sqrt{\dfrac{n_1 n_2(n_1 + n_2 + 1)}{12}}} = \frac{578 - \dfrac{24(24 + 24 + 1)}{2}}{\sqrt{\dfrac{24(24)(24 + 24 + 1)}{12}}} = \frac{-10}{48.4974} = -.21$$

Since the observed value of the test statistic does not fall in the rejection region ($z = -.21 \not< -1.28$), H_0 is not rejected. There is insufficient evidence to indicate the relational intimacy scores for the participants in the CMC group are lower than the relational intimacy scores for the participants in the FTF group at $\alpha = .10$.

7.77   a.   Since the data can take on a very limited number of values, the data are probably not from a normal distribution.

b.   Using SAS, the output for the Wilcoxon test is:

```
                      The NPAR1WAY Procedure

           Wilcoxon Scores (Rank Sums) for Variable GSHWS
                    Classified by Variable COND

                      Sum of      Expected       Std Dev         Mean
      COND      N     Scores      Under H0       Under H0        Score
      --------------------------------------------------------------------
      ATIPS     98    8039.0      11123.0        476.284142      82.030612
      TIPS      128   17612.0     14528.0        476.284142      137.593750

              Average scores were used for ties.


                    Wilcoxon Two-Sample Test

          Statistic                 8039.0000

          Normal Approximation
          Z                           -6.4741
          One-Sided Pr <  Z           <.0001
          Two-Sided Pr > |Z|          <.0001

          t Approximation
          One-Sided Pr <  Z           <.0001
          Two-Sided Pr > |Z|          <.0001

      Z includes a continuity correction of 0.5.


                    Kruskal-Wallis Test

          Chi-Square                 41.9273
          DF                               1
          Pr > Chi-Square             <.0001
```

To determine if there was a difference in the level of involvement in science homework assignments between TIPS and ATIPS students, we test:

H_0: The distributions of the involvement scores for science homework for the two groups have the same location

H_a: The distributions of the involvement scores for science homework for the TIPS students is shifted to the right or left of that for the ATIPS students

From the printout, the test statistic is $z = -6.4741$ and the p-value is $p < .0001$. Since the p-value is less than α ($p < .0001 < .05$), H_0 is rejected. There is sufficient evidence to indicate that there is a difference in the level of involvement in science homework assignments between TIPS and ATIPS students at $\alpha = .05$.

c.   Using SAS, the output for the Wilcoxon test is:

```
                    The NPAR1WAY Procedure

         Wilcoxon Scores (Rank Sums) for Variable MTHHWS
                   Classified by Variable COND

                      Sum of      Expected       Std Dev        Mean
      COND      N     Scores      Under H0       Under H0       Score
      ---------------------------------------------------------------
      ATIPS    98     10936.0      11123.0      472.485707    111.591837
      TIPS    128     14715.0      14528.0      472.485707    114.960938

             Average scores were used for ties.


                    Wilcoxon Two-Sample Test

         Statistic                  10936.0000

         Normal Approximation
         Z                             -0.3947
         One-Sided Pr <  Z              0.3465
         Two-Sided Pr > |Z|             0.6930

         t Approximation
         One-Sided Pr <  Z              0.3467
         Two-Sided Pr > |Z|             0.6934

      Z includes a continuity correction of 0.5.


                    Kruskal-Wallis Test

         Chi-Square                     0.1566
         DF                                  1
         Pr > Chi-Square                0.6923
```

To determine if there was a difference in the level of involvement in mathematics homework assignments between TIPS and ATIPS students, we test:

H_0: The distributions of the involvement scores for mathematics homework for the two groups have the same location
H_a: The distributions of the involvement scores for mathematics homework for the TIPS students is shifted to the right or left of that for the ATIPS students

From the printout, the test statistic is $z = -.3947$ and the p-value is $p = .6930$. Since the p-value is not less than α ($p = .6930 > .05$), H_0 is not rejected. There is insufficient evidence to indicate that there is a difference in the level of involvement in mathematics homework assignments between TIPS and ATIPS students at $\alpha = .05$.

d.   Using SAS, the output for the Wilcoxon test is:

```
                    The NPAR1WAY Procedure

          Wilcoxon Scores (Rank Sums) for Variable LAHWS
                    Classified by Variable COND

                      Sum of      Expected      Std Dev        Mean
     COND      N      Scores      Under H0      Under H0       Score
     ------------------------------------------------------------------
     ATIPS    98     10496.0      11123.0     464.844149    107.102041
     TIPS    128     15155.0      14528.0     464.844149    118.398438

               Average scores were used for ties.


                    Wilcoxon Two-Sample Test

          Statistic                  10496.0000


          Normal Approximation
          Z                             -1.3478
          One-Sided Pr <   Z             0.0889
          Two-Sided Pr > |Z|             0.1777


          t Approximation
          One-Sided Pr <   Z             0.0895
          Two-Sided Pr > |Z|             0.1791

        Z includes a continuity correction of 0.5.


                    Kruskal-Wallis Test

          Chi-Square                     1.8194
          DF                                  1
          Pr > Chi-Square                0.1774
```

To determine if there was a difference in the level of involvement in language arts homework assignments between TIPS and ATIPS students, we test:

H_0: The distributions of the involvement scores for language arts homework for the two groups have the same location

H_a: The distributions of the involvement scores for language arts homework for the TIPS students is shifted to the right or left of that for the ATIPS students

From the printout, the test statistic is $z = -1.3478$ and the p-value is $p = .1777$. Since the p-value is not less than α ($p = .1777 > .05$), H_0 is not rejected. There is insufficient evidence to indicate that there is a difference in the level of involvement in language arts homework assignments between TIPS and ATIPS students at $\alpha = .05$.

7.79   We assume that the probability distribution of differences is continuous so that the absolute differences will have unique ranks. Although tied (absolute) differences can be assigned average ranks, the number of ties should be small relative to the number of observations to assure validity.

7.81   a.   The hypotheses are:

H_0: The two sampled populations have identical probability distributions

H_a: The probability distributions for population A is shifted to the right of that for population B

b.   Some preliminary calculations are:

| Treatment | | Difference | Rank of Absolute |
A	B	A − B	Difference
54	45	9	5
60	45	15	10
98	87	11	7
43	31	12	9
82	71	11	7
77	75	2	2.5
74	63	11	7
29	30	−1	1
63	59	4	4
80	82	−2	2.5
			$T_- = 3.5$

The test statistic is $T_- = 3.5$

The rejection region is $T_- \le 8$, from Table VII, Appendix A, with $n = 10$ and $\alpha = .025$.

Since the observed value of the test statistic falls in the rejection region ($T_- = 3.5 \le 8$), H_0 is rejected. There is sufficient evidence to indicate the responses for A tend to be larger than those for B at $\alpha = .025$.

7.83   a. H_0:   The two sampled populations have identical probability distributions
H_a: The probability distribution for population A is located to the right of that for population B

b.   The test statistic is:

$$z = \frac{T_+ - \dfrac{n(n+1)}{4}}{\sqrt{\dfrac{n(n+1)(2n+1)}{24}}} = \frac{354 - \dfrac{30(30+1)}{4}}{\sqrt{\dfrac{30(30+1)(60+1)}{24}}} = \frac{121.5}{48.6184} = 2.499$$

The rejection region requires $\alpha = .05$ in the upper tail of the z distribution. From Table III, Appendix A, $z = 1.645$. The rejection region is $z > 1.645$.

Since the observed value of the test statistic falls in the rejection region ($z = 2.499 > 1.645$), H_0 is rejected. There is sufficient evidence to indicate population A is located to the right of that for population B at $\alpha = .05$.

c.   The p-value $= P(z \ge 2.499) = .5 - .4938 = .0062$ (using Table III, Appendix A).

d.   The necessary assumptions are:

1.   The sample of differences is randomly selected from the population of differences.
2.   The probability distribution from which the sample of paired differences is drawn is continuous.

7.85   a.   From the printout, the test statistic is $z = -4.638$.

b.   To determine if handling a museum object has a positive impact on a sick patient's well being, we test:

H_0: The distributions of patient's health status are identical before and after handling museum objects

H_a: The distribution of patient's health status after handling museum objects is shifted to the right of that before

The test statistic is $z = -4.638$ and the p-value is $p = .000/2 = .000$.

Since the p-value is less than α ($p < .000 < .01$), H_0 is rejected. There is sufficient evidence to indicate handling a museum object has a positive impact on a sick patient's well-being at $\alpha = .01$.

7.87   a.   To determine if the distributions of the FSI scores for good and average readers differ in location, we test:

H_0: The distribution of the FSI scores of good readers is identical to the distribution of the FSI scores of average readers

H_a: The distribution of the of the FSI scores of good readers is shifted to the right or left of the distribution of the FSI scores of average readers

b, c.

| | FSI Scores | | | |
STRATEGY	Good Readers	Average Readers	Difference	Rank of Absolute Difference
Word meaning	.38	.32	.06	2
Words in context	.29	.25	.04	1
Literal comprehension	.42	.25	.17	5
Draw inference from single string	.60	.26	.34	8
Draw inference from multiple string	.45	.31	.14	4
Interpretation of metaphor	.32	.14	.18	6.5
Find salient or main idea	.21	.03	.18	6.5
Form judgment	.73	.80	−.07	3

d.   Positive rank sum $T_+ = 33$. Negative rank sum $T_- = 3$. The test statistic is the smaller of of T_+ and T_- and is $T_- = 3$.

e.   The rejection region is $T_- \leq 4$, from Table VII, Appendix A, with $n = 8$ and $\alpha = .05$ (two-tails).

f.   Since the observed value of the test statistic falls in the rejection region ($T_- = 3 \leq 4$), H_0 is rejected. There is sufficient evidence to indicate the distributions of the FSI scores for good and average readers differ in location at $\alpha = .05$.

7.89   Some preliminary calculations:

Teacher	Pre-test	Post-test	Difference Pre-Post	Rank of Absolute Differences
1	53	74	-21	9
2	73	80	-7	8
3	70	94	-24	10
4	72	78	-6	6.5
5	77	78	-1	1.5
6	81	84	-3	5
7	73	71	2	3.5
8	87	88	-1	1.5
9	61	63	-2	3.5
10	76	83	-6	6.5

Positive rank sum $T_+ = 3.5$

To determine if the program was effective, we test:

H_0: The distribution of racial tolerance before the program has the same location as the distribution of racial tolerance after program

H_a: The distribution of racial tolerance before the program is shifted to the left of that for racial tolerance after the program

The test statistic is $T_+ = 3.5$.

Reject H_0 if $T_+ \le T_0$ where T_0 is based on $\alpha = .01$ and $n = 10$ (one-tailed):

Reject H_0 if $T_+ \le 5$ (from Table VII, Appendix A)

Since the observed value of the test statistic falls in the rejection region ($T_+ = 3.5 \le 5$), H_0 is rejected. There is sufficient evidence to indicate that the program was effective at $\alpha = .01$.

7.91   Some preliminary calculations are:

Day	Change in Temp.	Field Meas.	3D Model	Diff.	Rank of Absolute Difference
		Change in Transverse Strain			
Oct. 24	-6.3	-58	-52	-6	3
Dec. 3	13.2	69	59	10	5
Dec. 15	3.3	35	32	3	2
Feb. 2	-14.8	-32	-24	-8	4
Mar. 25	1.7	-40	-39	-1	1
May 24	-.2	-83	-71	-12	6

$T_+ = 7$

To determine if there is a shift in the change in transverse strain distributions between field measurements and the 3D model, we test:

H_0: The change in transverse strain distribution for field measurements is identical to the change in transverse strain distribution for the 3D model
H_a: The change in transverse strain distribution for field measurements is shifted to the right or left of the change in transverse strain distribution for the 3D model

The test statistic is $T =$ smaller of T_- and T_+ which is $T_+ = 7$.

The rejection region is $T_+ \leq 1$, from Table VII, Appendix A, with $n = 6$ and $\alpha = .05$ (two-tails).

Since the observed value of the test statistic does not fall in the rejection region $(T_+ = 7 \not\leq 1)$, H_0 is not rejected. There is insufficient evidence to indicate there is a shift in the change in transverse strain distributions between field measurements and the 3D model at $\alpha = .05$.

7.93   Some preliminary calculations are:

Patient	Before Thickness	After Thickness	Difference	Rank of Absolute difference
1	11.0	11.5	−0.5	6
2	4.0	6.4	−2.4	20
3	6.3	6.1	0.2	3
4	12.0	10.0	2.0	15.5
5	18.2	14.7	3.5	24
6	9.2	7.3	1.9	14
7	7.5	6.1	1.4	12
8	7.1	6.4	0.7	8.5
9	7.2	5.7	1.5	13
10	6.7	6.5	0.2	3
11	14.2	13.2	1.0	10
12	7.3	7.5	−0.2	3
13	9.7	7.4	2.3	18.5
14	9.5	7.2	2.3	18.5
15	5.6	6.3	−0.7	8.5
16	8.7	6.0	2.7	21
17	6.7	7.3	−0.6	7
18	10.2	7.0	3.2	23
19	6.6	5.3	1.3	11
20	11.2	9.0	2.2	17
21	8.6	6.6	2.0	15.5
22	6.1	6.3	−0.2	3
23	10.3	7.2	3.1	22
24	7.0	7.2	−0.2	3
25	12.0	8.0	4.0	25

Negative rank sum $T_- = 50.5$
Positive rank sum $T_+ = 274.5$

To determine if the treatment for tendonitis tends to reduce the thickness of tendons, we test:

H_0: The distribution of the thickness before treatment has the same location as the distribution of the thickness after the treatment

H_a: The distribution of the thickness before the treatment is shifted to the right of that for the thickness after the treatment

The test statistics is $T =$ smaller of T_- and T_+ which is $T_- = 50.5$.

Reject H_0 if $T_- \leq T_0$ where T_0 is based on $\alpha = .10$ and $n = 25$ (one-tailed).

Reject H_0 if $T_- \leq 101$ (From Table VII, Appendix A)

(Note: There is no value for $\alpha = .10$ for a one-tailed test. However, if we reject H_0 for $\alpha = .05$, we would also reject H_0 for $\alpha = .10$.)

Since the observed value of the test statistic falls in the rejection region ($T_- = 50.5 \leq 101$), H_0 is rejected. There is sufficient evidence to indicate the treatment for tendonitis tends to reduce the thickness of tendons at $\alpha = .10$.

7.95   For a one-way ANOVA, independent random samples are selected for each treatment, or treatments are randomly assigned to experimental units.

7.97   The statement "The ANOVA method is robust when the assumption of normality is not exactly satisfied" is true. That is, moderate departures from normality do not have much effect on the significance level of the ANOVA F-test or on confidence coefficients.

7.99   a.   In the second dot diagram **B**, the difference between the sample means is small relative to the variability within the sample observations. In the first dot diagram **A**, the values in each of the samples are grouped together with a range of 4, while in the second diagram **B**, the range of values is 8.

b.   For diagram **A**,

$$\bar{x}_1 = \frac{\sum x_1}{n} = \frac{7+8+9+9+10+11}{6} = \frac{54}{6} = 9$$

$$\bar{x}_2 = \frac{\sum x_2}{n} = \frac{12+13+14+14+15+16}{6} = \frac{84}{6} = 14$$

For diagram **B**,

$$\bar{x}_1 = \frac{\sum x_1}{n} = \frac{5+5+7+11+13+13}{6} = \frac{54}{6} = 9$$

$$\bar{x}_2 = \frac{\sum x_2}{n} = \frac{10+10+12+16+18+18}{6} = \frac{84}{6} = 14$$

c.　For diagram **A**,

$$\text{SST} = \sum_{i=1}^{2} n_i (\bar{x}_i - \bar{x})^2 = 6(9 - 11.5)^2 + 6(14 - 11.5)^2 + = 75$$

$$\left(\bar{x} = \frac{\sum x}{n} = \frac{54 + 84}{12} = 11.5 \right)$$

For diagram **B**,

$$\text{SST} = \sum_{i=1}^{2} n_i (\bar{x}_i - \bar{x})^2 = 6(6 - 11.5)^2 + 6(14 - 11.5)^2 = 75$$

d.　For diagram **A**,

$$s_1^2 = \frac{\sum x_1^2 - \dfrac{\left(\sum x_1\right)^2}{n_1}}{n_1 - 1} = \frac{496 - \dfrac{54^2}{6}}{6 - 1} = 2$$

$$s_2^2 = \frac{\sum x_2^2 - \dfrac{\left(\sum x_2\right)^2}{n_2}}{n_2 - 1} = \frac{1186 - \dfrac{84^2}{6}}{6 - 1} = 2$$

$$\text{SSE} = (n_1 - 1)s_1^2 + (n_2 - 1)s_2^2 = (6 - 1)^2(2) + (6 - 1)^2(2) = 20$$

For diagram **B**,

$$s_1^2 = \frac{\sum x_1^2 - \dfrac{\left(\sum x_1\right)^2}{n_1}}{n_1 - 1} = \frac{558 - \dfrac{54^2}{6}}{6 - 1} = 14.4$$

$$s_2^2 = \frac{\sum x_2^2 - \dfrac{\left(\sum x_2\right)^2}{n_2}}{n_2 - 1} = \frac{1248 - \dfrac{84^2}{6}}{6 - 1} = 14.4$$

$$\text{SSE} = (n_1 - 1)s_1^2 + (n_2 - 1)s_2^2 = (6 - 1)14.4 + (6 - 1)14.4 = 144$$

e.　For diagram **A**, SS(Total) = SST + SSE = 75 + 20 = 95

$$\text{SST is } \frac{\text{SST}}{\text{SS(Total)}} \times 100\% = \frac{75}{95} \times 100\% = 78.95\% \text{ of SS(Total)}$$

For diagram **B**, SS(Total) = SST + SSE = 75 + 144 = 219

$$\text{SST is } \frac{\text{SST}}{\text{SS(Total)}} \times 100\% = \frac{75}{219} \times 100\% = 34.25\% \text{ of SS(Total)}$$

f.　For diagram **A**, $\text{MST} = \dfrac{\text{SST}}{k - 1} = \dfrac{75}{2 - 1} = 75$

$$\text{MSE} = \frac{\text{SSE}}{n - k} = \frac{20}{12 - 2} = 2 \qquad F = \frac{\text{MST}}{\text{MSE}} = \frac{75}{2} = 37.5$$

For diagram **B**, $\text{MST} = \dfrac{\text{SST}}{k-1} = \dfrac{75}{2-1} = 75$

$$\text{MSE} = \frac{\text{SSE}}{n-k} = \frac{144}{12-2} = 14.4 \qquad F = \frac{\text{MST}}{\text{MSE}} = \frac{75}{14.4} = 5.21$$

g. The rejection region for both diagrams requires $\alpha = .05$ in the upper tail of the F distribution with numerator df $= k - 1 = 2 - 1 = 1$ and denominator df $= n - k = 12 - 2 = 10$. From Table IX, Appendix A, $F_{.05} = 4.96$. The rejection region is $F > 4.96$.

For diagram **A**, since the observed value of the test statistic falls in the rejection region ($F = 37.5 > 4.96$), H_0 is rejected. There is sufficient evidence to indicate the samples were drawn from populations with different means at $\alpha = .05$.

For diagram **B**, since the observed value of the test statistic falls in the rejection region ($F = 5.21 > 4.96$), H_0 is rejected. There is sufficient evidence to indicate the samples were drawn from populations with different means at $\alpha = .05$.

h. We must assume both populations are normally distributed with common variances.

i. For each dot diagram, we want to test:

$$H_0 : \mu_1 - \mu_2 = 0$$
$$H_a : \mu_1 - \mu_2 \neq 0$$

Some preliminary calculations:

Diagram A	**Diagram B**

$$s_p^2 = \frac{s_1^2 + s_2^2}{2} \qquad\qquad s_p^2 = \frac{s_1^2 + s_2^2}{2}$$

$$= \frac{2+2}{2} = 2 \quad (n_1 = n_2) \qquad\qquad = \frac{14.4 + 14.4}{2} = 14.4 \quad (n_1 = n_2)$$

The test statistics for the two diagrams are:

Diagram A	**Diagram B**

$$t = \frac{\bar{x}_1 - \bar{x}_2}{\sqrt{s_p^2\left(\dfrac{1}{n_1} + \dfrac{1}{n_2}\right)}} = \frac{9-14}{\sqrt{2\left(\dfrac{1}{6} + \dfrac{1}{6}\right)}} \qquad t = \frac{\bar{x}_1 - \bar{x}_2}{\sqrt{s_p^2\left(\dfrac{1}{n_1} + \dfrac{1}{n_2}\right)}} = \frac{9-14}{\sqrt{14.4\left(\dfrac{1}{6} + \dfrac{1}{6}\right)}}$$

$$= -6.12 \qquad\qquad\qquad = -2.28$$

For the each test, the rejection region requires $\alpha / 2 = .05 / 2 = .025$ in each tail of the t distribution with df $= n_1 + n_2 - 2 = 6 + 6 - 2 = 10$. From Table IV, Appendix A, $t_{.025} = 2.228$. The rejection region is $t < -2.228$ or $t > 2.228$.

For the Diagram A test, since the observed value of the test statistic falls in the rejection region ($t = -6.12 < -2.228$), H_0 is rejected. There is sufficient evidence to indicate the samples were drawn from populations with different means at $\alpha = .05$.

For the Diagram B test, since the observed value of the test statistic falls in the rejection region ($t = -2.28 < -2.228$), H_0 is rejected. There is sufficient evidence to indicate the samples were drawn from populations with different means at $\alpha = .05$.

For the t tests, the test statistics are $t = -6.12$ and $t = -2.28$. For the F tests, the test statistics are $F = 37.5$ and $F = 5.21$. If we square the t values (($-6.12)^2 = 37.5$ and $(-2.28)^2 = 5.20$), we get the F values. In addition, if we square the t value from the rejection region ($t^2 = 2.228^2 = 4.96$) we get the F value from the rejection region ($F = 4.96$).

Assumptions for the t test:

1.   Both populations have relative frequency distributions that are approximately normal.
2.   The two population variances are equal.
3.   Samples are selected randomly and independently from the populations.

Assumptions for the F test:

1.   Both population probability distributions are normal.
2.   The two population variances are equal.
3.   Samples are selected randomly and independently from the respective populations.

The assumptions are the same for both tests.

7.101   a.   Answers will vary. Using MINITAB, the numbers selected to receive the usual follow-up care are:

2, 3, 7, 9, 11, 12, 13, 14, 15, 16, 17, 18, 19, 20, 22, 23, 25, 26, 27, 31, 32, 33, 35, 42, 44, 45, 47, 48, 49, 51, 52, 56, 57, 63, 64, 65, 66, 67, 69, 70, 71, 72, 75, 78, 79, 80, 82, 83, 86, 87, 88, 89, 93, 94, 95, 96, 99, 101, 102, 103, 104, 105, 106, 107, 108, 110, 111, 112, 114, 115, 116, 117, 118, 120, 121, 123, 125, 126, 128, 130, 138, 140, 142, 143, 144, 145, 147, 152, 154, 155, 156, 159, 161, 163, 165, 166, 168, 170, 171, 173, 174, 175, 176, 177, 179, 181, 182, 183, 186, 188, 190, 191, 192, 193, 194, 196, 198, 199, 202, 204, 207, 208, 244, 212, 218, 219, 221, 224, 228, 229, 231, 232, 233, 234, 235, 240, 243, 244, 246, 252, 259, 262, 264, 267, 269, 270, 273, 274, 275, 276, 277, 278, 279, 283, 285, 286, 288, 292, 298, 300, 302, 306, 311, 314, 315, 317, 321, 323, 329, 330, 332, 334, 335, 338, 340, 342, 344, 345, 347, 354, 356, 359, 360, 362, 363, 369, 370, 371, 372, 373, 375, 376, 377, 378, 381, 383, 384, 385, 386, 391, 392, 394, 401, 402, 405

All other numbers will be assigned to the group attending yoga classes.

b.   The results from this assignment would not be valid. Those patients with the most severe cancer would probably be more fatigued and sleepy than those with less severe cancer, regardless of which group they were assigned. To assign all the most severe cancer patients to one group will skew the results.

7.103   a.   The experimental units are the NCAA tennis coaches. The dependent variable is the rating on a 7-point scale of how important the coaches think the web site was. The treatments are Division I, Division II and Division III.

b.   To determine if the mean ratings of the web sites are different for the different levels of coaches, we test:

$$H_0:\ \mu_{\mathrm{I}} = \mu_{\mathrm{II}} = \mu_{\mathrm{III}}$$
$$H_a:\ \text{At least one mean differs}$$

c.   Since the p-value is less than $\alpha = .05$ ($p < .003 < .05$), H_0 is rejected. There is sufficient evidence to indicate a difference in mean responses among the different divisions of the colleges and universities at $\alpha = .05$.

7.105   a.   The experimental design used is a completely randomized design.

b.   There are 4 treatments in this experiment. The four treatments are the 4 "colonies" to which the robots were assigned – 3, 6, 9, or 12 robots per colony. The dependent variable is the energy expended per robot.

c.   To determine if the mean energy expended (per robot) of the four different colony sizes differed, we test:

$$H_0:\ \mu_1 = \mu_2 = \mu_3 = \mu_4$$
$$H_a:\ \text{At least two treatment means differ}$$

d.   The test statistic is $F = 7.70$ and the p-value is $p < .001$. Since the p-value is less than $\alpha = .05$, H_0 is rejected. There is sufficient evidence to indicate that mean energy expended (per robot) in the four different colony sizes differed at $\alpha = .05$.

7.107   a.   The experimental units are the participants in the study.

b.   The dependent variable is the brand recall score.

c.   There is one factor in this study – TV viewing group. Since there is only one factor, the treatments correspond to the factor levels of this variable. Thus, the treatments are the same as the three levels of TV viewer group. These 3 levels are violent content code, sex content code, and neutral TV.

d.   The means given are only sample means. If new samples were selected and sample means computed, the values and order of the sample means could change. In addition, the variances are not taken into account.

e.   The test statistic is $F = 20.45$ and the p-value is p-value $= 0.000$.

f.   Since the p-value is less than α ($p = 0.000 < .01$), H_o is rejected. There is sufficient evidence to indicate differences in the mean recall scores among the three viewing groups at $\alpha = .01$. The researchers can conclude that the content of the TV show affects the recall of imbedded commercials.

7.109   a.   This experiment is a one-way ANOVA. The 3 treatments are honey dosage, DM dosage, and no dosage.

b.   Using MINITAB, the ANOVA is:

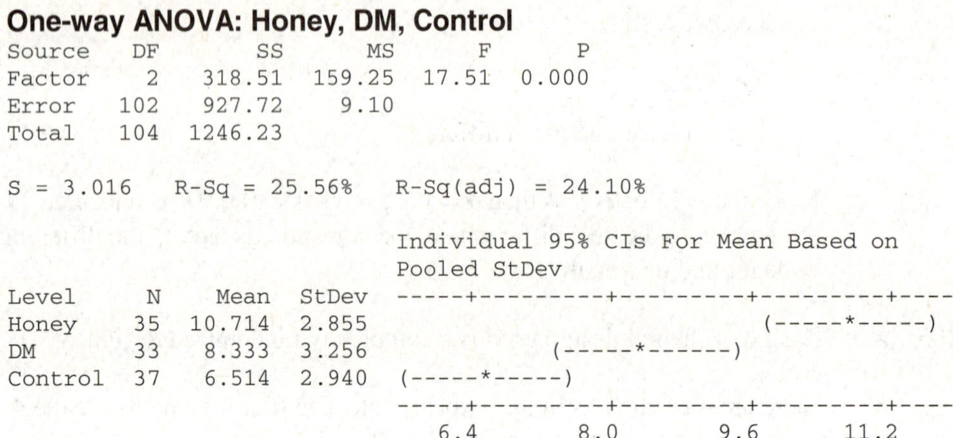

```
One-way ANOVA: Honey, DM, Control
Source    DF        SS       MS       F       P
Factor     2    318.51   159.25   17.51   0.000
Error    102    927.72     9.10
Total    104   1246.23


S = 3.016    R-Sq = 25.56%    R-Sq(adj) = 24.10%

                                    Individual 95% CIs For Mean Based on
                                    Pooled StDev
Level       N    Mean   StDev   -----+---------+---------+---------+----
Honey      35  10.714   2.855                                 (-----*-----)
DM         33   8.333   3.256                 (-----*------)
Control    37   6.514   2.940   (-----*-----)
                                -----+---------+---------+---------+----
                                   6.4       8.0       9.6      11.2

Pooled StDev = 3.016
```

To determine if there are differences in the mean improvement scores among the 3 groups, we test:

H_0: $\mu_1 = \mu_2 = \mu_3$
H_a: At least two treatment means differ

From the printout, the test statistic is $F = 17.51$ and the p-value is $p = 0.000$.

Since the p-value is so small $(p = 0.000)$, H_0 is rejected. There is sufficient evidence to indicate a difference in mean improvement scores among the 3 groups for any reasonable value of α.

7.111   Using MINITAB, the results of the analysis are:

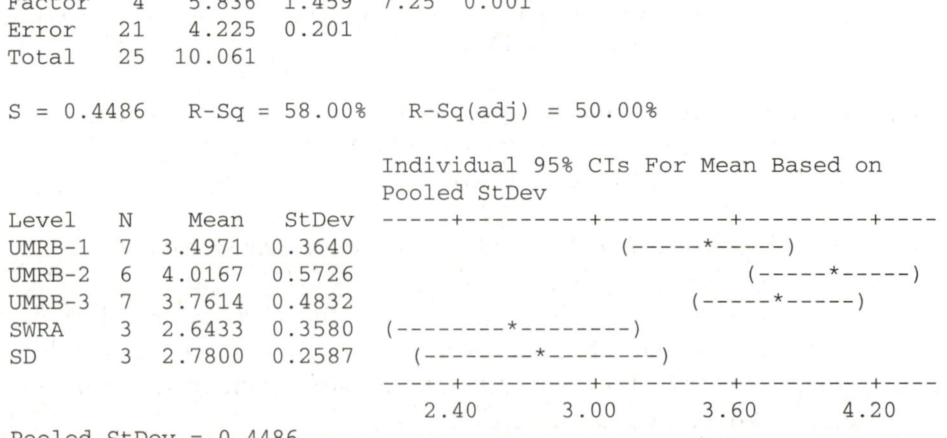

```
One-way ANOVA: UMRB-1, UMRB-2, UMBR-3, SWRA, SD
Source   DF       SS      MS      F       P
Factor    4    5.836   1.459   7.25   0.001
Error    21    4.225   0.201
Total    25   10.061

S = 0.4486   R-Sq = 58.00%   R-Sq(adj) = 50.00%

                                  Individual 95% CIs For Mean Based on
                                  Pooled StDev
Level    N     Mean    StDev   -----+---------+---------+---------+----
UMRB-1   7   3.4971   0.3640                      (-----*-----)
UMRB-2   6   4.0167   0.5726                            (-----*-----)
UMRB-3   7   3.7614   0.4832                        (-----*-----)
SWRA     3   2.6433   0.3580   (--------*--------)
SD       3   2.7800   0.2587     (--------*--------)
                               -----+---------+---------+---------+----
                                  2.40      3.00      3.60      4.20
Pooled StDev = 0.4486
```

To determine if there are differences among the mean Al/Be ratios for the 5 boreholes, we test:

$$H_0: \ \mu_1 = \mu_2 = \mu_3 = \mu_4 = \mu_5$$
$$H_a: \ \text{At least 2 means differ}$$

From the printout, the test statistic is $F = 7.25$ and the p-value is $p = 0.001$. Since the p-value is less than $\alpha = .10$ $(p = .001 < .10)$, H_0 is rejected. There is sufficient evidence to indicate a difference in the mean Al/Be ratios among the 5 boreholes at $\alpha = .10$.

7.113　a.　To compare average SAT scores of males and females, the target parameter is $\mu_1 - \mu_2$.

　　　　b.　To compare the difference between mean waiting times at two supermarket checkout lines, the target parameter is $\mu_1 - \mu_2$.

　　　　c.　To compare the mean starting salaries of mathematics, psychology, and engineering students, the target parameters are $\mu_1, \mu_2,$ and μ_3.

7.115　a.　
$$s_p^2 = \frac{(n_1 - 1)s_1^2 + (n_1 - 1)s_2^2}{n_1 + n_2 - 2} = \frac{11(74.2) + 13(60.5)}{12 + 14 - 2} = 66.7792$$

$$H_0: \ \mu_1 - \mu_2 = 0$$
$$H_a: \ \mu_1 - \mu_2 > 0$$

The test statistic is $t = \dfrac{(\bar{x}_1 - \bar{x}_2) - 0}{\sqrt{s_p^2\left(\dfrac{1}{n_1} + \dfrac{1}{n_2}\right)}} = \dfrac{(17.8 - 15.3) - 0}{\sqrt{66.7792\left(\dfrac{1}{12} + \dfrac{1}{14}\right)}} = .78$

The rejection region requires $\alpha = .05$ in the upper tail of the t distribution with df $= n_1 + n_2 - 2 = 12 + 14 - 2 = 24$. From Table IV, Appendix A, for df $= 24$, $t_{.05} = 1.711$. The rejection region is $t > 1.711$.

Since the observed value of the test statistic does not fall in the rejection region ($t = 0.78$ $\not> 1.711$), H_0 is not rejected. There is insufficient evidence to indicate that $\mu_1 > \mu_2$ at $\alpha = .05$.

　　　　b.　For confidence coefficient .99, $\alpha = .01$ and $\alpha / 2 = .01 / 2 = .005$. From Table IV, Appendix A, with df $= n_1 + n_2 - 2 = 12 + 14 - 2 = 24$, $t_{.005} = 2.797$. The confidence interval is:

$$(\bar{x}_1 - \bar{x}_2) \pm t_{.005}\sqrt{s_p^2\left(\frac{1}{n_1} + \frac{1}{n_2}\right)} \Rightarrow (17.8 - 15.3) \pm 2.797\sqrt{66.7792\left(\frac{1}{12} + \frac{1}{14}\right)}$$

$$\Rightarrow 2.50 \pm 8.99 \Rightarrow (-6.49, 11.49)$$

c. For confidence coefficient .99, $\alpha = .01$ and $\alpha/2 = .01/2 = .005$. From Table III, Appendix A, $z_{.005} = 2.58$.

$$n_1 = n_2 = \frac{(z_{\alpha/2})^2 \left(\sigma_1^2 + \sigma_2^2\right)}{(SE)^2} = \frac{(2.58)^2 (74.2 + 60.5)}{2^2} = 224.15 \approx 225$$

7.117   a. For confidence coefficient .90, $\alpha = .10$ and $\alpha/2 = .05$. From Table III, Appendix A, $z_{.05} = 1.645$. The confidence interval is:

$$(\bar{x}_1 - \bar{x}_2) \pm z_{.05} \sqrt{\frac{s_1^2}{n_1} + \frac{s_2^2}{n_2}} \Rightarrow (12.2 - 8.3) \pm 1.645 \sqrt{\frac{2.1}{135} + \frac{3.0}{148}} \Rightarrow 3.90 \pm .31 \Rightarrow (3.59,\ 4.21)$$

b. H_0: $\mu_1 - \mu_2 = 0$
H_a: $\mu_1 - \mu_2 \neq 0$

The test statistic is $z = \dfrac{(\bar{x}_1 - \bar{x}_2) - 0}{\sqrt{\dfrac{s_1^2}{n_1} + \dfrac{s_2^2}{n_2}}} = \dfrac{(12.2 - 8.3) - 0}{\sqrt{\dfrac{2.1}{135} + \dfrac{3.0}{148}}} = 20.60$

The rejection region requires $\alpha/2 = .01/2 = .005$ in each tail of the z distribution. From Table III, Appendix A, $z_{.005} = 2.58$. The rejection region is $z < -2.58$ or $z > 2.58$.

Since the observed value of the test statistic falls in the rejection region ($z = 20.60 > 2.58$), H_0 is rejected. There is sufficient evidence to indicate that $\mu_1 \neq \mu_2$ at $\alpha = .01$.

c. For confidence coefficient .90, $\alpha = .10$ and $\alpha/2 = .05$. From Table III, Appendix A, $z_{.05} = 1.645$.

$$n_1 = n_2 = \frac{(z_{\alpha/2})^2 \left(\sigma_1^2 + \sigma_2^2\right)}{(SE)^2} = \frac{(1.645)^2 (2.1 + 3.0)}{.2^2} = 345.02 \approx 346$$

7.119   Use the Wilcoxon signed rank test. Some preliminary calculations are:

Pair	X	Y	Difference	Rank of Absolute Difference
1	19	12	7	3
2	27	19	8	4.5
3	15	7	8	4.5
4	35	25	10	6
5	13	11	2	1.5
6	29	10	19	8
7	16	16	0	(eliminated)
8	22	10	12	7
9	16	18	−2	1.5
				$T_- = 1.5$

To determine if the probability distribution of x is shifted to the right of that for y, we test:

H_0:   The probability distributions are identical for the two variables
H_a: The probability distribution of x is shifted to the right of the probability
distribution of y

The test statistic is $T = T_- = 1.5$

Reject H_0 if $T \leq T_0$ where T_0 is based on $\alpha = .05$ and $n = 8$ (one-tailed):

Reject H_0 if $T \leq 6$ (from Table VII, Appendix A).

Since the observed value of the test statistic falls in the rejection region ($T = 1.5 \leq 6$), H_0 is rejected. There is sufficient evidence to conclude that the probability distribution of x is shifted to the right of that for y at $\alpha = .05$.

7.121   From the printout, the 95% confidence interval for the difference in mean seabird densities of oiled and unoiled transects is (−2.93, 2.49). We are 95% confident that the true difference in mean seabird densities of oiled and unoiled transects is between −2.93 and 2.49. Since 0 is contained in the interval, there is no evidence to indicate that the mean seabird densities are different for the oiled and unoiled transects.

7.123   a.   The variable measured for this experiment is the time needed to match the picture with the sentence.

   b.   The experimental units are the climbers.

   c.   Since the same climbers were timed at the base camp and at a camp 5 miles above sea level, the data should be analyzed as a paired difference experiment.

7.125   a.   Let $\mu_1 =$ mean height of Australian boys who repeated a grade and $\mu_2 =$ mean height of Australian boys who never repeated a grade.

   To determine if the average height of Australian boys who repeated a grade is less than the average height of boys who never repeated, we test:

$H_0: \mu_1 - \mu_2 = 0$

$H_a: \mu_1 - \mu_2 < 0$

b.   The test statistic is $z = \dfrac{\bar{x}_1 - \bar{x}_2}{\sqrt{\dfrac{s_1^2}{n_1} + \dfrac{s_2^2}{n_2}}} = \dfrac{-.04 - .30}{\sqrt{\dfrac{1.17^2}{86} + \dfrac{.97^2}{1346}}} = -2.64$

The rejection region requires $\alpha = .05$ in the lower tail of the z distribution. From Table III, Appendix A, $z_{.05} = 1.645$. The rejection region is $z < -1.645$.

Since the observed value of the test statistic falls in the rejection region ($z = -2.64 < -1.645$), H_0 is rejected. There is sufficient evidence to indicate that the average height of Australian boys who repeated a grade is less than the average height of boys who never repeated at $\alpha = .05$.

c.   Let μ_1 = mean height of Australian girls who repeated a grade and μ_2 = mean height of Australian girls who never repeated a grade.

To determine if the average height of Australian girls who repeated a grade is less than the average height of girls who never repeated, we test:

$H_0: \mu_1 - \mu_2 = 0$

$H_a: \mu_1 - \mu_2 < 0$

The test statistic is $z = \dfrac{\bar{x}_1 - \bar{x}_2}{\sqrt{\dfrac{s_1^2}{n_1} + \dfrac{s_2^2}{n_2}}} = \dfrac{.26 - .22}{\sqrt{\dfrac{.94^2}{43} + \dfrac{1.04^2}{1366}}} = .27$

The rejection region requires $\alpha = .05$ in the lower tail of the z distribution. From Table III, Appendix A, $z_{.05} = 1.645$. The rejection region is $z < -1.645$.

Since the observed value of the test statistic does not fall in the rejection region ($z = .27 \not< -1.645$), H_0 is not rejected. There is insufficient evidence to indicate that the average height of Australian girls who repeated a grade is less than the average height of girls who never repeated at $\alpha = .05$.

7.127   For confidence coefficient .95, $\alpha = .05$ and $\alpha / 2 = .05 / 2 = .025$. From Table IV, Appendix A, with df = $n_d - 1 = 26 - 1 = 25$, $t_{.025} = 2.060$. The 95% confidence interval is:

$$\bar{x}_d \pm t_{.025} \dfrac{s_d}{\sqrt{n_d}} \Rightarrow 10.5 \pm 2.060 \dfrac{7.6}{\sqrt{26}} \Rightarrow 10.5 \pm 3.07 \Rightarrow (7.43,\ 13.57)$$

We are 95% confident that the difference in the mean anxiety levels between before and after the visit is between 7.43 and 13.57. Since all of these values are positive, we can conclude that mean anxiety level after the visit is significantly less than the anxiety level before the visit.

7.129   a.   Let μ_1 = mean reading response times for the tongue-twister list and μ_2 = mean reading response times for the control list. Then $\mu_d = \mu_1 - \mu_2$.

To compare the mean reading response times for the tongue-twister and control lists, we test:

$$H_0: \ \mu_d = 0$$
$$H_a: \ \mu_d \neq 0$$

b.   The test statistic is $z = \dfrac{\bar{x}_d - 0}{s_d / \sqrt{n_d}} = \dfrac{.25 - 0}{.78 / \sqrt{42}} = 2.08$

The p-value is $P(z \leq -2.08) + P(z \geq 2.08) = (.5000 - .4812) + (.5000 - .4812) = .0188 + .0188 = .0376$. (Table III, Appendix A.)

c.   Since the observed p-value is less than α ($p = .0376 < .05$), H_0 is rejected. There is sufficient evidence to conclude that the mean reading response times differ for the tongue-twister and control lists at $\alpha = .05$.

7.131   a.   Let μ_1 = mean DIQ score for SLI children and μ_2 = mean DIQ score for YND children.

Some preliminary calculations are:

$$\bar{x}_1 = \frac{\sum x_1}{n_1} = \frac{936}{10} = 93.6 \qquad s_1^2 = \frac{\sum x_1^2 - \dfrac{\left(\sum x_1\right)^2}{n_1}}{n_1 - 1} = \frac{88352 - \dfrac{(936)^2}{10}}{10 - 1} = 82.4889$$

$$\bar{x}_2 = \frac{\sum x_2}{n_2} = \frac{953}{10} = 95.3 \qquad s_2^2 = \frac{\sum x_2^2 - \dfrac{\left(\sum x_2\right)^2}{n_2}}{n_1 - 1} = \frac{91329 - \dfrac{(953)^2}{10}}{10 - 1} = 56.4556$$

$$s_p^2 = \frac{(n_1 - 1)s_1^2 + (n_2 - 1)^2 s_2^2}{n_1 + n_2 - 2} = \frac{(10 - 1)82.4889 + (10 - 1)56.4556}{10 + 10 - 2} = 69.4723$$

To determine if the mean DIQ scores differ for the two groups, we test:

$$H_0: \ \mu_1 - \mu_2 = 0$$
$$H_a: \ \mu_1 - \mu_2 \neq 0$$

The test statistic is $t = \dfrac{\bar{x}_1 - \bar{x}_2 - 0}{\sqrt{s_p^2\left(\dfrac{1}{n_1} + \dfrac{1}{n_2}\right)}} = \dfrac{(93.6 - 95.3) - 0}{\sqrt{69.4723\left(\dfrac{1}{10} + \dfrac{1}{10}\right)}} = -0.46$

The rejection region requires $\alpha/2 = .10/2 = .05$ in each tail of the t distribution with df $= n_1 + n_2 - 2 = 10 + 10 - 2 = 18$. From Table IV, Appendix A, $t_{.05} = 1.734$. The rejection region is $t < -1.734$ or $t > 1.734$.

Since the observed value of the test statistic does not fall in the rejection region ($t = -.46 \not< -1.734$), H_0 is not rejected. There is insufficient evidence to indicate that the mean DIQ scores differ for the two groups at $\alpha = .10$.

b.   Some preliminary calculations are:

SLI Children	Rank	YND Children	Rank
86	3	110	19.5
94	12	92	10
89	6	86	3
110	19.5	90	7.5
87	5	90	7.5
86	3	92	10
98	15	100	16
84	1	105	17
107	18	96	14
95	13	92	10
	$T_1 = 95.5$		$T_2 = 114.5$

To determine if there is a shift in location for the two distributions, we test:

H_0: The two sampled populations have identical probability distributions
H_a: The probability distribution for SLI children is shifted to the left or to the right
       of that for YND children

The test statistic is $T_1 = 95.5$ since both samples have the same number of measurements.

The null hypothesis will be rejected if $T_1 \leq T_L$ or $T_1 \geq T_U$ where T_L and T_U correspond to $\alpha = .10$ (two-tailed), $n_1 = 10$ and $n_2 = 10$. From Table VI, Appendix A, $T_L = 83$ and $T_U = 127$.

   Reject H_0 if $T_1 \leq 83$ or $T_1 \geq 127$.

Since the observed value of the test statistic does not fall in the rejection region ($T_1 = 95.5 \not\geq 127$ and $T_1 \not\leq 83$), H_0 is not rejected. There is insufficient evidence to indicate the probability distribution for SLI children is shifted to the left or to the right of that for YND children at $\alpha = .10$.

7.133   Some preliminary calculations:

	Panic Attacks per Week				Panic Attacks per Week		
Patient	Placebo	Inositol	$D = Pl - In$	Patient	Placebo	Inositol	$D = Pl - In$
1	0	0	0	12	0	2	−2
2	2	1	1	13	3	1	2
3	0	3	−3	14	3	1	2
4	1	2	−1	15	3	4	−1
5	0	0	0	16	4	2	2
6	10	5	5	17	6	4	2
7	2	0	2	18	15	21	−6
8	6	4	2	19	28	8	20
9	1	1	0	20	30	0	30
10	1	0	1	21	13	0	13
11	1	3	−2				

$$\overline{x}_d = \frac{\sum x_d}{n_d} = \frac{67}{21} = 3.19 \qquad s_d^2 = \frac{\sum x_d^2 - \frac{\left(\sum x_d\right)^2}{n_d}}{n_d - 1} = \frac{1{,}575 - \frac{67^2}{21}}{21 - 1} = 68.0619$$

To determine if there is a difference in the mean number of panic attacks between the placebo and the drug Inositol, we test:

H_0: $\mu_d = 0$

H_a: $\mu_d \neq 0$

The test statistic is $t = \dfrac{\overline{x}_d - 0}{s_d / \sqrt{n_d}} = \dfrac{3.19 - 0}{\sqrt{68.0619} / \sqrt{21}} = 1.77$

Since no α was given, we will use $\alpha = .05$. The rejection region requires $\alpha / 2 = .05 / 2 = .025$ in each tail of the t distribution with df $= n_d - 1 = 21 - 1 = 20$. From Table IV, Appendix A, $t_{.025} = 2.086$. The rejection region is $t < -2.086$ or $t > 2.086$.

Since the observed value of the test statistic does not fall in the rejection region ($t = 1.77 \not> 2.086$), H_0 is not rejected. There is insufficient evidence to indicate that there is a difference in the mean number of panic attacks between the placebo and the drug Inositol at $\alpha = .05$. There is no evidence that the drug is effective.

7.135   Using MINITAB, the ANOVA table is:

One-way ANOVA: y versus FACES

```
Source   DF       SS      MS       F        P
FACES     5    23.09    4.62    3.96    0.007
Error    30    34.99    1.17
Total    35    58.07

S = 1.080    R-Sq = 39.75%    R-Sq(adj) = 29.71%
```

To determine if differences exist among the mean dominance ratings of the six facial expressions, we test:

$$H_0: \ \mu_1 = \mu_2 = \mu_3 = \mu_4 = \mu_5 = \mu_6$$
$$H_a: \ \text{At least two treatment means differ}$$

where μ_i represents the mean dominance rating for facial expression i.

The test statistic is $F = 3.96$.

The rejection region requires $\alpha = .10$ in the upper tail of the F distribution with $v_1 = k - 1 = 6 - 1 = 5$ and $v_2 = n - k = 36 - 6 = 30$. Using Table VIII, Appendix A, $F_{.10} = 2.05$. The rejection region is $F > 2.05$.

Since the observed value of the test statistic falls in the rejection region ($F = 3.96 > 2.05$), H_0 is rejected. There is sufficient evidence to indicate that differences exist among the mean dominance ratings among the facial expressions at $\alpha = .10$.

7.137   Let $\mu_1 =$ mean rating for music majors and $\mu_2 =$ mean rating of nonmusic majors.

For confidence coefficient .95, $\alpha = .05$ and $\alpha / 2 = .05 / 2 = .025$. From Table III, Appendix A, $z_{.025} = 1.96$. The confidence interval is:

$$(\bar{x}_1 - \bar{x}_2) \pm z_{.025}\sqrt{\frac{s_1^2}{n_1} + \frac{s_2^2}{n_2}} \Rightarrow (4.26 - 4.59) \pm 1.96\sqrt{\frac{.81^2}{100} + \frac{.78^2}{100}}$$

$$\Rightarrow -.33 \pm .220 \Rightarrow (-.550, -.110)$$

We are 95% confident that the difference in the mean rating between music majors and nonmusic majors is between $-.55$ and $-.11$.

7.139   Some preliminary calculations are:

Fat	Plasma	Difference	Rank of Absolute Difference
4.9	2.5	2.4	11.5
6.9	3.5	3.4	16
10.0	6.8	3.2	15
4.4	4.7	-0.3	4
4.6	4.6	0	(eliminated)
1.1	1.8	-0.7	8
2.3	2.5	-0.2	2.5
5.9	3.1	2.8	14
7.0	3.1	3.9	17
5.5	3.0	2.5	13
7.0	6.9	0.1	1
1.4	1.6	-0.2	2.5
11.0	20.0	-9.0	19
2.5	4.1	-1.6	9
4.4	2.1	2.3	10
4.2	1.8	2.4	11.5
41.0	36.0	5.0	18
2.9	3.3	-0.4	5
7.7	7.2	0.5	6.5
2.5	2.0	0.5	6.5

$$T_- = 50$$
$$T_+ = 140$$

To determine if the distribution of TCDD levels in fat is shifted above or below the distribution of TCDD levels in plasma, we test:

H_0: The distribution of TCDD levels in fat is identical the distribution of TCDD levels in plasma

H_a: The distribution of TCDD levels in fat is shifted above or below the distribution of TCDD levels in plasma

The test statistic is T = smaller of T_- and T_+ which is $T_- = 50$.

The rejection region is $T_- \le 46$, from Table VII, Appendix A, with $n = 19$ and $\alpha = .05$ (two-tails).

Since the observed value of the test statistic does not fall in the rejection region ($T_- = 50 \not< 46$), H_0 is not rejected. There is insufficient evidence to indicate the distribution of TCDD levels in fat is shifted above or below the distribution of TCDD levels in plasma at $\alpha = .05$.

7.141   Using MINITAB, the descriptive statistics are:

Descriptive Statistics: CREATIVE, INFO, DECPERS, SKILLS, TASKID, AGE, EDYRS

```
Variable  GROUP      N    Mean   StDev  Minimum       Q1   Median       Q3  Maximum
CREATIVE  NOSPILL   67  4.4478  0.5304   3.0000   4.0000   4.0000   5.0000   5.0000
          SPILLOV   47  5.2553  0.4408   5.0000   5.0000   5.0000   6.0000   6.0000

INFO      NOSPILL   67   4.657   2.226    1.000    3.000    5.000    7.000    7.000
          SPILLOV   47   5.213   1.719    1.000    4.000    5.000    7.000    7.000

DECPERS   NOSPILL   67   2.746   1.964    1.000    1.000    2.000    5.000    7.000
          SPILLOV   47   3.319   2.023    1.000    1.000    3.000    5.000    7.000

SKILLS    NOSPILL   67   4.821   1.100    2.000    4.000    5.000    5.000    7.000
          SPILLOV   47   5.851   1.161    2.000    5.000    6.000    7.000    7.000

TASKID    NOSPILL   67   4.239   1.156    1.000    3.000    4.000    5.000    7.000
          SPILLOV   47   4.809   2.028    1.000    3.000    5.000    7.000    7.000

AGE       NOSPILL   67  45.343   7.971   21.000   41.000   45.000   50.000   63.000
          SPILLOV   47  46.298   5.767   30.000   44.000   47.000   51.000   54.000

EDYRS     NOSPILL   67  13.224   1.312   12.000   12.000   13.000   14.000   18.000
          SPILLOV   47  13.085   1.060   12.000   12.000   13.000   14.000   16.000
```

Let μ_1 = mean characteristic for the workers with positive spillover and μ_2 = mean characteristic for the workers with no spillover. To determine if there is a difference in the mean characteristic between the positive spillover group and the no spillover group, we test:

$$H_0: \ \mu_1 - \mu_2 = 0$$
$$H_a: \ \mu_1 - \mu_2 \neq 0$$

The test statistic is $z = \dfrac{\bar{x}_1 - \bar{x}_2}{\sqrt{\left(\dfrac{s_1^2}{n_1} + \dfrac{s_2^2}{n_2}\right)}}$

Since no α was given, we will run all tests at $\alpha = .05$. The rejection region requires $\alpha / 2 = .05 / 2 = .025$ in each tail of the z distribution. From Table III, Appendix A, $z_{.025} = 1.96$. The rejection region is $z < -1.96$ or $z > 1.96$.

Use of Creative Ideas:

The test statistic is $z = \dfrac{\bar{x}_1 - \bar{x}_2}{\sqrt{\left(\dfrac{s_1^2}{n_1} + \dfrac{s_2^2}{n_2}\right)}} = \dfrac{5.2553 - 4.4478}{\sqrt{\dfrac{.4408^2}{47} + \dfrac{.5304^2}{67}}} = 8.85$

Since the observed value of the test statistic falls in the rejection region ($z = 8.85 > 1.96$), H_0 is rejected. There is sufficient evidence to indicate a difference in the mean use of creative ideas between the positive spillover group and the no spillover group at $\alpha = .05$.

Utilization of Information:

The test statistic is $z = \dfrac{\overline{x}_1 - \overline{x}_2}{\sqrt{\left(\dfrac{s_1^2}{n_1} + \dfrac{s_2^2}{n_2}\right)}} = \dfrac{5.213 - 4.657}{\sqrt{\dfrac{1.719^2}{47} + \dfrac{2.226^2}{67}}} = 1.50$

Since the observed value of the test statistic does not fall in the rejection region ($z = 1.50 \ngtr 1.96$), H_o is not rejected. There is insufficient evidence to indicate a difference in the mean utilization of information between the positive spillover group and the no spillover group at $\alpha = .05$.

Decisions Regarding Personnel Matters:

The test statistic is $z = \dfrac{\overline{x}_1 - \overline{x}_2}{\sqrt{\left(\dfrac{s_1^2}{n_1} + \dfrac{s_2^2}{n_2}\right)}} = \dfrac{3.319 - 2.746}{\sqrt{\dfrac{2.023^2}{47} + \dfrac{1.964^2}{67}}} = 1.51$

Since the observed value of the test statistic does not fall in the rejection region ($z = 1.51 \ngtr 1.96$), H_0 is not rejected. There is insufficient evidence to indicate a difference in the mean participation in decisions regarding personnel matters between the positive spillover group and the no spillover group at $\alpha = .05$.

Good Use of Job Skills:

The test statistic is $z = \dfrac{\overline{x}_1 - \overline{x}_2}{\sqrt{\left(\dfrac{s_1^2}{n_1} + \dfrac{s_2^2}{n_2}\right)}} = \dfrac{5.851 - 4.821}{\sqrt{\dfrac{1.161^2}{47} + \dfrac{1.100^2}{67}}} = 4.76$

Since the observed value of the test statistic falls in the rejection region ($z = 4.76 > 1.96$), H_0 is rejected. There is sufficient evidence to indicate a difference in the mean good use of skills between the positive spillover group and the no spillover group at $\alpha = .05$.

Task Identity:

The test statistic is $z = \dfrac{\overline{x}_1 - \overline{x}_2}{\sqrt{\left(\dfrac{s_1^2}{n_1} + \dfrac{s_2^2}{n_2}\right)}} = \dfrac{4.809 - 4.239}{\sqrt{\dfrac{2.028^2}{47} + \dfrac{1.156^2}{67}}} = 1.74$

Since the observed value of the test statistic does not fall in the rejection region ($z = 1.74 \ngtr 1.96$), H_0 is not rejected. There is insufficient evidence to indicate a difference in the mean task identity between the positive spillover group and the no spillover group at $\alpha = .05$.

Of the 5 job-related characteristics, there are only 2 that have significantly different means for the 2 groups. They are "use of creative ideas" and "good use of skills". For both of these characteristics, the mean for the positive spillover group was significantly higher than the mean for the no spillover group.

Age:

The test statistic is $z = \dfrac{\overline{x}_1 - \overline{x}_2}{\sqrt{\left(\dfrac{s_1^2}{n_1} + \dfrac{s_2^2}{n_2}\right)}} = \dfrac{46.298 - 45.343}{\sqrt{\dfrac{5.767^2}{47} + \dfrac{7.971^2}{67}}} = 0.74$

Since the observed value of the test statistic does not fall in the rejection region ($z = 0.74 \not> 1.96$), H_0 is not rejected. There is insufficient evidence to indicate a difference in the mean age between the positive spillover group and the no spillover group at $\alpha = .05$.

Education:

The test statistic is $z = \dfrac{\overline{x}_1 - \overline{x}_2}{\sqrt{\left(\dfrac{s_1^2}{n_1} + \dfrac{s_2^2}{n_2}\right)}} = \dfrac{13.085 - 13.224}{\sqrt{\dfrac{1.060^2}{47} + \dfrac{1.312^2}{67}}} = -0.62$

Since the observed value of the test statistic does not fall in the rejection region ($z = -0.62 \not> 1.96$), H_0 is not rejected. There is insufficient evidence to indicate a difference in the mean years of education between the positive spillover group and the no spillover group at $\alpha = .05$.

Comparing Population Proportions

Chapter
8

8.1 The conditions required for a large-sample inference about $(p_1 - p_2)$ are:

1. The two samples are randomly selected in an independent manner from the two target populations.
2. The sample sizes, n_1 and n_2, are both large so that the sampling distribution of $(\hat{p}_1 - \hat{p}_2)$ will be approximately normal. (This condition will be satisfied if all of the following are true: $n_1\hat{p}_1 \geq 15$, $n_1\hat{q}_1 \geq 15$ and $n_2\hat{p}_2 \geq 15$, $n_2\hat{q}_2 \geq 15$.)

8.3 a. The distribution of x_1 is binomial with n_1 trials and the probability of success, p_1. The distribution of x_2 is binomial with n_2 trials and the probability of success, p_2.

b. For large samples, the sampling distribution of $(\hat{p}_1 - \hat{p}_2)$ is approximately normal with

$$\text{mean } (p_1 - p_2), \text{ and standard deviation } \sigma_{\hat{p}_1 - \hat{p}_2} = \sqrt{\frac{p_1 q_1}{n_1} + \frac{p_2 q_2}{n_2}}.$$

8.5 For confidence coefficient .95, $\alpha = 1 - .95 = .05$ and $\alpha/2 = .05/2 = .025$. From Table III, Appendix A, $z_{.025} = 1.96$. The 95% confidence interval for $p_1 - p_2$ is approximately:

a. $(\hat{p}_1 - \hat{p}_2) \pm z_{\alpha/2} \sqrt{\dfrac{\hat{p}_1\hat{q}_1}{n_1} + \dfrac{\hat{p}_2\hat{q}_2}{n_2}}$

$$\Rightarrow (.65 - .58) \pm 1.96\sqrt{\frac{.65(1-.65)}{400} + \frac{.58(1-.58)}{400}} \Rightarrow .07 \pm .067 \Rightarrow (.003, .137)$$

b. $(\hat{p}_1 - \hat{p}_2) \pm z_{\alpha/2} \sqrt{\dfrac{\hat{p}_1\hat{q}_1}{n_1} + \dfrac{\hat{p}_2\hat{q}_2}{n_2}}$

$$\Rightarrow (.31 - .25) \pm 1.96\sqrt{\frac{.31(1-.31)}{180} + \frac{.25(1-.25}{250}} \Rightarrow .06 \pm .086 \Rightarrow (-.026, .146)$$

c. $(\hat{p}_1 - \hat{p}_2) \pm z_{\alpha/2} \sqrt{\dfrac{\hat{p}_1\hat{q}_1}{n_1} + \dfrac{\hat{p}_2\hat{q}_2}{n_2}}$

$$\Rightarrow (.46 - .61) \pm 1.96\sqrt{\frac{.46(1-.46)}{100} + \frac{.61(1-.61)}{120}} \Rightarrow -.15 \pm .131 \Rightarrow (-.281, -.019)$$

8.7 H_0: $(p_1 - p_2) = .1$
H_a: $(p_1 - p_2) > .1$

Since D_0 is not equal to 0, the test statistic is:

$$z = \frac{(\hat{p}_1 - \hat{p}_2) - D_0}{\sqrt{\dfrac{p_1 q_1}{n_1} + \dfrac{p_2 q_2}{n_2}}} = \frac{(\hat{p}_1 - \hat{p}_2) - D_0}{\sqrt{\dfrac{\hat{p}_1 \hat{q}_1}{n_1} + \dfrac{\hat{p}_2 \hat{q}_2}{n_2}}} = \frac{(.4 - .2) - .1}{\sqrt{\dfrac{.4(1-.4)}{50} + \dfrac{.2(1-.2)}{60}}} = \frac{1}{.0864} = 1.16$$

The rejection region requires $\alpha = .05$ in the upper tail of the z distribution. From Table III, Appendix A, $z_{.05} = 1.645$. The rejection region is $z > 1.645$.

Since the observed value of the test statistic does not fall in the rejection region ($z = 1.16 \not> 1.645$), H_0 is not rejected. There is insufficient evidence to show $(p_1 - p_2) > .1$ at $\alpha = .05$.

8.9   a. $\hat{p}_1 = \dfrac{x_1}{n_1} = \dfrac{746}{1358} = .549$

b. $\hat{p}_2 = \dfrac{x_2}{n_2} = \dfrac{967}{1379} = .701$

c.   For confidence coefficient .90, $\alpha = .10$ and $\alpha / 2 = .10 / 2 = .05$. From Table III, Appendix A, $z_{.05} = 1.645$. The 90% confidence interval is:

$$(\hat{p}_1 - \hat{p}_2) \pm z_{.05} \sqrt{\frac{\hat{p}_1 \hat{q}_1}{n_1} + \frac{\hat{p}_2 \hat{q}_2}{n_2}} \Rightarrow (.549 - .701) \pm 1.645 \sqrt{\frac{.549(.451)}{1358} + \frac{.701(.299)}{1379}}$$

$$\Rightarrow -.152 \pm .030 \Rightarrow (-.182, \ -.122)$$

d.   We are 90% confident that the true value of the difference in the proportion of Dutch boys and Dutch girls who have never bullied another student is between $-.182$ and $-.122$.

e.   Since the 90% confidence interval contains only negative numbers, the proportion of girls who do not bully is larger than that for boys. Thus, the proportion of Dutch boys who bully is significantly greater than the proportion of Dutch girls who bully.

8.11   a.   The observed proportion of St. John's wort patients who were in remission is:

$$\hat{p}_1 = \frac{x_1}{n_1} = \frac{14}{98} = .143$$

b.   The observed proportion of placebo patients who were in remission is:

$$\hat{p}_2 = \frac{x_2}{n_2} = \frac{5}{102} = .049$$

c. Some preliminary calculations are:

$$\hat{p} = \frac{x_1 + x_2}{n_1 + n_2} = \frac{14 + 5}{98 + 102} = \frac{19}{200} = .095 \qquad \hat{q} = 1 - \hat{p} = 1 - .095 = .905$$

To determine if the proportion of St. John's wort patients in remission exceeds the proportion of placebo patients in remission, we test:

H_0: $p_1 - p_2 = 0$
H_a: $p_1 - p_2 > 0$

The test statistic is $z = \dfrac{\hat{p}_1 - \hat{p}_2}{\sqrt{\hat{p}\hat{q}\left(\dfrac{1}{n_1} + \dfrac{1}{n_2}\right)}} = \dfrac{.143 - .049}{\sqrt{.095(.905)\left(\dfrac{1}{98} + \dfrac{1}{102}\right)}} = 2.27$

The rejection region requires $\alpha = .01$ in the upper tail of the z distribution. From Table III, Appendix A, $z_{.01} = 2.33$. The rejection region is $z > 2.33$.

Since the observed value of the test statistic does not fall in the rejection region ($z = 2.27 \not> 2.33$), H_0 is not rejected. There is insufficient evidence to indicate the proportion of St. John's wort patients in remission exceeds the proportion of placebo patients in remission at $\alpha = .01$.

d. The hypotheses and test statistic are the same as in part **c**.

The rejection region requires $\alpha = .10$ in the upper tail of the z distribution. From Table III, Appendix A, $z_{.10} = 1.28$. The rejection region is $z > 1.28$.

Since the observed value of the test statistic falls in the rejection region ($z = 2.27 > 1.28$), H_0 is rejected. There is sufficient evidence to indicate the proportion of St. John's wort patients in remission exceeds the proportion of placebo patients in remission $\alpha = .10$.

e. By changing the value of α in this problem, the conclusion changes. When α is small ($\alpha = .01$), we did not reject H_0. When α increases ($\alpha = .10$), we rejected H_0. As α increases, the chance of making a Type I error increases, but the chance of a Type II error decreases.

8.13 a. Let p_1 = proportion of African-American drivers who were searched and p_2 = proportion of White drivers who were searched. The parameter of interest is $p_1 - p_2$, or the difference in the proportions of drivers searched between the two ethnic groups.

Some preliminary calculations are:

$$\hat{p}_1 = \frac{x_1}{n_1} = \frac{12,016}{61,688} = .195 \qquad\qquad \hat{p}_2 = \frac{x_2}{n_2} = \frac{5,312}{106,892} = .050$$

$$\hat{p} = \frac{x_1 + x_2}{n_1 + n_2} = \frac{12,016 + 5,312}{61,688 + 106,892} = .103$$

To determine if there is a difference in the proportions of African-American and White drivers who are searched by the LA police, we test:

$$H_0:\ p_1 - p_2 = 0$$
$$H_a:\ p_1 - p_2 \neq 0$$

The test statistic is $z = \dfrac{(\hat{p}_1 - \hat{p}_2) - 0}{\sqrt{\hat{p}\hat{q}\left(\dfrac{1}{n_1} + \dfrac{1}{n_2}\right)}} = \dfrac{(.195 - .050) - 0}{\sqrt{.103(.897)\left(\dfrac{1}{61,688} + \dfrac{1}{106,892}\right)}} = 94.35$

The rejection region requires $\alpha / 2 = .05 / 2 = .025$ in each tail of the z distribution. From Table III, Appendix A, $z_{.025} = 1.96$. The rejection region is $z < -1.96$ or $z > 1.96$.

Since the observed value of the test statistic falls in the rejection region ($z = 94.35 > 1.96$), H_0 is rejected. There is sufficient evidence to indicate a difference in the proportions of African-American and White drivers who are searched by the LA police at $\alpha = .05$.

b. Let p_1 = "hit rate" of African-American drivers and p_2 = "hit rate" of White drivers. The parameter of interest is $p_1 - p_2$, or the difference in the "hit rates" for the two ethnic groups.

Some preliminary calculations are:

$$\hat{p}_1 = \frac{x_1}{n_1} = \frac{5,134}{12,016} = .427 \qquad\qquad \hat{p}_2 = \frac{x_2}{n_2} = \frac{3,006}{5,312} = .566$$

For confidence coefficient .95, $\alpha = .05$ and $\alpha / 2 = .05 / 2 = .025$. From Table III, Appendix A, $z_{.025} = 1.96$. The 95% confidence interval is:

$$(.427 - .566) \pm 1.96\sqrt{\frac{.427(.573)}{12,016} + \frac{.566(.434)}{5,312}} \Rightarrow -.139 \pm .016 \Rightarrow (-.155,\ \ -.123)$$

We are 95% confident that the difference in "hit rates" between African-American and White drivers searched by the LA police is between $-.155$ and $-.123$. Since both end points of the interval are negative, there is evidence that the "hit rate" for African-American drives is less than that for White drivers.

8.15   Let p_1 = proportion of rice weevils found dead after 4 days of exposure to nitrogen gas and p_2 = proportion of rice weevils found dead after 3.5 days of exposure to nitrogen gas. The parameter of interest is $p_1 - p_2$, or the difference in the proportions of rice weevils found dead between the 2 different exposures to nitrogen gas.

Some preliminary calculations are:

$$\hat{p}_1 = \frac{x_1}{n_1} = \frac{31,386}{31,421} = .999 \qquad\qquad \hat{p}_2 = \frac{x_2}{n_2} = \frac{23,516}{23,676} = .993$$

$$\hat{p} = \frac{x_1 + x_2}{n_1 + n_2} = \frac{31,386 + 23,516}{31,421 + 23,676} = .996$$

To determine if there is a difference in the proportions of rice weevils found dead between the 2 exposure times, we test:

H_o: $p_1 - p_2 = 0$
H_a: $p_1 - p_2 \neq 0$

The test statistic is $z = \dfrac{(\hat{p}_1 - \hat{p}_2) - 0}{\sqrt{\hat{p}\hat{q}\left(\dfrac{1}{n_1} + \dfrac{1}{n_2}\right)}} = \dfrac{(.999 - .993) - 0}{\sqrt{.996(.004)\left(\dfrac{1}{31,421} + \dfrac{1}{23,676}\right)}} = 11.05$

The rejection region requires $\alpha / 2 = .10 / 2 = .05$ in each tail of the z distribution. From Table III, Appendix A, $z_{.05} = 1.645$. The rejection region is $z < -1.645$ or $z > 1.645$.

Since the observed value of the test statistic falls in the rejection region ($z = 11.05 > 1.645$), H_o is rejected. There is sufficient evidence to indicate a difference in the proportions of rice weevils found dead between the 2 exposure times at $\alpha = .10$.

8.17   Let p_1 = recall rate of the 60 to 72 year-old seniors and p_2 = recall rate of the 73 to 83 year-old seniors.

Some preliminary calculations are:

$$\hat{p}_1 = \frac{x_1}{n_1} = \frac{31}{40} = .775 \qquad \hat{p}_2 = \frac{x_2}{n_2} = \frac{22}{40} = .55 \qquad \hat{p} = \frac{x_1 + x_2}{n_1 + n_2} = \frac{31 + 22}{40 + 40} = .6625$$

To determine if the recall rates of the two groups differ, we test:

H_0: $p_1 - p_2 = 0$
H_a: $p_1 - p_2 \neq 0$

The test statistic is $z = \dfrac{(\hat{p}_1 - \hat{p}_2) - 0}{\sqrt{\hat{p}\hat{q}\left(\dfrac{1}{n_1} + \dfrac{1}{n_2}\right)}} = \dfrac{(.775 - .55) - 0}{\sqrt{.6625(.3375)\left(\dfrac{1}{40} + \dfrac{1}{40}\right)}} = 2.13$

The rejection region requires $\alpha / 2 = .05 / 2 = .025$ in each tail of the z distribution. From Table III, Appendix A, $z_{.025} = 1.96$. The rejection region is $z < -1.96$ or $z > 1.96$.

Since the observed value of the test statistic falls in the rejection region ($z = 2.13 > 1.96$), H_0 is rejected. There is sufficient evidence to indicate the recall rates of the two groups differ at $\alpha = .05$.

8.19   Let p_1 = proportion of volunteers who figured out the third rule in the group that slept and p_2 = proportion of volunteers who figured out the third rule in the group that stayed awake all night. The parameter of interest is $p_1 - p_2$, or the difference in the proportions of volunteers who figured out the third rule between the 2 groups.

Some preliminary calculations are:

$$\hat{p}_1 = \frac{x_1}{n_1} = \frac{39}{50} = .78 \qquad\qquad \hat{p}_2 = \frac{x_2}{n_2} = \frac{15}{50} = .30$$

For confidence coefficient .90, $\alpha = .10$ and $\alpha/2 = .10/2 = .05$. From Table III, Appendix A, $z_{.05} = 1.645$. The 90% confidence interval is:

$$(.78 - .30) \pm 1.645\sqrt{\frac{.78(.22)}{50} + \frac{.30(.70)}{50}} \Rightarrow .48 \pm .144 \Rightarrow (.336, \ .624)$$

We are 90% confident that the difference in the proportions of volunteers who figured out the third rule between the 2 groups is between .336 and .624. Since both end points are greater than 0, there is evidence that the proportion of those who slept who figured out the third rule is greater than the proportion of those who stayed awake all night who figured out the third rule.

8.21   Let p_1 = proportion of Yayoi farmers with LEH defects, p_2 = proportion of Eastern Jomon foragers with LEH defects, and p_3 = proportion of Western Jomon foragers with LEH defects.

First, we will test Theory 1.

Some preliminary calculations are:

$$\hat{p} = \frac{n_1\hat{p}_1 + n_2\hat{p}_2}{n_1 + n_2} = \frac{182(.631) + 164(.482)}{182 + 164} = .560$$

To determine if foragers with a broad-based economy (Eastern Jomon) have a lower prevalence of LEH defects than early agriculturists (Yayoi), we test:

H_0: $p_1 - p_2 = 0$
H_a: $p_1 - p_2 > 0$

The test statistic is $z = \dfrac{\hat{p}_1 - \hat{p}_2}{\sqrt{\hat{p}\hat{q}\left(\dfrac{1}{n_1} + \dfrac{1}{n_2}\right)}} = \dfrac{.631 - .482}{\sqrt{.560(.440)\left(\dfrac{1}{182} + \dfrac{1}{164}\right)}} = 2.79$

The rejection region requires $\alpha = .01$ in the upper tail of the z-distribution. From Table III, Appendix A, $z_{.01} = 2.33$. The rejection region is $z > 2.33$.

Since the observed value of the test statistic falls in the rejection region ($z = 2.79 > 2.33$), H_0 is rejected. There is sufficient evidence to indicate that foragers with a broad-based economy (Eastern Jomon) have a lower prevalence of LEH defects than early agriculturists (Yayoi) at $\alpha = .01$.

Next, we test Theory 2.

Some preliminary calculations are:

$$\hat{p} = \frac{n_1 \hat{p}_1 + n_3 \hat{p}_3}{n_1 + n_3} = \frac{182(.631) + 122(.648)}{182 + 122} = .638$$

To determine if foragers with a wet rice economy (Western Jomon) have a different prevalence of LEH defects than early agriculturists (Yayoi), we test:

H_0: $p_1 - p_2 = 0$
H_a: $p_1 - p_2 \neq 0$

The test statistic is $z = \dfrac{\hat{p}_1 - \hat{p}_2}{\sqrt{\hat{p}\hat{q}\left(\dfrac{1}{n_1} + \dfrac{1}{n_2}\right)}} = \dfrac{.631 - .648}{\sqrt{.638(.362)\left(\dfrac{1}{182} + \dfrac{1}{122}\right)}} = -.30$

The rejection region requires $\alpha = .01 / 2 = .005$ in each tail of the z-distribution. From Table III, Appendix A, $z_{.005} = 2.58$. The rejection region is $z > 2.58$ or $z < -2.58$.

Since the observed value of the test statistic does not fall in the rejection region ($z = -.30 \not< -2.58$), H_0 is not rejected. There is insufficient evidence to indicate that foragers with a wet rice economy (Western Jomon) have a different prevalence of LEH defects than early agriculturists (Yayoi) at $\alpha = .01$.

8.23   If the sample size calculation yields a value of n that is too large to be practical, we might decide to use a large sampling error (SE) in order to reduce the sample size, or we might decrease the confidence coefficient.

8.25   For confidence coefficient .90, $\alpha = 1 - .90 = .10$ and $\alpha/2 = .10/2 = .05$. From Table III, Appendix A, $z_{.05} = 1.645$. For width = .1, the standard error is $SE = .1/2 = .05$.

$$n_1 = n_2 = \frac{\left(z_{\alpha/2}\right)^2 (p_1 q_1 + p_2 q_2)}{(SE)^2} = \frac{1.645^2 (.6(.4) + .6(.4))}{.05^2} = 519.6 \approx 520$$

In order to estimate the difference in proportions using a 90% confidence interval of width .1, we would need to sample 520 observations from each population. Since only enough money was budgeted to sample 100 observations from each population, insufficient funds have been allocated.

8.27   a.   For confidence coefficient .95, $\alpha = 1 - .95 = .05$ and $\alpha / 2 = .05 / 2 = .025$. From Table III, Appendix A, $z_{.05} = 1.96$. The standard error is $SE = .015$. From Exercise 8.14, $\hat{p}_1 = .184$ and $\hat{p}_2 = .177$.

$$n_1 = n_2 = \frac{\left(z_{\alpha/2}\right)^2 (p_1 q_1 + p_2 q_2)}{(SE)^2} = \frac{1.96^2 (.184(.816) + .177(.823))}{.015^2} = 5050.7 \approx 5051$$

In order to estimate the difference in proportions using a 95% confidence interval to within .015, we would need to sample 5,051 observations from each population.

b. It would be extremely difficult to obtain information on over 10,000 patients. In addition, the cost would be extremely high.

c. A difference of .015 is very small. This difference would be a difference of only 15 out of every 1,000 patients. In practice, this difference would be almost meaningless.

8.29 For confidence coefficient .90, $\alpha = 1 - .90 = .10$ and $\alpha/2 = .10/2 = .05$. From Table III, Appendix A, $z_{.05} = 1.645$. We are given that the proportion of sets that need repair is about .2, so we will use .2 to estimate both p_1 and p_2. For a width of .05, the standard error is SE = .05. We also are given that $n_1 = 2n_2$.

$$SE = z_{\alpha/2}\sqrt{\frac{p_1 q_1}{n_1} + \frac{p_2 q_2}{n_2}} \Rightarrow SE = z_{\alpha/2}\sqrt{\frac{p_1 q_1}{2n_2} + \frac{p_2 q_2}{n_2}}$$

$$\Rightarrow (SE)^2 = (z_{\alpha/2})^2\left(\frac{p_1 q_1}{2n_2} + \frac{p_2 q_2}{n_2}\right) \Rightarrow (SE)^2 = (z_{\alpha/2})^2\left(\frac{p_1 q_1 + 2p_2 q_2}{2n_2}\right)$$

$$\Rightarrow 2n_2 = \frac{(z_{\alpha/2})^2(p_1 q_1 + 2p_2 q_2)}{(SE)^2} \Rightarrow n_2 = \frac{(z_{\alpha/2})^2(p_1 q_1 + 2p_2 q_2)}{2(SE)^2}$$

$$\Rightarrow n_2 = \frac{1.645^2(.2(.8) + 2(.2)(.8))}{.05^2} = 259.8 \approx 260$$

Thus, the manufacturer would sample 2(260) = 520 buyers of its sets and 260 buyers of its competitors sets.

8.31 The characteristics of the multinomial experiment are:

1. The experiment consists of n identical trials.
2. There are k possible outcomes to each trial.
3. The probabilities of the k outcomes, denoted $p_1, p_2, \ldots, p_k$, remain the same from trial to trial, where $p_1 + p_2 + \cdots + p_k = 1$.
4. The trials are independent.
5. The random variables of interest are the counts $n_1, n_2, \ldots, n_k$ in each of the k cells.

The characteristics of the binomial are the same as those for the multinomial with $k = 2$.

8.33 a. With df = 10, $\chi^2_{.05} = 18.3070$

b. With df = 50, $\chi^2_{.990} = 29.7067$

c. With df = 16, $\chi^2_{.10} = 23.5418$

d. With df = 50, $\chi^2_{.005} = 79.4900$

8.35 a. The rejection region requires $\alpha = .05$ in the upper tail of the χ^2 distribution with df $= k - 1 = 3 - 1 = 2$. From Table V, Appendix A, $\chi^2_{.05} = 5.99147$. The rejection region is $\chi^2 > 5.99147$.

b. The rejection region requires $\alpha = .10$ in the upper tail of the χ^2 distribution with $df = k - 1 = 5 - 1 = 4$. From Table V, Appendix A, $\chi^2_{.10} = 7.77944$. The rejection region is $\chi^2 > 7.77944$.

c. The rejection region requires $\alpha = .01$ in the upper tail of the χ^2 distribution with $df = k - 1 = 4 - 1 = 3$. From Table V, Appendix A, $\chi^2_{.01} = 11.3449$. The rejection region is $\chi^2 > 11.3449$.

8.37   Some preliminary calculations are:

If the probabilities are the same, $p_{1,0} = p_{2,0} = p_{3,0} = p_{4,0} = .25$

$$E_1 = np_{1,0} = 205(.25) = 51.25$$
$$E_2 = E_3 = E_4 = 205(.25) = 51.25$$

a. To determine if the multinomial probabilities differ, we test:

H_0: $p_1 = p_2 = p_3 = p_4 = .25$
H_a: At least one of the probabilities differs from .25

The test statistic is

$$\chi^2 = \sum \frac{[n_i - E_i]^2}{E_i} = \frac{(43 - 51.25)^2}{51.25} + \frac{(56 - 51.25)^2}{51.25} + \frac{(59 - 51.25)^2}{51.25} + \frac{(47 - 51.25)^2}{51.25} = 3.293$$

The rejection region requires $\alpha = .05$ in the upper tail of the χ^2 distribution with $df = k - 1 = 4 - 1 = 3$. From Table V, Appendix A, $\chi^2_{.05} = 7.81473$. The rejection region is $\chi^2 > 7.81473$.

Since the observed value of the test statistic does not fall in the rejection region $(\chi^2 = 3.293 \not> 7.81473)$, H_0 is not rejected. There is insufficient evidence to indicate the multinomial probabilities differ at $\alpha = .05$.

b. The Type I error is concluding the multinomial probabilities differ when, in fact, they do not.

The Type II error is concluding the multinomial probabilities are equal, when, in fact, they are not.

c. For confidence coefficient .95, $\alpha = .05$ and $\alpha / 2 = .05 / 2 = .025$. From Table III, Appendix A, $z_{.025} = 1.96$.

$$\hat{p}_3 = 59 / 205 = .288$$

The confidence interval is:

$$\hat{p} \pm z_{.025}\sqrt{\frac{\hat{p}\hat{q}}{n}} \Rightarrow .288 \pm 1.96\sqrt{\frac{.288(.712)}{205}} \Rightarrow .288 \pm .062 \Rightarrow (.226, .350)$$

8.39 a. The qualitative variable of interest is nonfunctional jaw habits. It has 4 levels: bruxism, teeth clenching, both bruxism and clenching, and neither.

b. A one-way table of the data is:

Cell

	Bruxism	Clench	Bruxism & Clench	Neither
n_i	3	11	30	16

c. To determine whether the percentages associated with the admitted habits differ, we test:

H_0: $p_1 = p_2 = p_3 = p_4 = .25$
H_a: At least one p_i differs from .25, $i = 1, 2, 3, 4$

d. $E_1 = E_2 = E_3 = E_4 = np_{i,0} = 60(.25) = 15$.

e. The test statistic is

$$\chi^2 = \sum \frac{[n_i - E_i]^2}{E_i} = \frac{[3-15]^2}{15} + \frac{[11-15]^2}{15} + \frac{[30-15]^2}{15} + \frac{[16-15]^2}{15} = 25.733$$

f. The rejection region requires $\alpha = .05$ in the upper tail of the χ^2 distribution with df = $k - 1 = 4 - 1 = 3$. From Table V, Appendix A, $\chi^2_{.05} = 7.81473$. The rejection region is $\chi^2 > 7.81473$.

g. Since the observed value of the test statistic falls in the rejection region $(\chi^2 = 25.733 > 7.81473)$, H_0 is rejected. There is sufficient evidence to indicate the percentages associated with the admitted habits differ at $\alpha = .05$.

h. $\hat{p}_3 = \dfrac{30}{60} = .5$

For confidence coefficient .95, $\alpha = .05$ and $\alpha / 2 = .05 / 2 = .025$. From Table III, Appendix A, $z_{.025} = 1.96$. The 95% confidence interval is:

$$\hat{p}_i \pm z_{.025}\sqrt{\frac{\hat{p}_i\hat{q}_i}{n}} \Rightarrow .5 \pm 1.96\sqrt{\frac{.5(.5)}{60}} \Rightarrow .5 \pm .127 \Rightarrow (.373, .627)$$

We are 95% confident that the true proportion of dental patients who admit to both habits is between .373 and .627.

8.41   a.   The qualitative variable of interest in this problem is the type of pottery found. There are 4 levels of the variable: burnished, monochrome, painted, and other.

       b.   If all 4 types of pottery occur with equal probabilities, the values of p_1, p_2, p_3, and p_4 are all .25.

       c.   To determine if one type of pottery is more likely to occur at the site than any other, we test:

$$H_0: \ p_1 = p_2 = p_3 = p_4 = .25$$
$$H_a: \ \text{At least one } p_i \neq .25 \ \text{ for } i = 1, 2, 3, 4$$

       d.   Some preliminary calculations are:

$$E_1 = E_2 = E_3 = E_4 = np_{i,0} = 837(.25) = 209.25.$$

$$\chi^2 = \sum \frac{[n_i - E_i]^2}{E_i} = \frac{[133 - 209.25]^2}{209.25} + \frac{[460 - 209.25]^2}{209.25} + \frac{[183 - 209.25]^2}{209.25} + \frac{[61 - 209.25]^2}{209.25}$$

$$= 436.591$$

       e.   The p-value is $p = P(\chi^2 \geq 436.591)$. Using Table V, Appendix A, with df $= k - 1$ $= 4 - 1 = 3$, $p = P(\chi^2 \geq 436.591) < .005$.

Since the p-value is less than α ($p < .005 < .10$), H_0 is rejected. There is sufficient evidence to indicate at least one type of pottery is more likely to occur at the site than another at $\alpha = .10$.

8.43   a.   If it is just as likely to have a boy as a girl, then for any child, $P(G) = \frac{1}{2}$ and $P(B) = \frac{1}{2}$. Since the gender of one child is independent of the next, $P(GG) = \frac{1}{2} \left(\frac{1}{2} \right) = \frac{1}{4} = .25$, $P(GB) = \frac{1}{2} \left(\frac{1}{2} \right) = \frac{1}{4} = .25$, $P(BG) = \frac{1}{2} \left(\frac{1}{2} \right) = \frac{1}{4} = .25$, and $P(BB) = \frac{1}{2} \left(\frac{1}{2} \right) = \frac{1}{4} = .25$.

       b.   Since each probability is .25, the expected number of families of each combination is $np_i = 42,888 \, (\, .25 \,) = 10,722$.

       c.   The test statistics is

$$\chi^2 = \sum \frac{[n_i - E_i]^2}{E_i} = \frac{[9,523 - 10,722]^2}{10,722} + \frac{[11,118 - 10,722]^2}{10,722} + \frac{[10,913 - 10,722]^2}{10,722}$$

$$+ \frac{[11,334 - 10,722]^2}{10,722} = 187.04$$

       d.   The rejection region requires $\alpha = .10$ in the upper tail of the χ^2 distribution with $df = k - 1 = 4 - 1 = 3$. From Table V, Appendix A, $\chi^2_{.10} = 6.25139$. The rejection region is $\chi^2 > 6.25139$.

Since the observed value of the test statistic falls in the rejection region $(\chi^2 = 187.04 > 6.25139)$, H_0 is rejected. There is sufficient evidence to indicate that at least one probability of gender configuration for a two-child family is not .25 at $\alpha = .10$.

e.   The expected numbers of families for each gender configuration are:

GG: $np_1 = 42,888(.23795) = 10,205.2$
GB: $np_2 = 42,888(.24985) = 10,715.6$
BG: $np_3 = 42,888(.24985) = 10,715.6$
BB: $np_4 = 42,888(.26235) = 11,251.7$

The test statistics is

$$\chi^2 = \sum \frac{[n_i - E_i]^2}{E_i} = \frac{[9,523 - 10,205.2]^2}{10,205.2} + \frac{[11,118 - 10,715.6]^2}{10,715.6} + \frac{[10,913 - 10,715.6]^2}{10,715.6}$$

$$+ \frac{[11,334 - 11,251.7]^2}{11,251.7} = 64.95$$

Since the observed value of the test statistic falls in the rejection region $(\chi^2 = 64.95 > 6.25139)$, H_0 is rejected. There is sufficient evidence to indicate that at least one probability of gender configuration for a two-child family is not the proposed probability at $\alpha = .10$.

8.45   a.   There were 1,470 responses that were missing. In addition, 14 responses were 8 = Don't know and 7 responses were 9 = Missing. Those responding with 8 or 9 were not included. The frequency table is:

Response	Frequency
1	450
2	627
3	219
4	23
Totals	**1319**

b.   To determine if the true proportions in each category are equal, we test:

H_0: $p_1 = p_2 = p_3 = p_4 = .25$
H_a: At least one p_i differs

c. $E_1 = E_2 = E_3 = E_4 = np_{i,0} = 1,319(.25) = 329.75$

d.   The test statistic is

$$\chi^2 = \sum \frac{[n_i - E_i]^2}{E_i}$$

$$= \frac{[450 - 329.75]^2}{329.75} + \frac{[627 - 329.75]^2}{329.75} + \frac{[219 - 329.75]^2}{329.75} + \frac{[23 - 329.75]^2}{329.75} = 634.36$$

e.   The rejection region requires $\alpha = .10$ in the upper tail of the χ^2 distribution with df = $k - 1 = 4 - 1 = 3$. From Table V, Appendix A, $\chi^2_{.10} = 6.25139$. The rejection region is $\chi^2 > 6.25139$.

Since the observed value of the test statistic falls in the rejection region $(\chi^2 = 634.36 > 6.25139)$, H_0 is rejected. There is sufficient evidence to indicate at least one proportion differs at $\alpha = .10$.

f.   To determine if the true proportions follow the proportions given, we test:

H_0: $p_1 = .30, p_2 = .50, p_3 = .15, p_4 = .05$
H_a: At least one p_i differs from its hypothesized value

$E_1 = np_{1,0} = 1,319(.30) = 395.7$ $E_2 = np_{2,0} = 1,319(.50) = 659.5$
$E_3 = np_{3,0} = 1,319(.15) = 197.85$ $E_4 = np_{4,0} = 1,319(.05) = 65.95$

The test statistic is

$$\chi^2 = \sum \frac{[n_i - E_i]^2}{E_i}$$

$$= \frac{[450 - 395.7]^2}{395.7} + \frac{[627 - 659.5]^2}{659.5} + \frac{[219 - 197.85]^2}{197.85} + \frac{[23 - 65.95]^2}{65.95} = 39.29$$

The rejection region requires $\alpha = .10$ in the upper tail of the χ^2 distribution with df = $k - 1 = 4 - 1 = 3$. From Table V, Appendix A, $\chi^2_{.10} = 6.25139$. The rejection region is $\chi^2 > 6.25139$.

Since the observed value of the test statistic falls in the rejection region $(\chi^2 = 39.29 > 6.25139)$, H_0 is rejected. There is sufficient evidence to indicate at least one proportion differs from its hypothesized value at $\alpha = .10$.

8.47   Some preliminary calculations are:

$E_1 = np_{1,0} = 2,097(.02) = 41.94$
$E_2 = np_{2,0} = 2,097(.25) = 524.25$
$E_3 = np_{3,0} = 2,097(.73) = 1,530.81$

To determine if the distribution of the E4/E4 genotypes for the population of young adults differs from the norm, we test:

H_0: $p_1 = .02$; $p_2 = .25$; $p_3 = .73$
H_a: At least one p_i differs from its hypothesized value

The test statistic is

$$\chi^2 = \sum \frac{[n_i - E_i]^2}{E_i} = \frac{[56 - 41.94]^2}{41.94} + \frac{[517 - 524.25]^2}{524.25} + \frac{[1524 - 1530.81]^2}{1530.81} = 4.84$$

The rejection region requires $\alpha = .05$ in the upper tail of the χ^2 distribution with $df = k - 1$ $= 3 - 1 = 2$. From Table V, Appendix A, $\chi^2_{.05} = 5.99146$. The rejection region is $\chi^2 > 5.99146$.

Since the observed value of the test statistic does not fall in the rejection region $(\chi^2 = 4.84 \ngtr 5.99146)$, H_0 is not rejected. There is insufficient evidence to indicate the distribution of the E4/E4 genotypes for the population of young adults differs from the norm at $\alpha = .05$.

8.49   a.   Some preliminary calculations are:

$E_A = np_{A,0} = 700(.09) = 63$	$E_B = np_{B,0} = 700(.02) = 14$
$E_C = np_{C,0} = 700(.02) = 14$	$E_D = np_{D,0} = 700(.04) = 28$
$E_E = np_{E,0} = 700(.12) = 84$	$E_F = np_{F,0} = 700(.02) = 14$
$E_G = np_{G,0} = 700(.03) = 21$	$E_H = np_{H,0} = 700(.02) = 14$
$E_I = np_{I,0} = 700(.09) = 63$	$E_J = np_{J,0} = 700(.01) = 7$
$E_K = np_{K,0} = 700(.01) = 7$	$E_L = np_{L,0} = 700(.04) = 28$
$E_M = np_{M,0} = 700(.02) = 14$	$E_N = np_{N,0} = 700(.06) = 42$
$E_O = np_{O,0} = 700(.08) = 56$	$E_P = np_{P,0} = 700(.02) = 14$
$E_Q = np_{Q,0} = 700(.01) = 7$	$E_R = np_{R,0} = 700(.06) = 42$
$E_S = np_{S,0} = 700(.04) = 28$	$E_T = np_{T,0} = 700(.06) = 42$
$E_U = np_{U,0} = 700(.04) = 28$	$E_V = np_{V,0} = 700(.02) = 14$
$E_W = np_{W,0} = 700(.02) = 14$	$E_X = np_{X,0} = 700(.01) = 7$
$E_Y = np_{Y,0} = 700(.02) = 14$	$E_Z = np_{Z,0} = 700(.01) = 7$
$E_{BLANK} = np_{BL,0} = 700(.02) = 14$	

To determine if the ScrabbleExpress™ presents the player with "unfair word selection opportunities" that are different from the Scrabble™ board game, we test:

H_0: $p_A = .09$, $p_B = .02$, $p_C = .02$, $p_D = .04$, $p_E = .12$, $p_F = .02$, $p_G = .03$, $p_H = .02$, $p_I = .09$, $p_J = .01$, $p_K = .01$, $p_L = .04$, $p_M = .02$, $p_N = .06$, $p_O = .08$, $p_P = .02$, $p_Q = .01$, $p_R = .06$, $p_S = .04$, $p_T = .06$, $p_U = .04$, $p_V = .02$, $p_W = .02$, $p_X = .01$, $p_Y = .02$, $p_Z = .01$, $p_{BLANK} = .02$

H_a: At least one p_i differs from its hypothesized value, i = A, B, …, BLANK

The test statistic is

$$\chi^2 = \sum \frac{[n_i - E_i]^2}{E_i} = \frac{[39-63]^2}{63} + \frac{[18-14]^2}{14} + \frac{[30-14]^2}{14} + \cdots + \frac{[34-14]^2}{14} = 360.48$$

The rejection region requires $\alpha = .05$ in the upper tail of the χ^2 distribution with df $= k - 1 = 27 - 1 = 26$. From Table V, Appendix A, $\chi^2_{.05} = 38.8852$. The rejection region is $\chi^2 > 38.8852$.

Since the observed value of the test statistic falls in the rejection region ($\chi^2 = 360.48 > 38.8852$), H_0 is rejected. There is sufficient evidence to indicate the ScrabbleExpress^{TM} presents the player with "unfair word selection opportunities" that are different from the Scrabble^{TM} board game at $\alpha = .05$.

b.   The true proportion of letters drawn from the board game is

$$p_A + p_E + p_I + p_O + p_U = .09 + .12 + .09 + .08 + .04 = .42$$

The number of vowels drawn by the electronic game is $39 + 31 + 25 + 20 + 21 = 136$. The proportion of vowels drawn is $136 / 700 = .194$.

For confidence coefficient .95, $\alpha = .05$ and $\alpha / 2 = .05 / 2 = .025$. From Table III, Appendix A, $z_{.025} = 1.96$. The 95% confidence interval is:

$$\hat{p}_i \pm z_{.025} \sqrt{\frac{\hat{p}_i \hat{q}_i}{n}} \Rightarrow .194 \pm 1.96 \sqrt{\frac{.194(.806)}{700}} \Rightarrow .194 \pm .029 \Rightarrow (.165, \ .223)$$

We are 95% confident that the true proportion of vowels drawn by the electronic game is between .165 and .223. The actual proportion of vowels drawn in the board game is .42. Since this value is not contained in the 95% confidence interval, one can conclude that the ScrabbleExpress^{TM} game produces too few vowels in the 7-letter draws.

8.51   A contingency table with fixed marginals is a table where the column totals for one qualitative variable are known in advance. For example, we could ask a sample of 150 males and a sample of 150 females their preference to a particular brand. The column totals for gender are known in advance.

8.53   The conditions that are required for a valid chi-square test of data from a contingency table are:

1.   The n observed counts are a random sample from the population of interest.
2.   The sample size, n, will be large enough so that, for every cell, the expected count, E_{ij}, will be equal to 5 or more.

8.55   a. H_0: The row and column classifications are independent
H_a: The row and column classifications are dependent

b. The test statistic is $\chi^2 = \sum\sum \dfrac{[n_{ij} - \hat{E}_{ij}]^2}{\hat{E}_{ij}}$

The rejection region requires $\alpha = .01$ in the upper tail of the χ^2 distribution with df = $(r-1)(c-1) = (2-1)(3-1) = 2$. From Table V, Appendix A, $\chi^2_{.01} = 9.21034$. The rejection region is $\chi^2 > 9.21034$.

c. The expected cell counts are:

$$\hat{E}_{11} = \frac{R_1 C_1}{n} = \frac{96(25)}{167} = 14.37 \qquad \hat{E}_{21} = \frac{R_2 C_1}{n} = \frac{71(25)}{167} = 10.63$$

$$\hat{E}_{12} = \frac{R_1 C_2}{n} = \frac{96(64)}{167} = 36.79 \qquad \hat{E}_{22} = \frac{R_2 C_2}{n} = \frac{71(64)}{167} = 27.21$$

$$\hat{E}_{13} = \frac{R_1 C_3}{n} = \frac{96(78)}{167} = 44.84 \qquad \hat{E}_{23} = \frac{R_2 C_3}{n} = \frac{71(78)}{167} = 33.16$$

d. The test statistic is

$$\chi^2 = \sum\sum \frac{[n_{ij} - \hat{E}_{ij}]^2}{\hat{E}_{ij}} = \frac{(9-14.37)^2}{14.37} + \frac{(34-36.79)^2}{36.79} + \frac{(53-44.84)^2}{44.84} + \frac{(16-10.63)^2}{10.63}$$

$$+ \frac{(30-27.21)^2}{27.21} + \frac{(25-33.16)^2}{33.16} = 8.71$$

Since the observed value of the test statistic does not fall in the rejection region $(\chi^2 = 8.71 \not> 9.21034)$, H_0 is not rejected. There is insufficient evidence to indicate the row and column classifications are dependent at $\alpha = .01$.

8.57   Some preliminary calculations are:

$$\hat{E}_{11} = \frac{R_1 C_1}{n} = \frac{154(134)}{439} = 47.007 \qquad \hat{E}_{21} = \frac{186(134)}{439} = 56.774$$

$$\hat{E}_{12} = \frac{154(163)}{439} = 57.180 \qquad \hat{E}_{22} = \frac{186(163)}{439} = 69.062$$

$$\hat{E}_{13} = \frac{154(142)}{439} = 49.813 \qquad \hat{E}_{23} = \frac{186(142)}{439} = 60.164$$

$$\hat{E}_{31} = \frac{99(134)}{439} = 30.219 \qquad \hat{E}_{33} = \frac{99(142)}{439} = 32.023$$

$$\hat{E}_{32} = \frac{99(163)}{439} = 36.759$$

To determine if the row and column classifications are dependent, we test:

H_0: The row and column classifications are independent
H_a: The row and column classifications are dependent

The test statistic is

$$\chi^2 = \sum\sum \frac{[n_{ij} - \hat{E}_{ij}]^2}{\hat{E}_{ij}} = \frac{(40 - 47.007)^2}{47.007} + \frac{(72 - 57.180)^2}{57.180} + \frac{(42 - 49.813)^2}{49.813} + \frac{(63 - 56.774)^2}{56.774}$$

$$+ \frac{(53 - 69.062)^2}{69.062} + \frac{(70 - 60.164)^2}{60.164} + \frac{(31 - 32.023)^2}{32.023} + \frac{(38 - 36.759)^2}{36.759} + \frac{(30 - 32.023)^2}{32.023}$$

$$= 12.33$$

The rejection region requires $\alpha = .05$ in the upper tail of the χ^2 distribution with df = $(r - 1)(c - 1) = (3 - 1)(3 - 1) = 4$. From Table V, Appendix A, $\chi^2_{.05} = 9.48773$. The rejection region is $\chi^2 > 9.48773$.

Since the observed value of the test statistic falls in the rejection region $(\chi^2 = 12.33 > 9.48773)$, H_0 is rejected. There is sufficient evidence to indicate the row and column classifications are dependent at $\alpha = .05$.

8.59   a. $\hat{p}_1 = \dfrac{6}{77} = .078$

     b. $\hat{p}_2 = \dfrac{11}{70} = .157$

     c.   The proportion of girls drawing a dog is almost 2 times the proportion of boys drawing a dog. It is possible that the likelihood of drawing a dog depends on gender.

     d.   To determine if the likelihood of drawing a dog depends on gender, we test:

          H_0: Presence of a dog and gender are independent
          H_a: Presence of a dog and gender are dependent

     e.   From the printout, the test statistic is $\chi^2 = 2.250$ and the p-value is $p = 0.134$.

        Since the p value is not less than $\alpha = .05$ $(p = .134 \nless .05)$, H_0 is not rejected. There is insufficient evidence to indicate the likelihood of drawing a dog depends on gender at $\alpha = .05$.

f.   Using MINITAB, the results are:

Tabulated statistics: TV, GENDER

```
Using frequencies in NUMBER


Rows: TV   Columns: GENDER

            Boy    Girl     All

No           66      61     127
           66.52  60.48  127.00

Yes          11       9      20
           10.48   9.52   20.00

All          77      70     147
           77.00  70.00  147.00

Cell Contents:        Count
                      Expected count


Pearson Chi-Square = 0.064, DF = 1, P-Value = 0.801
Likelihood Ratio Chi-Square = 0.064, DF = 1, P-Value = 0.801
```

To determine if the likelihood of drawing a TV in the bedroom depends on gender, we test:

H_0: Presence of TV and gender are independent
H_a: Presence of TV and gender are dependent

From the printout, the test statistic is $\chi^2 = .064$ and the *p*-value is $p = 0.801$.

Since the *p*-value is not less than $\alpha = .05$ $(p = .801 \not< .05)$, H_0 is not rejected. There is insufficient evidence to indicate the likelihood of drawing TV depends on gender at $\alpha = .05$.

8.61   a. $\hat{p}_1 = \dfrac{111}{170} = .653$

b. $\hat{p}_2 = \dfrac{61}{110} = .555$

c. $\hat{p}_3 = \dfrac{11}{30} = .355$

d.   The sample proportions range from .367 to .653. It appears that the proportion of news stories that are deceptive depends on story tone.

e.   To determine whether the authenticity of a news story depends on tone, we test:

H_0: Authenticity of news story and tone are independent
H_a: Authenticity of news story and tone are dependent

f.   From the printout, the test statistic is $\chi^2 = 10.427$ and the p-value is $p = .005$. Since the p-value is less than $\alpha\,(p = .005 < .05)$, H_0 is rejected. There is sufficient evidence to indicate that the authenticity of a news story depends on tone at $\alpha = .05$.

8.63 a.   The two categorical variables are Risk of masculinity and type of event. Risk of masculinity has two levels: High and Low. Type of event has two levels: Violent event and Avoided-Violent event.

b.   The experimental units are 1,507 newly incarcerated men.

c. $E_{11} = \dfrac{R_1 C_1}{n} = \dfrac{379(1,037)}{1,507} = 260.798$

d. $E_{12} = \dfrac{R_1 C_2}{n} = \dfrac{379(470)}{1,507} = 118.202$ $E_{21} = \dfrac{R_2 C_1}{n} = \dfrac{1,128(1,037)}{1,507} = 776.202$

$E_{22} = \dfrac{R_2 C_2}{n} = \dfrac{1,128(470)}{1,507} = 351.798$

e. $\chi^2 = \sum\sum \dfrac{[n_{ij} - \hat{E}_{ij}]^2}{\hat{E}_{ij}} = \dfrac{[236 - 260.798]^2}{260.798} + \dfrac{[143 - 118.202]^2}{118.202}$

$+ \dfrac{[801 - 776.202]^2}{776.202} + \dfrac{[327 - 351.798]^2}{351.798} = 10.101$

f.   To determine whether event type depends on high/low risk masculinity, we test:

H_0: Event Type and High/Low Risk Masculinity are independent
H_a: Event Type and High/Low Risk Masculinity are dependent

The test statistic is $\chi^2 = 10.101$.

The rejection region requires $\alpha = .05$ in the upper tail of the χ^2 distribution with $df = (r - 1)(c - 1) = (2 - 1)(2 - 1) = 1$. From Table V, Appendix A, $\chi^2_{.05} = 3.84146$. The rejection region is $\chi^2 > 3.84146$.

Since the observed value of the test statistic falls in the rejection region $(\chi^2 = 10.101 > 3.84146)$, H_0 is rejected. There is sufficient evidence that event type depends on high/low risk masculinity at $\alpha = .05$.

8.65   Some preliminary calculations are:

$\hat{E}_{11} = \dfrac{R_1 C_1}{n} = \dfrac{192(115)}{526} = 41.98$ $\hat{E}_{12} = \dfrac{R_1 C_2}{n} = \dfrac{192(411)}{526} = 150.02$

$$\hat{E}_{21} = \frac{R_2 C_1}{n} = \frac{157(115)}{526} = 34.33 \qquad \hat{E}_{22} = \frac{R_2 C_2}{n} = \frac{157(411)}{526} = 122.67$$

$$\hat{E}_{31} = \frac{R_3 C_1}{n} = \frac{177(115)}{526} = 38.70 \qquad \hat{E}_{32} = \frac{R_3 C_2}{n} = \frac{177(411)}{526} = 138.30$$

To determine if the proportion of students diagnosed with MR depends on the IQ test/retest method, we test:

H_0: MR diagnosis and Test/Retest are independent
H_a: MR diagnosis and Test/Retest are dependent

The test statistic is

$$\chi^2 = \sum\sum \frac{[n_{ij} - \hat{E}_{ij}]^2}{E_{ij}} = \frac{[25 - 41.98]^2}{41.98} + \frac{[54 - 34.33]^2}{34.33} + \frac{[36 - 38.70]^2}{38.70} + \frac{[167 - 150.02]^2}{150.02}$$

$$+ \frac{[103 - 122.67]^2}{122.67} + \frac{[141 - 138.30]^2}{138.30} = 23.46$$

The rejection region requires $\alpha = .01$ in the upper tail of the χ^2 distribution with df $= (r-1)(c-1) = (3-1)(2-1) = 2$. From Table V, Appendix A, $\chi^2_{.01} = 9.21034$. The rejection region is $\chi^2 > 9.21034$.

Since the observed value of the test statistic falls in the rejection region $(\chi^2 = 23.46 > 9.21034)$, H_0 is rejected. There is sufficient evidence to indicate the proportion of students diagnosed with MR depends on the IQ test/retest method at $\alpha = .01$.

8.67   Some preliminary calculations are:

$$\hat{E}_{11} = \frac{R_1 C_1}{n} = \frac{2097(149)}{6560} = 47.630 \qquad \hat{E}_{12} = \frac{R_1 C_2}{n} = \frac{2097(1647)}{6560} = 526.488$$

$$\hat{E}_{13} = \frac{R_1 C_3}{n} = \frac{2097(4764)}{6560} = 1522.882 \qquad \hat{E}_{21} = \frac{R_2 C_1}{n} = \frac{2182(149)}{6560} = 49.561$$

$$\hat{E}_{22} = \frac{R_2 C_2}{n} = \frac{2182(1647)}{6560} = 547.828 \qquad \hat{E}_{23} = \frac{R_2 C_3}{n} = \frac{2182(4764)}{6560} = 1584.611$$

$$\hat{E}_{31} = \frac{R_3 C_1}{n} = \frac{2281(149)}{6560} = 51.809 \qquad \hat{E}_{32} = \frac{R_3 C_2}{n} = \frac{2281(1647)}{6560} = 572.684$$

$$\hat{E}_{33} = \frac{R_3 C_3}{n} = \frac{2281(4764)}{6560} = 1656.507$$

To determine if there are significant genotype differences across the three age groups, we test:

H_0: Genotype and age group are independent

H_a: Genotype and age group are dependent

The test statistic is

$$\chi^2 = \sum \frac{\left[n_{ij} - \hat{E}_{ij}\right]^2}{\hat{E}_{ij}} = \frac{[56-47.630]^2}{47.630} + \frac{[517-526.488]^2}{526.488} + \frac{[1524-1522.882]^2}{1522.882}$$

$$+ \frac{[45-49.561]^2}{49.561} + \frac{[566-547.828]^2}{547.828} + \frac{[1571-1584.611]^2}{1584.611}$$

$$+ \frac{[48-51.809]^2}{51.809} + \frac{[564-572.684]^2}{572.684} + \frac{[1669-1656.507]^2}{1656.507} = 3.288$$

The rejection region requires $\alpha = .05$ in the upper tail of the χ^2 distribution with $df = (r-1)(c-1) = (3-1)(3-1) = 4$. From Table V, Appendix A, $\chi^2_{.05} = 9.48773$. The rejection region is $\chi^2 > 9.48773$.

Since the observed value of the test statistics does not fall in the rejection region $(\chi^2 = 3.288 \not> 9.48773)$, H_0 is not rejected. There is insufficient evidence to indicate there are significant genotype differences across the three age groups at $\alpha = .05$.

8.69   Using MINITAB, the results of the analyses are:

Results for: AIRTHR~1.MTP

Tabulated statistics: Instruction, Strategy

```
Rows: Instruction    Columns: Strategy

                Guess     Other      TTBC       All

Cue                5         6        13        24
               7.000     8.500     8.500    24.000

Pattern            9        11         4        24
               7.000     8.500     8.500    24.000

All               14        17        17        48
              14.000    17.000    17.000    48.000

Cell Contents:        Count
                      Expected count

Pearson Chi-Square = 7.378, DF = 2, P-Value = 0.025
Likelihood Ratio Chi-Square = 7.668, DF = 2, P-Value = 0.022
```

To determine if the choice of heuristic strategy depends on type of instruction provided, we test:

H_0: Choice of heuristic strategy and instruction provided are independent
H_0: Choice of heuristic strategy and instruction provided are dependent

From the printout, the test statistic is $\chi^2 = 7.378$ and the p-value is $p = .025$.

Since the p-value is less than $\alpha = .05$ $(p = .025 < .05)$, H_0 is rejected. There is sufficient evidence to indicate the choice of heuristic strategy depends on type of instruction provided at $\alpha = .05$.

If $\alpha = .01$, then the p-value is not less than $\alpha (p = .025 \not< .01)$. H_0 would not be rejected at $\alpha = .01$.

8.71   a.   Using MINITAB, the results of the analyses are:

Tabulated statistics: SHSS, CAHS

```
Using frequencies in Fr

Rows: SHSS    Columns: CAHS

               L       M       H       V       All

 L            32      14       2       0        48
           18.09   16.25   11.45    2.22     48.00

 M            11      14       6       0        31
           11.68   10.49    7.39    1.43     31.00

 H             6      14      19       3        42
           15.83   14.22   10.02    1.94     42.00

 V             0       2       4       3         9
            3.39    3.05    2.15    0.42      9.00

 All          49      44      31       6       130
           49.00   44.00   31.00    6.00    130.00

Cell Contents:       Count
                     Expected count


Pearson Chi-Square = 60.103, DF = 9
Likelihood Ratio Chi-Square = 59.638, DF = 9

* WARNING * 1 cells with expected counts less than 1
* WARNING * Chi-Square approximation probably invalid

* NOTE * 7 cells with expected counts less than 5
```

From the printout, 7 cells have expected counts less than 5. In order for the test to be valid, all of the cells should have expected counts greater than 5. Thus, we should not proceed with the analysis.

b.   Combining the High and Very High categories, the new table is:

<table>
<tr><td rowspan="2"></td><td rowspan="2"></td><td colspan="3" align="center">CAHS LEVEL</td></tr>
<tr><td>Low</td><td>Medium</td><td>High/Very High</td></tr>
<tr><td rowspan="3">SHSS: C LEVEL</td><td>Low</td><td>32</td><td>14</td><td>2</td></tr>
<tr><td>Medium</td><td>11</td><td>14</td><td>6</td></tr>
<tr><td>High/Very High</td><td>6</td><td>16</td><td>29</td></tr>
</table>

c.   The expected cell counts are:

$$\hat{E}_{11} = \frac{R_1 C_1}{n} = \frac{48(49)}{130} = 18.0923 \qquad \hat{E}_{21} = \frac{R_2 C_1}{n} = \frac{31(49)}{130} = 11.6846$$

$$\hat{E}_{12} = \frac{R_1 C_2}{n} = \frac{48(44)}{130} = 16.2462 \qquad \hat{E}_{22} = \frac{R_2 C_2}{n} = \frac{31(44)}{130} = 10.4923$$

$$\hat{E}_{13} = \frac{R_1 C_3}{n} = \frac{48(37)}{130} = 13.6615 \qquad \hat{E}_{23} = \frac{R_2 C_3}{n} = \frac{31(37)}{130} = 8.8231$$

$$\hat{E}_{31} = \frac{R_3 C_1}{n} = \frac{51(49)}{130} = 19.2231 \qquad \hat{E}_{32} = \frac{R_3 C_2}{n} = \frac{51(44)}{130} = 17.2615$$

$$\hat{E}_{33} = \frac{R_3 C_3}{n} = \frac{51(37)}{130} = 14.5154$$

Since all of the expected cell counts are now 5 or greater, the assumption is met.

d.   To determine whether CAHS Levels and SHSS:C Levels are dependent, we test:

H_0: CAHS Levels and SHSS:C Levels are independent
H_a: CAHS Levels and SHSS:C Levels are dependent

The test statistic is

$$\chi^2 = \sum\sum \frac{[n_{ij} - \hat{E}_{ij}]^2}{\hat{E}_{ij}} = \frac{(32-18.0923)^2}{18.0923} + \frac{(14-16.2462)^2}{16.2462} + \frac{(2-13.6615)^2}{13.6615} + \frac{(11-11.6846)^2}{11.6846}$$

$$+ \frac{(14-10.4923)^2}{10.4923} + \frac{(6-8.8231)^2}{8.8231} + \frac{(6-19.2231)^2}{19.2231} + \frac{(16-17.2615)^2}{17.2615} + \frac{(29-14.5154)^2}{14.5154} = 46.71$$

The rejection region requires $\alpha = .05$ in the upper tail of the χ^2 distribution with df $=$
$(r - 1)(c - 1) = (3 - 1)(3 - 1) = 4$. From Table V, Appendix A, $\chi^2_{.05} = 9.48773$.
The rejection region is $\chi^2 > 9.48773$.

Since the observed value of the test statistic falls in the rejection region
$(\chi^2 = 46.70 > 9.48773)$, H_0 is rejected. There is sufficient evidence to indicate that
CAHS Levels and SHSS:C Levels are dependent at $\alpha = .05$.

8.73   a.   Some preliminary calculations are:

$$\hat{E}_{11} = \frac{R_1 C_1}{n} = \frac{31(24)}{38} = 19.58 \qquad \hat{E}_{12} = \frac{R_1 C_2}{n} = \frac{31(14)}{38} = 11.42$$

$$\hat{E}_{21} = \frac{R_2 C_1}{n} = \frac{7(24)}{38} = 4.42 \qquad \hat{E}_{22} = \frac{R_2 C_2}{n} = \frac{7(14)}{38} = 2.58$$

To determine whether the vaccine is effective in treating the MN strain of HIV, we test:

H_0: Vaccine and MN strain are independent
H_a: Vaccine and MN strain are dependent

The test statistic is

$$\chi^2 = \sum\sum \frac{[n_{ij} - \hat{E}_{ij}]^2}{\hat{E}_{ij}} = \frac{(22-19.58)^2}{19.58} + \frac{(9-11.42)^2}{11.42} + \frac{(2-4.42)^2}{4.42} + \frac{(5-2.58)^2}{2.58} = 4.407$$

The rejection region requires $\alpha = .05$ in the upper tail of the χ^2 distribution with df = $(r-1)(c-1) = (2-1)(2-1) = 1$. From Table V, Appendix A, $\chi^2_{.05} = 3.84146$. The rejection region is $\chi^2 > 3.84146$.

Since the observed value of the test statistic falls in the rejection region $(\chi^2 = 4.407 > 3.84146)$, H_0 is rejected. There is sufficient evidence to indicate that the vaccine is effective in treating the MN strain of HIV at $\alpha = .05$.

b.   The necessary assumptions are:

1. The n observed counts are a random sample from the population of interest.

2. The sample size, n, will be large enough so that, for every cell, the expected count, $E(n_{ij})$, will be equal to 5 or more.

For this example, the second assumption is violated. Two of the expected cell counts are less than 5. What we have computed for the test statistic may not have a χ^2 distribution.

c.   For this contingency table:

$$p = \frac{\binom{7}{2}\binom{31}{22}}{\binom{38}{24}} = \frac{\dfrac{7!}{2!5!}\dfrac{31!}{22!9!}}{\dfrac{38!}{24!14!}} = .0438$$

d.   If vaccine and MN strain are independent, then the proportion of positive results should be relatively the same for both patient groups. In the two tables presented, the proportion of positive results for the vaccinated group is smaller than the proportion for the original table.

For the first table,

$$p = \frac{\binom{7}{1}\binom{31}{23}}{\binom{38}{24}} = \frac{\dfrac{7!}{1!6!}\dfrac{31!}{23!8!}}{\dfrac{38!}{24!14!}} = .0057$$

For the second table:

$$p = \frac{\binom{7}{0}\binom{31}{24}}{\binom{38}{24}} = \frac{\dfrac{7!}{0!7!}\dfrac{31!}{24!7!}}{\dfrac{38!}{24!14!}} = .0003$$

e. The p-value for Fisher's exact test is $p = .0438 + .0057 + .0003 = .0498$. Since the p-value is small, there is evidence to reject H_0. There is sufficient evidence to indicate that the vaccine is effective in treating the MN strain of HIV at $\alpha > .0498$.

8.75 The statement "Rejecting the null hypothesis in a chi-square test for independence implies that a causal relationship between the two categorical variables exists." is false. If the variables are dependent (reject H_0), it simply implies that they are related. One cannot infer causal relationships from contingency table analysis.

8.77 Some preliminary calculations are:

$$\hat{p}_1 = \frac{x_1}{n_1} = \frac{110}{200} = .55; \quad \hat{p}_2 = \frac{x_2}{n_2} = \frac{130}{200} = .65; \quad \hat{p} = \frac{x_1 + x_2}{n_1 + n_2} = \frac{110 + 130}{200 + 200} = \frac{240}{400} = .60$$

a. H_0: $p_1 - p_2 = 0$
   H_a: $p_1 - p_2 < 0$

The test statistic is $z = \dfrac{(\hat{p}_1 - \hat{p}_2) - 0}{\sqrt{\hat{p}\hat{q}\left(\dfrac{1}{n_1} + \dfrac{1}{n_2}\right)}} = \dfrac{(.55 - .65) - 0}{\sqrt{.6(1-.6)\left(\dfrac{1}{200} + \dfrac{1}{200}\right)}} = \dfrac{-.10}{.049} = -2.04$

The rejection region requires $\alpha = .10$ in the lower tail of the z distribution. From Table III, Appendix A, $z_{.10} = 1.28$. The rejection region is $z < -1.28$.

Since the observed value of the test statistic falls in the rejection region ($z = -2.04 < -1.28$), H_0 is rejected. There is sufficient evidence to conclude ($p_1 - p_2 < 0$) at $\alpha = .10$.

b. For confidence coefficient .95, $\alpha = 1 - .95 = .05$ and $\alpha/2 = .05/2 = .025$. From Table III, Appendix A, $z_{.025} = 1.96$. The 95% confidence interval for ($p_1 - p_2$) is approximately:

$$(\hat{p}_1 - \hat{p}_2) \pm z_{\alpha/2}\sqrt{\frac{\hat{p}_1\hat{q}_1}{n_1} + \frac{\hat{p}_2\hat{q}_2}{n_2}} \Rightarrow (.55 - .65) \pm 1.96\sqrt{\frac{.55(1-.55)}{200} + \frac{.65(1-.65)}{200}}$$

$$\Rightarrow -.10 \pm .096 \Rightarrow (-.196, -.004)$$

c.　From part **b**, $z_{.025} = 1.96$. Using the information from our samples, we can use $p_1 = .55$ and $p_2 = .65$. For a width of .01, the bound is .005.

$$n_1 = n_2 = \frac{(z_{\alpha/2})^2 (p_1 q_1 + p_2 q_2)}{(SE)^2} = \frac{(1.96)^2 (.55(1-.55) + .65(1-.65))}{.005^2}$$

$$= \frac{1.82476}{.000025} = 72{,}990.4 \approx 72{,}991$$

8.79　a.　Some preliminary calculations are:

$$\hat{E}_{11} = \frac{R_1 C_1}{n} = \frac{50(50)}{250} = 10 \qquad \hat{E}_{21} = \frac{R_2 C_1}{n} = \frac{100(50)}{250} = 20$$

$$\hat{E}_{12} = \frac{R_1 C_2}{n} = \frac{50(90)}{250} = 18 \qquad \hat{E}_{22} = \frac{R_2 C_2}{n} = \frac{100(90)}{250} = 36$$

$$\hat{E}_{13} = \frac{R_1 C_3}{n} = \frac{50(110)}{250} = 22 \qquad \hat{E}_{23} = \frac{R_2 C_3}{n} = \frac{100(100)}{250} = 44$$

$$\hat{E}_{31} = \frac{R_3 C_1}{n} = \frac{100(50)}{250} = 20 \qquad \hat{E}_{32} = \frac{R_3 C_2}{n} = \frac{100(90)}{250} = 36$$

$$\hat{E}_{33} = \frac{R_3 C_3}{n} = \frac{100(110)}{250} = 44$$

To determine if the rows and columns are dependent, we test:

H_0:　Rows and columns are independent
H_a:　Rows and columns are dependent

The test statistic is $\chi^2 = \sum\sum \dfrac{[n_{ij} - E_{ij}]^2}{\hat{E}_{ij}} = \dfrac{(20-10)^2}{10} + \cdots + \dfrac{(30-44)^2}{44} = 54.14$

The rejection region requires $\alpha = .05$ in the upper tail of the χ^2 distribution with df = $(r-1)(c-1) = (3-1)(3-1) = 4$. From Table V, Appendix A, $\chi^2_{.05} = 9.48773 = 9.48773$. The rejection region is $\chi^2 > 9.48773$.

Since the observed value of the test statistic falls in the rejection region $(\chi^2 = 54.14 > 9.48773)$, H_0 is rejected. There is sufficient evidence to indicate a dependence between rows and columns at $\alpha = .05$.

b.　No, the analysis remains identical.

c.　Yes, the assumptions on the sampling differ.

d.   The percentages are in the table below.

	Column			
	1	**2**	**3**	**Totals**
1	$\dfrac{20}{50} \times 100\% = 40\%$	$\dfrac{20}{90} \times 100\% = 22.2\%$	$\dfrac{10}{110} \times 100\% = 9.1\%$	$\dfrac{50}{250} \times 100\% = 20\%$
Row 2	$\dfrac{10}{50} \times 100\% = 20\%$	$\dfrac{20}{90} \times 100\% = 22.2\%$	$\dfrac{70}{110} \times 100\% = 63.6\%$	$\dfrac{100}{250} \times 100\% = 40\%$
3	$\dfrac{20}{50} \times 100\% = 40\%$	$\dfrac{50}{90} \times 100\% = 55.6\%$	$\dfrac{30}{110} \times 100\% = 37.3\%$	$\dfrac{100}{250} \times 100\% = 40\%$

e.   The graph is:

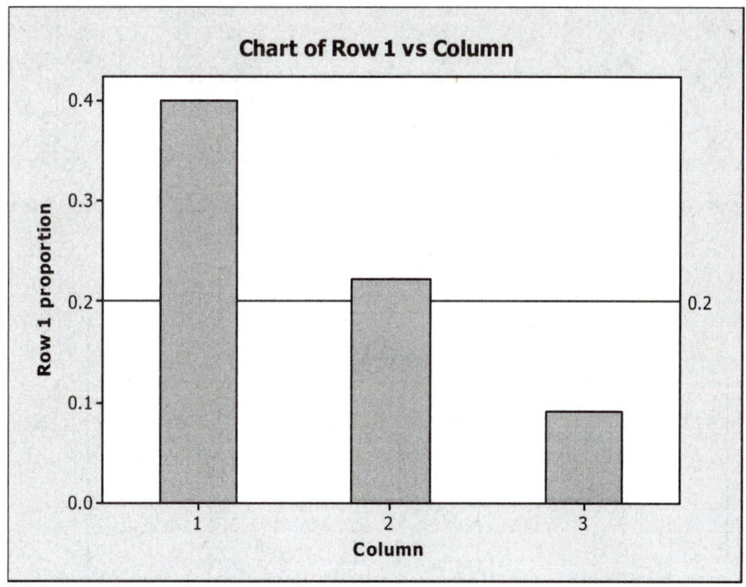

If the rows and columns are independent, then the proportion of observations for each column should be relatively the same. In this graph, the percentages are quite different, supporting the results of the test in part **a**.

f.   The graph for row 2 is:

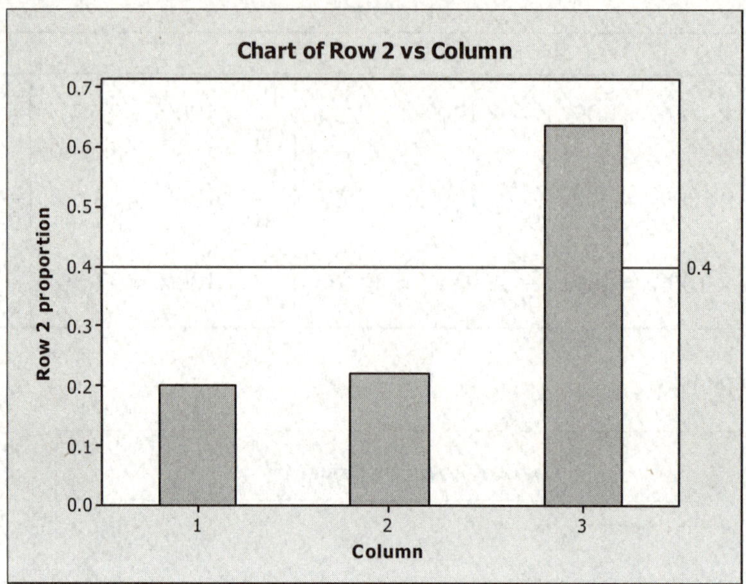

The percentages are quite different, supporting the results of the test in part **a.**

g.   The graph for row 3 is:

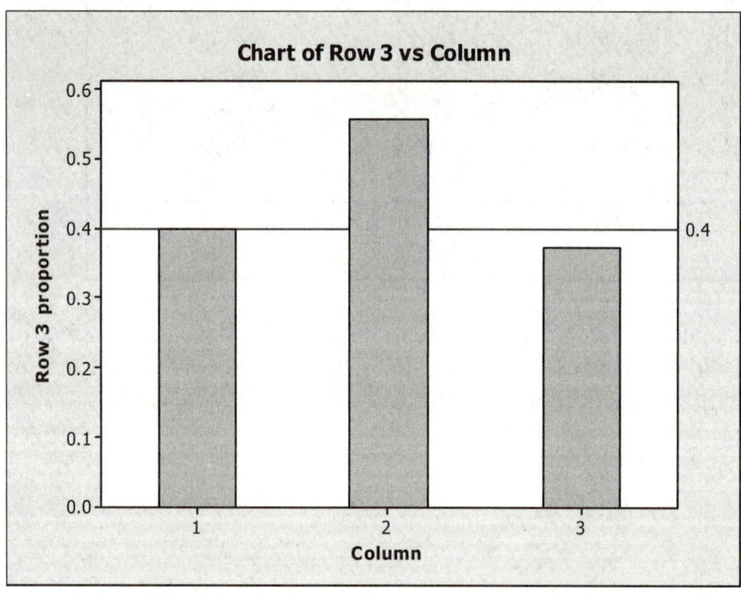

The percentages are quite different, supporting the results of the test in part **a.**

8.81   a.   The qualitative variable of interest in the study is location of major sports franchises. There are three levels of the variable – downtown, central city, and suburban areas.

b.   Let p_1 = proportion of facilities in downtown, p_2 = proportion if facilities in central cities, and p_3 = proportion of facilities in suburban areas. To determine whether the proportions of major sports facilities in downtown, central city, and suburban areas are the same in 1997 as they were in 1985, we test:

H_0: $p_1 = .40$, $p_2 = .30$, $p_3 = .30$
H_a: At least one of the probabilities differs from the hypothesized value

c.   If H_0 is true, we would expect the following number of sports facilities in the 3 locations:

$$E_1 = n\, p_{1,0} = 113\,(.40) = 45.2$$

$$E_2 = n\, p_{2,0} = 113\,(.30) = 33.9$$

$$E_3 = n\, p_{3,0} = 113\,(.30) = 33.9$$

d.   The test statistic is $\chi^2 = \sum \dfrac{[n_i - E_i]^2}{E_i} = \dfrac{(58-45.2)^2}{45.2} + \dfrac{(26-33.9)^2}{33.9} + \dfrac{(29-33.9)^2}{33.9} = 6.17$

e.   Using Table V, Appendix A, with df $= k - 1 = 3 - 1 = 2$, $P(\chi^2 > 5.99147) = .05$ and $P(\chi^2 > 7.37776) = .025$. Thus, $.025 < P(\chi^2 > 6.17) < .05$.

Since the p-value is less than α $(p < .05)$, H_0 is rejected. There is sufficient evidence to indicate the proportions of major sports facilities in downtown, central city, and suburban areas are different in 1997 from what they were in 1985 at $\alpha = .05$.

8.83   a.   The 2×2 contingency table is:

LLD

Genetic Trait	Yes	No	Total
Yes	21	15	36
No	150	306	456
Total	171	321	492

b.   Some preliminary calculations are:

$$\hat{E}_{11} = \frac{R_1 C_1}{n} = \frac{36(171)}{492} = 12.51 \qquad \hat{E}_{12} = \frac{R_1 C_2}{n} = \frac{36(321)}{492} = 23.49$$

$$\hat{E}_{21} = \frac{R_2 C_1}{n} = \frac{456(171)}{492} = 158.49 \qquad \hat{E}_{22} = \frac{R_2 C_2}{n} = \frac{456(321)}{492} = 297.51$$

To determine if the genetic trait occurs at a higher rate in LDD patients than in the controls, we test:

H_0: Genetic trait and LDD group are independent
H_a: Genetic trait and LDD group are dependent

The test statistic is

$$\chi^2 = \sum\sum \frac{[n_{ij} - \hat{E}_{ij}]^2}{\hat{E}_{ij}} = \frac{(21-12.51)^2}{12.51} + \frac{(15-23.49)^2}{23.49} + \frac{(150-158.49)^2}{158.49}$$

$$+ \frac{(306-297.51)^2}{297.51} = 9.52$$

The rejection region requires $\alpha = .01$ in the upper tail of the χ^2 distribution with df $= (r-1)(c-1) = (2-1)(2-1) = 1$. From Table V, Appendix A, $\chi^2_{.01} = 6.63490$. The rejection region is $\chi^2 > 6.63490$.

Since the observed value of the test statistic falls in the rejection region $(\chi^2 = 9.52 > 6.63490)$, H_0 is rejected. There is sufficient evidence to indicate that the genetic trait occurs at a higher rate in LDD patients than in the controls at $\alpha = .01$.

c.   To construct a bar graph, we will first compute the proportion of LDD/Control patients that have the genetic trait. Of the LLD patients, $21 / 171 = .123$ have the trait. For the Controls, $15 / 321 = .043$. For the row totals, $36 / 492 = .073$. The bar graph is:

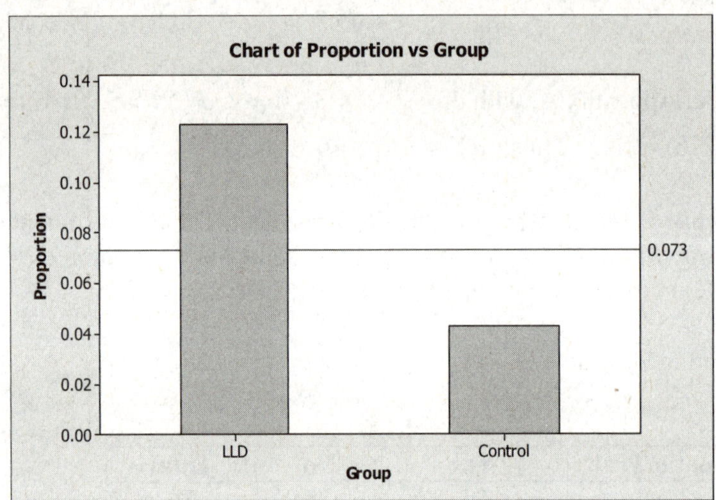

Because the bars are not close to being the same height, it indicates that the genetic trait appears at a higher rate among LLD patients than among the controls. This supports the conclusion of the test in part **b**.

8.85   Some preliminary calculations are:

$$\hat{E}_{11} = \frac{R_1 C_1}{n} = \frac{62(106)}{275} = 23.90 \qquad \hat{E}_{12} = \frac{R_1 C_2}{n} = \frac{62(112)}{275} = 25.25$$

$$\hat{E}_{13} = \frac{R_1 C_3}{n} = \frac{62(57)}{275} = 12.85 \qquad \hat{E}_{21} = \frac{R_2 C_1}{n} = \frac{59(106)}{275} = 22.74$$

$$\hat{E}_{22} = \frac{R_2 C_2}{n} = \frac{59(112)}{275} = 24.03 \qquad \hat{E}_{23} = \frac{R_2 C_3}{n} = \frac{59(57)}{275} = 12.23$$

$$\hat{E}_{31} = \frac{R_3 C_1}{n} = \frac{59(106)}{275} = 22.74 \qquad \hat{E}_{32} = \frac{R_3 C_2}{n} = \frac{59(112)}{275} = 24.03$$

$$\hat{E}_{33} = \frac{R_3 C_3}{n} = \frac{59(57)}{275} = 12.23 \qquad \hat{E}_{41} = \frac{R_4 C_1}{n} = \frac{95(106)}{275} = 36.62$$

$$\hat{E}_{42} = \frac{R_4 C_2}{n} = \frac{95(112)}{275} = 38.69 \qquad \hat{E}_{43} = \frac{R_4 C_3}{n} = \frac{95(57)}{275} = 19.69$$

To determine if political strategy of ethnic groups depends on world region, we test:

H_0: Political strategy and world region are independent
H_a: Political strategy and world region are dependent

The test statistic is

$$\chi^2 = \sum\sum \frac{[n_{ij} - \hat{E}_{ij}]^2}{\hat{E}_{ij}} = \frac{(24 - 23.90)^2}{23.90} + \frac{(31 - 25.25)^2}{25.25} + \frac{(7 - 12.85)^2}{12.85} + \frac{(32 - 22.74)^2}{22.74}$$

$$+ \frac{(23 - 24.03)^2}{24.03} + \frac{(4 - 12.23)^2}{12.23} + \frac{(11 - 22.74)^2}{22.74} + \frac{(22 - 24.03)^2}{24.03}$$

$$+ \frac{(26 - 12.23)^2}{12.23} + \frac{(39 - 36.62)^2}{36.62} + \frac{(36 - 38.69)^2}{38.69} + \frac{(20 - 19.69)^2}{19.69} = 35.409$$

The rejection region requires $\alpha = .10$ in the upper tail of the χ^2 distribution with df = $(r - 1)(c - 1) = (4 - 1)(3 - 1) = 6$. From Table V, Appendix A, $\chi^2_{.10} = 10.6446$. The rejection region is $\chi^2 > 10.6446$.

Since the observed value of the test statistic falls in the rejection region $(\chi^2 = 35.409 > 10.6446)$, H_0 is rejected. There is sufficient evidence to indicate that the political strategy of the ethnic groups depends on world region at $\alpha = .10$.

To graph the data, we will first compute the percent of observations in each category of Political Strategy for each World Region. To do this, we divide each cell frequency by the row total and then multiply by 100%. Using MINITAB, a graph of the data is:

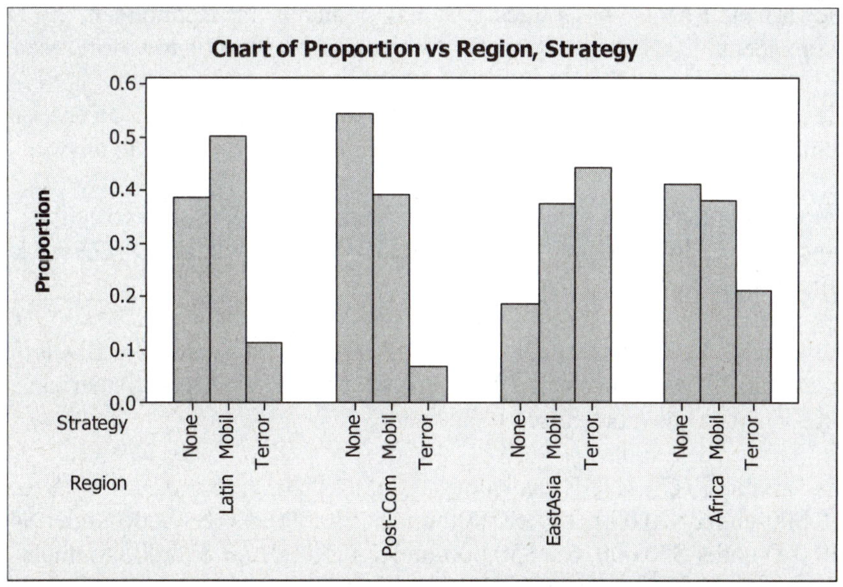

The graph supports the results of the test above. The percents in each level of Political Strategy differ for the different values of World Region.

8.87   a. $\hat{p}_1 = \dfrac{x_1}{n_1} = \dfrac{29}{189} = .153$

     b. $\hat{p}_2 = \dfrac{x_2}{n_2} = \dfrac{32}{149} = .215$

     c.   For confidence coefficient .90, $\alpha = .10$ and $\alpha / 2 = .10 / 2 = .05$. From Table III, Appendix A, $z_{.05} = 1.645$. The 90% confidence interval is:

$$(\hat{p}_1 - \hat{p}_2) \pm z_{.05}\sqrt{\dfrac{\hat{p}_1\hat{q}_1}{n_1} + \dfrac{\hat{p}_2\hat{q}_2}{n_2}} \Rightarrow (.153 - .215) \pm 1.645\sqrt{\dfrac{.153(.847)}{189} + \dfrac{.215(.785)}{149}}$$

$$\Rightarrow -.062 \pm .070 \Rightarrow (-.132, .008)$$

     d.   We are 90% confident that the difference in the true proportion of super-experienced bidders who fall prey to the winner's curse and the true proportion of less-experienced bidders who fall prey to the winner's curse is between -0.132 and 0.008. Since 0 is contained in the interval, there is no evidence to indicate that experience in bidding affects the likelihood of the winner's curse occurring.

8.89   First, the values of the variables will be defined.

    1.   The variable WAR has values 1, 2, and 9. 1 = Support, 2 = Oppose, and 9 = Don't know or refused to answer.

    2.   The variable INTERNET has values 1, 2, and 9. 1 = Yes, 2 = No, and 9 = Don't know or refused to answer.

    3.   The variable PARTY has values 1, 2, 3, 4, 5, and 9. 1 = Republican, 2 = Democrat, 3 = Independent, 4 = No preference, 5 = Other, and 9 = Don't know or refused to answer.

    4.   The variable VET has values 1, 2, 3, 4, and 9. 1 = Yes, I have, 2 = Yes, other household member, 3 = Yes, both, 4 = No, and 9 = Don't know or refused to answer.

    5.   The variable IDEO has values 1, 2, 3, 4, 5, and 9. 1 = Very conservative, 2 = Conservative, 3 = Moderate, 4 = Liberal, 5 = Very liberal, and 9 = Don't know or refused to answer.

    6.   The variable RACE has values 1, 2, 3, 4, 5, and 9. 1 = White, 2 = Black or African American, 3 = Asian or Pacific Islander, 4 = Mixed, 5 = Native American, 6 = Other and 9 = Don't know or refused to answer.

    7.   The variable INCRANGE has values 1, 2, 3, 4, 5, 6, 7, 8, and 9. 1 = <\$10,000, 2 = \$10,000-under \$20,000, 3 = \$20,000-under \$30,000, 4 = \$30,000-under \$40,000, 5 = \$40,000-under \$50,000, 6 = \$50,000-under \$75,000, 7 = \$75,000 to under \$100,000, 8 = Over \$100,000, and 9 = Don't know or refused to answer.

    8.   The variable CUMMUNIT has values 1, 2, 3, and 9. 1 = Urban, 2 = Suburban, 3 = Rural, and 9 = Don't know or refused to answer.

To do the analysis, all values of 9 were converted to missing values and are not included in the analysis. For all tests, $\alpha = .05$ was used.

WAR versus INTERNET. Using MINITAB, the results of the analyses are:

Tabulated statistics: War, Internet

```
Rows: War    Columns: Internet

                 1        2  Missing       All

1              703      399        4      1102
             77.94    76.29        *     77.33

2              199      124        1       323
             22.06    23.71        *     22.67

Missing         27       38        0         *
                 *        *        *         *

All            902      523        *      1425
            100.00   100.00        *    100.00


Cell Contents:         Count
                       % of Column


Pearson Chi-Square = 0.512, DF = 1, P-Value = 0.474
Likelihood Ratio Chi-Square = 0.510, DF = 1, P-Value = 0.475
```

To determine if Support of Iraq War and access of internet are dependent, we test:

H_0: Support of the Iraq War and access of internet are independent
H_a: Support of the Iraq War and access of internet are dependent

The test statistic is $\chi^2 = .512$ and the p-value is $p = .474$. Since the p-value is not less than $\alpha = .05$, H_0 is not rejected. There is insufficient evidence to indicate that Support of Iraq war and access of internet are dependent at $\alpha = .05$.

WAR versus PARTY. Since only 3 respondents selected "Other" I combined the "Other" group with the "No preference" group" so that the expected counts in all cells are at least 5. Using MINITAB, the results of the analyses are:

Tabulated statistics: War, Party

```
Rows: War    Columns: Party

                    1       2       3       4  Missing      All

1                 451     262     315      53       25     1081
                95.55   60.23   73.77   80.30        *    77.21

2                  21     173     112      13        5      319
                 4.45   39.77   26.23   19.70        *    22.79

Missing             8      27      21       5        4        *
                    *       *       *       *        *        *

All               472     435     427      66        *     1400
               100.00  100.00  100.00  100.00        *   100.00

Cell Contents:         Count
                       % of Column


Pearson Chi-Square = 164.761, DF = 3, P-Value = 0.000
Likelihood Ratio Chi-Square = 189.289, DF = 3, P-Value = 0.000
```

To determine if Support of Iraq War and Party are dependent, we test:

H_0: Support of the Iraq War and Party are independent
H_a: Support of the Iraq War and Party are dependent

The test statistic is $\chi^2 = 164.761$ and the p-value is $p = 0.000$. Since the p-value is less than $\alpha = .05$, H_0 is rejected. There is sufficient evidence to indicate that Support of Iraq war and Party are dependent at $\alpha = .05$.

If you look at the column percents in the above table, we see that 95.55% of the Republicans support the Iraq War, while only 60.23% of Democrats support the Iraq War. In addition, 73.33% of the Independents support the war and 80.3% of those with no preference or other support the war.

WAR versus VET. Using MINITAB, the results of the analyses are:

Tabulated statistics: War, vet

```
Rows: War    Columns: vet

                   1        2        3        4  Missing       All

1                209      237       27      631        2      1104
               84.27    76.45    77.14    75.66        *     77.37

2                 39       73        8      203        1       323
               15.73    23.55    22.86    24.34        *     22.63

Missing            6       15        0       44        0         *
                   *        *        *        *        *         *

All              248      310       35      834        *      1427
              100.00   100.00   100.00   100.00        *    100.00

Cell Contents:        Count
                      % of Column
```

```
Pearson Chi-Square = 8.295, DF = 3, P-Value = 0.040
Likelihood Ratio Chi-Square = 8.858, DF = 3, P-Value = 0.031
```

To determine if Support of Iraq War and Veteran status are dependent, we test:

H_0: Support of the Iraq War and Veteran status are independent
H_a: Support of the Iraq War and Veteran status are dependent

The test statistic is $\chi^2 = 8.295$ and the p-value is $p = 0.040$. Since the p-value is less than $\alpha = .05$, H_0 is rejected. There is sufficient evidence to indicate that Support of Iraq War and Veteran status are dependent at $\alpha = .05$.

If you look at the column percents in the above table, we see that 84.27% of the veterans support the Iraq War, while 76.45% of those who have a family member who is a veteran support the Iraq War. In addition, 77.14% of those who are vets and have a family member who is a vet support the war and 75.66% of those who are not veterans support the war.

WAR versus IDEO. Using MINITAB, the results of the analyses are:

Tabulated statistics: War, ideo

```
Rows: War    Columns: ideo

                 1        2        3        4        5  Missing      All

1               52      453      415      102       21       63     1043
             75.36    89.88    78.75    52.85    34.43        *    77.03

2               17       51      112       91       40       13      311
             24.64    10.12    21.25    47.15    65.57        *    22.97

Missing          2       17       19       11        2       14        *
                 *        *        *        *        *        *        *

All             69      504      527      193       61        *     1354
            100.00   100.00   100.00   100.00   100.00        *   100.00


Cell Contents:      Count
                    % of Column


Pearson Chi-Square = 174.386, DF = 4, P-Value = 0.000
Likelihood Ratio Chi-Square = 161.296, DF = 4, P-Value = 0.000
```

To determine if Support of Iraq War and Political views are dependent, we test:

H_0: Support of the Iraq War and Political views are independent
H_a: Support of the Iraq War and Political views are dependent

The test statistic is $\chi^2 = 174.386$ and the p-value is $p = 0.000$. Since the p-value is less than $\alpha = .05$, H_0 is rejected. There is sufficient evidence to indicate that Support of Iraq War and Political views are dependent at $\alpha = .05$.

If you look at the column percents in the above table, we see that 75.36% of the Very Conservatives support the Iraq War, while 89.88% of the Conservatives support the Iraq War. In addition, 78.75% of Moderates support the war, 52.85% of Liberals support the war, and 34.43% of the Very Liberals support the war.

WAR versus RACE. Because there were less than 24 respondents for the race categories Asian/Pacific Islander, Mixed, Native American, and Other, these 4 categories were combined. Using MINITAB, the results of the analyses are:

Tabulated statistics: War, Race

```
Rows: War    Columns: Race

                   1         2         3   Missing       All

1                989        51        54        12      1094
               81.00     46.79     72.97         *     77.92

2                232        58        20        14       310
               19.00     53.21     27.03         *     22.08

Missing           50         9         4         2         *
                   *         *         *         *         *

All             1221       109        74         *      1404
              100.00    100.00    100.00         *    100.00


Cell Contents:         Count
                       % of Column


Pearson Chi-Square = 69.181, DF = 2, P-Value = 0.000
Likelihood Ratio Chi-Square = 57.984, DF = 2, P-Value = 0.000
```

To determine if Support of Iraq War and Race are dependent, we test:

H_0: Support of the Iraq War and Race are independent
H_a: Support of the Iraq War and Race are dependent

The test statistic is $\chi^2 = 69.181$ and the p-value is $p = 0.000$. Since the p-value is less than $\alpha = .05$, H_0 is rejected. There is sufficient evidence to indicate that Support of Iraq War and Race are dependent at $\alpha = .05$.

If you look at the column percents in the above table, we see that 81.00% of the Whites support the Iraq War, while 46.79% of the Blacks support the Iraq War. In addition, 72.96% of the Others support the war.

WAR versus INCRANGE Using MINITAB, the results of the analyses are:

Tabulated statistics: War, incrange

```
Rows: War    Columns: incrange

                      1         2         3         4         5         6         7         8

1                    37        79       122       112       103       196       132       134
                  61.67     73.15     77.71     77.78     83.06     82.35     77.19     74.03

2                    23        29        35        32        21        42        39        47
                  38.33     26.85     22.29     22.22     16.94     17.65     22.81     25.97

Missing              12        10         4         9         2         5         2         3
                      *         *         *         *         *         *         *         *

All                  60       108       157       144       124       238       171       181
                 100.00    100.00    100.00    100.00    100.00    100.00    100.00    100.00

              Missing       All

1                 191       915
                    *     77.35

2                  56       268
                    *     22.65

Missing            18         *
                    *         *

All                 *      1183
                    *    100.00

Cell Contents:        Count
                      % of Column


Pearson Chi-Square = 16.387, DF = 7, P-Value = 0.022
Likelihood Ratio Chi-Square = 15.684, DF = 7, P-Value = 0.028
```

To determine if Support of Iraq War and Income range are dependent, we test:

H_0: Support of the Iraq War and Income range are independent
H_a: Support of the Iraq War and Income range are dependent

The test statistic is $\chi^2 = 16.387$ and the p-value is $p = 0.022$. Since the p-value is less than $\alpha = .05$, H_0 is rejected. There is sufficient evidence to indicate that Support of Iraq War and Income range are dependent at $\alpha = .05$.

If you look at the column percents in the above table, we see that those in the income ranges of \$40,000 to under \$50,000 and from \$50,000 to under \$75,000 support the Iraq War the most. Those in the lowest income range and the highest income range tend to not support the Iraq War as much as those in the middle income brackets.

WAR versus COMMUNIT Using MINITAB, the results of the analyses are:

Tabulated statistics: War, communit

```
Rows: War    Columns: communit

               1        2        3       All

 1                235      605      266     1106
               69.73    78.27    83.13    77.34

 2                102      168       54      324
               30.27    21.73    16.88    22.66

 Missing           15       35       15        *
                    *        *        *        *

 All              337      773      320     1430
              100.00   100.00   100.00   100.00


 Cell Contents:          Count
                         % of Column


 Pearson Chi-Square = 17.618, DF = 2, P-Value = 0.000
 Likelihood Ratio Chi-Square = 17.309, DF = 2, P-Value = 0.000
```

To determine if Support of Iraq War and Community are dependent, we test:

H_0: Support of the Iraq War and Community are independent
H_a: Support of the Iraq War and Community are dependent

The test statistic is $\chi^2 = 17.618$ and the p-value is $p = 0.000$. Since the p-value is less than $\alpha = .05$, H_0 is rejected. There is sufficient evidence to indicate that Support of Iraq War and Community are dependent at $\alpha = .05$.

If you look at the column percents in the above table, we see that 69.73% of those who live in the urban areas support the Iraq War, while 78.27% of those who live in suburban areas support the Iraq War. In addition, 83.13% of those who live in Rural areas support the war.

8.91   a.   Some preliminary calculations are:

$E_1 = np_{1,0} = 85(.26) = 22.1$
$E_2 = np_{2,0} = 85(.30) = 25.5$
$E_3 = np_{3,0} = 85(.11) = 9.35$
$E_4 = np_{4,0} = 85(.14) = 11.9$
$E_5 = np_{5,0} = 85(.19) = 16.15$

To determine of probabilities differ from the hypothesized values, we test:

H_0: $p_1 = .26, p_2 = .30, p_3 = .11, p_4 = .14, p_5 = .19$
H_a: At least one of the probabilities differs from its hypothesized value

The test statistic is

$$\chi^2 = \sum \frac{[n_i - E_i]^2}{E_i} = \frac{(32-22.1)^2}{22.1} + \frac{(26-25.5)^2}{25.5} + \frac{(15-9.35)^2}{9.35} + \frac{(6-11.9)^2}{11.9} + \frac{(6-16.15)^2}{16.15} = 17.16$$

The rejection region requires $\alpha = .05$ in the upper tail of the χ^2 distribution with $df = k - 1 = 5 - 1 = 4$. From Table V, Appendix A, $\chi^2_{.05} = 9.48773$. The rejection region is $\chi^2 > 9.48773$.

Since the observed value of the test statistic falls in the rejection region $(\chi^2 = 17.16 > 9.48773)$, reject H_0. There is sufficient evidence to indicate the probabilities differ from their hypothesized values at $\alpha = .05$.

b. $\hat{p}_1 = \dfrac{32}{85} = .376$

For confidence coefficient .95, $\alpha = .05$ and $\alpha / 2 = .05 / 2 = .025$. From Table III, Appendix A, $z_{.025} = 1.96$. The 95% confidence interval is:

$$\hat{p}_i \pm z_{.025} \sqrt{\frac{\hat{p}_i \hat{q}_i}{n}} \Rightarrow .376 \pm 1.96 \sqrt{\frac{.376(.624)}{85}} \Rightarrow .376 \pm .103 \Rightarrow (.273, \ .479)$$

We are 95% confident that the true proportion of Avonex MS patients who are exacerbation-free during a 2-year period is between .273 and .479.

c.   From previous studies, it is known that 26% of the MS patients on a placebo experienced no exacerbations in a 2-year period. Since .26 does not fall in the 95% confidence interval, there is evidence that Avonex patients are more likely to have no exacerbations than the placebo patients at $\alpha = .05$.

8.93   Some preliminary calculations are:

$$\hat{E}_{11} = \frac{R_1 C_1}{n} = \frac{12(10)}{24} = 5 \qquad\qquad \hat{E}_{12} = \frac{R_1 C_2}{n} = \frac{12(14)}{24} = 7$$

$$\hat{E}_{21} = \frac{R_2 C_1}{n} = \frac{12(10)}{24} = 5 \qquad\qquad \hat{E}_{22} = \frac{R_2 C_2}{n} = \frac{12(14)}{24} = 7$$

To determine if a relationship exists between food choice and whether or not chickadees fed on gypsy moth eggs, we test:

H_0: Food choice and whether or not chickadees fed on gypsy moth eggs are independent
H_a: Food choice and whether or not chickadees fed on gypsy moth eggs are dependent

The test statistic is $\chi^2 = \sum\sum \dfrac{[n_{ij} - \hat{E}_{ij}]^2}{\hat{E}_{ij}} = \dfrac{(2-5)^2}{5} + \dfrac{(10-7)^2}{7} + \dfrac{(8-5)^2}{5} + \dfrac{(4-7)^2}{7} = 6.171$

The rejection region requires $\alpha = .10$ in the upper tail of the χ^2 distribution with df = $(r-1)(c-1) = (2-1)(2-1) = 1$. From Table V, Appendix A, $\chi^2_{.10} = 2.70554$. The rejection region is $\chi^2 > 2.70554$.

Since the observed value of the test statistic falls in the rejection region $(\chi^2 = 6.171 > 2.70554)$, H_0 is rejected. There is sufficient evidence to indicate that a relationship exists between food choice and whether or not chickadees fed on gypsy moth eggs at $\alpha = .10$.

8.95   a.   Some preliminary calculations are:

$E_1 = np_{1,0} = 1,000(.50) = 500$ $E_2 = np_{2,0} = 1,000(.22) = 220$
$E_3 = np_{3,0} = 1,000(.11) = 110$ $E_4 = np_{4,0} = 1,000(.17) = 170$

To determine if the data disagree with the percentages reported by Nielson/NetRatings, we test:

H_0: $p_1 = .50, p_2 = .22, p_3 = .11$, and $p_4 = .17$
H_a: At least one p_i differs from it hypothesized value

The test statistic is

$$\chi^2 = \sum \frac{[n_i - E_i]^2}{E_i}$$

$$= \frac{[487-500]^2}{500} + \frac{[245-220]^2}{220} + \frac{[121-110]^2}{110} + \frac{[147-170]^2}{170} = 7.39$$

The rejection region requires $\alpha = .05$ in the upper tail of the χ^2 distribution with df = $k - 1 = 4 - 1 = 3$. From Table V, Appendix A, $\chi^2_{.05} = 7.81473$. The rejection region is $\chi^2 > 7.81473$.

Since the observed value of the test statistic does not fall in the rejection region $(\chi^2 = 7.39 \not> 7.81473)$, H_0 is not rejected. There is insufficient evidence to indicate the data disagree with the percentages reported by Nielson/NetRatings at $\alpha = .05$.

b. $\hat{p}_1 = \dfrac{487}{1000} = .487$

For confidence coefficient .95, $\alpha = .05$ and $\alpha/2 = .05/2 = .025$. From Table III, Appendix A, $z_{.025} = 1.96$. The 95% confidence interval is:

$$\hat{p}_1 \pm z_{.025}\sqrt{\frac{\hat{p}_1\hat{q}_1}{n}} \Rightarrow .487 \pm 1.96\sqrt{\frac{.487(.513)}{1000}} \Rightarrow .487 \pm .031 \Rightarrow (.456, .518)$$

We are 95% confident that the true proportion of all Internet searches that use the Google Search engine is between .456 and .518.

8.97   a.   The observed frequencies for the 3 groups are:

Mail-only – 262
Internet-only – 43
Both – 135

Some preliminary calculations are:

$$E_1 = E_2 = E_3 = np_{1,0} = 440(1/3) = 146.67$$

To determine if the proportions of mail-only, internet-only and both users are different, we test:

H_0: $p_1 = p_2 = p_3 = 1/3$
H_a: At least one p_i differs from it hypothesized value

The test statistic is

$$\chi^2 = \sum \frac{[n_i - E_i]^2}{E_i} = \frac{[262 - 146.67]^2}{146.67} + \frac{[43 - 146.67]^2}{146.67} + \frac{[135 - 146.67]^2}{146.67} = 164.89$$

The rejection region requires $\alpha = .05$ in the upper tail of the χ^2 distribution with df = $k - 1 = 3 - 1 = 2$. From Table V, Appendix A, $\chi^2_{.05} = 5.99147$. The rejection region is $\chi^2 > 5.99147$.

Since the observed value of the test statistic falls in the rejection region $(\chi^2 = 164.89 > 5.99147)$, H_0 is rejected. There is sufficient evidence to indicate the proportions of mail-only, internet-only and both users are different at $\alpha = .05$.

Using MINITAB, a graph of the categories is:

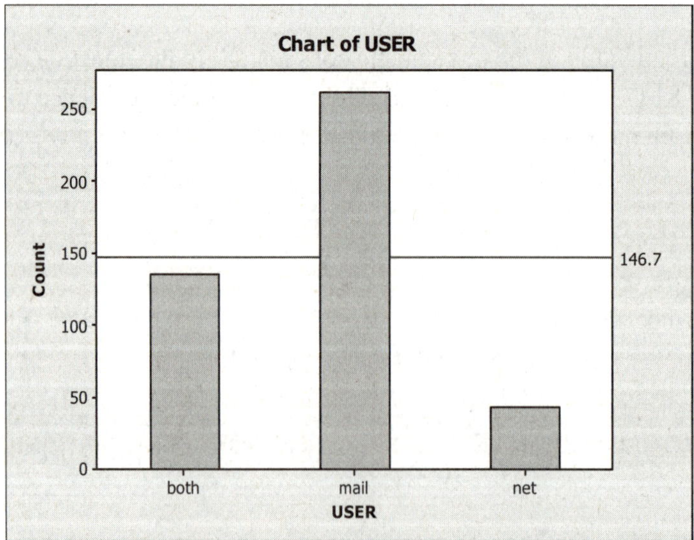

b.   **Type of User vs Gender**:

Using MINITAB, the contingency table is:

Tabulated statistics: USER, GENDER

```
Rows: USER    Columns: GENDER

           Female     Male      All

both          104       31      135
            77.04    22.96   100.00
            32.70    25.41    30.68

mail          178       84      262
            67.94    32.06   100.00
            55.97    68.85    59.55

net            36        7       43
            83.72    16.28   100.00
            11.32     5.74     9.77

All           318      122      440
            72.27    27.73   100.00
           100.00   100.00   100.00


Cell Contents:        Count
                      % of Row
                      % of Column


Pearson Chi-Square = 6.797, DF = 2, P-Value = 0.033
Likelihood Ratio Chi-Square = 7.105, DF = 2, P-Value = 0.029
```

To determine if coupon user type is related to gender, we test:

H_0: Type of coupon user and gender are independent
H_0: Type of coupon user and gender are dependent

From the printout, the test statistic is $\chi^2 = 6.797$ and the p-value is $p = .003$.

Since the observed p-value is less than $\alpha = .05$ ($p = .003 < .05$), H_0 is rejected. There is sufficient evidence to indicate the type of coupon user is related to gender at $\alpha = .05$.

Using MINITAB, a graph of the results is:

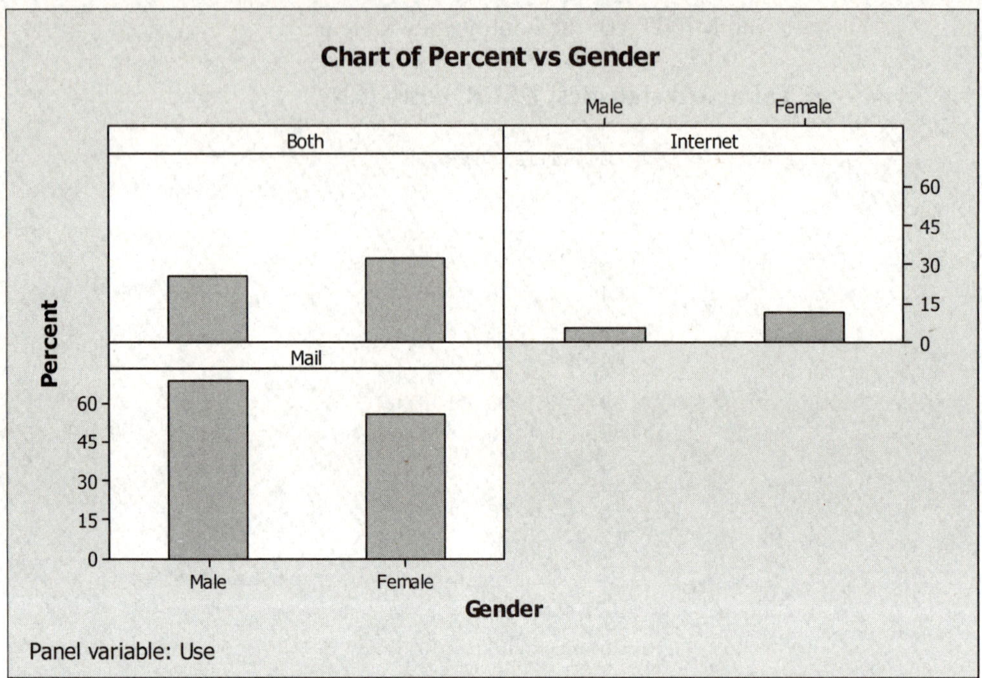

The majority of male and female users use coupons distributed through the mail. However, the proportion of male users who use coupons distributed through the mail is larger than the proportion of females. Very few users use coupons distributed via the Internet, but a higher proportion of females use coupons distributed via the Internet than males.

Type of User vs Education:

Using MINITAB, the contingency table is:

Tabulated statistics: USER, EDUC

```
Rows: USER    Columns: EDUC

              COLL      GRAD        HS      PROF       All

both            62        38        19        16       135
              45.93     28.15     14.07     11.85    100.00
              34.83     28.79     31.67     22.86     30.68

mail            96        85        34        47       262
              36.64     32.44     12.98     17.94    100.00
              53.93     64.39     56.67     67.14     59.55

net             20         9         7         7        43
              46.51     20.93     16.28     16.28    100.00
              11.24      6.82     11.67     10.00      9.77

All            178       132        60        70       440
              40.45     30.00     13.64     15.91    100.00
             100.00    100.00    100.00    100.00    100.00

Cell Contents:        Count
                      % of Row
                      % of Column


Pearson Chi-Square = 6.587, DF = 6,  P-Value = 0.361
Likelihood Ratio Chi-Square = 6.786, DF = 6,  P-Value = 0.341
```

To determine if coupon user type is related to education, we test:

H_0: Type of coupon user and education are independent
H_0: Type of coupon user and education are dependent

From the printout, the test statistic is $\chi^2 = 6.587$ and the p-value is $p = .361$.

Since the observed p-value is not less than $\alpha = .05$ ($p = .361 \not< .05$), H_0 is not rejected. There is insufficient evidence to indicate the type of coupon user is related to education at $\alpha = .05$.

Using MINITAB, a graph of the results is:

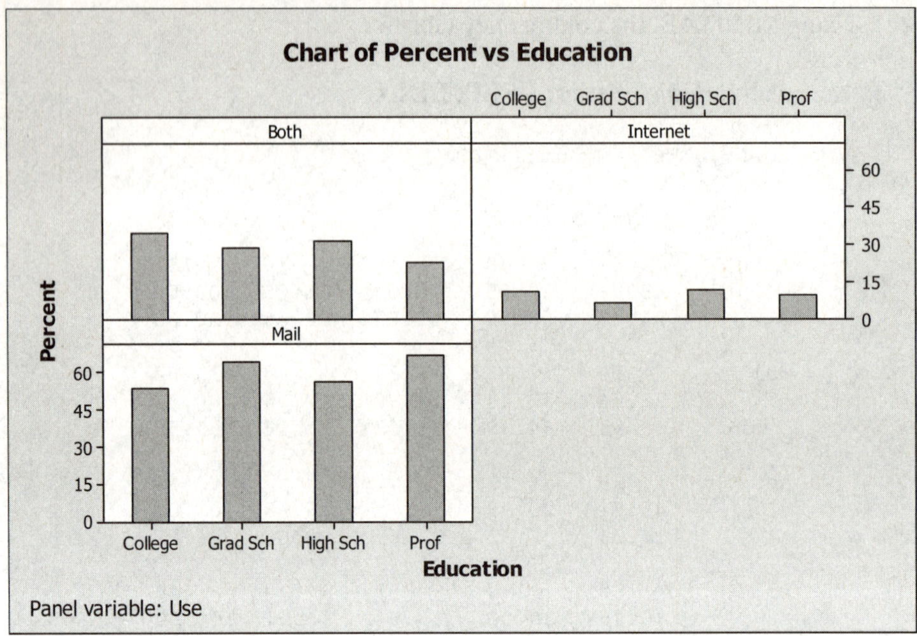

For each category of use, the proportions of each education level are about the same. This corresponds to the conclusion above that coupon use is not related to education level.

Type of User vs Work Status:

Using MINITAB, the contingency table is:

Tabulated statistics: USER, WORK

```
Rows: USER    Columns: WORK

                 1        2        3        4      All

both            90       13       17       15      135
              66.67     9.63    12.59    11.11   100.00
              33.71    25.00    33.33    21.43    30.68

mail           148       31       31       52      262
              56.49    11.83    11.83    19.85   100.00
              55.43    59.62    60.78    74.29    59.55

net             29        8        3        3       43
              67.44    18.60     6.98     6.98   100.00
              10.86    15.38     5.88     4.29     9.77

All            267       52       51       70      440
              60.68    11.82    11.59    15.91   100.00
             100.00   100.00   100.00   100.00   100.00

Cell Contents:        Count
                      % of Row
                      % of Column


Pearson Chi-Square = 11.687, DF = 6, P-Value = 0.069
Likelihood Ratio Chi-Square = 12.208, DF = 6, P-Value = 0.057

* NOTE * 1 cells with expected counts less than 5
```

To determine if coupon user type is related to work status, we test:

H_0: Type of coupon user and work status are independent
H_a: Type of coupon user and work status are dependent

From the printout, the test statistic is $\chi^2 = 11.687$ and the p-value is $p = .069$.

Since the observed p-value is not less than $\alpha = .05$ $(p = .069 \not< .05)$, H_0 is not rejected. There is insufficient evidence to indicate the type of coupon user is related to work status at $\alpha = .05$.

Using MINITAB, a graph of the results is:

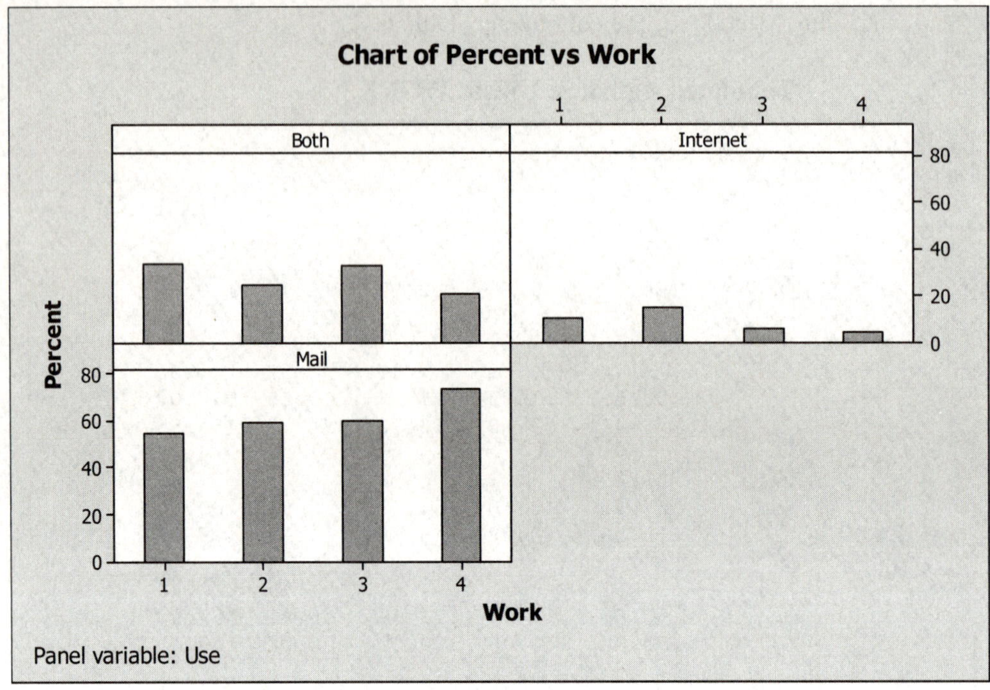

Again, for each category of use, the proportions of each work level are about the same. This corresponds to the conclusion above that coupon use is not related to work level.

Type of User vs Coupon satisfaction:

Using MINITAB, the contingency table is:

Tabulated statistics: USER, SATISF

```
Rows: USER    Columns: SATISF

             No      Some      Yes      All

both          3         9      123      135
           2.22      6.67    91.11   100.00
           8.57     11.25    37.85    30.68

mail         28        62      172      262
          10.69     23.66    65.65   100.00
          80.00     77.50    52.92    59.55

net           4         9       30       43
           9.30     20.93    69.77   100.00
          11.43     11.25     9.23     9.77

All          35        80      325      440
           7.95     18.18    73.86   100.00
         100.00    100.00   100.00   100.00

Cell Contents:        Count
                      % of Row
                      % of Column


Pearson Chi-Square = 30.418, DF = 4, P-Value = 0.000
Likelihood Ratio Chi-Square = 34.934, DF = 4, P-Value = 0.000

* NOTE * 1 cells with expected counts less than 5
```

To determine if coupon user type is related to coupon satisfaction, we test:

H_0: Type of coupon user and coupon satisfaction are independent
H_0: Type of coupon user and coupon satisfaction are dependent

From the printout, the test statistic is $\chi^2 = 30.418$ and the p-value is $p = .000$.

Since the observed p-value is less than $\alpha = .05$ ($p = .000 < .05$), H_0 is rejected. There is sufficient evidence to indicate the type of coupon user is related to coupon satisfaction at $\alpha = .05$.

Using MINITAB, a graph of the results is:

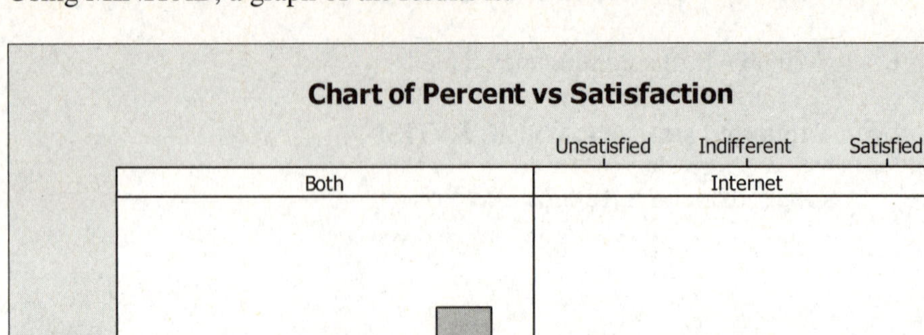

Again, very few users use coupons distributed via the Internet. The proportions of the different levels of satisfaction are about the same for this category of coupon use. For coupons distributed through the mail, the proportions of users who were unsatisfied or indifferent are larger than the proportion satisfied. However, for users of coupons distributed through both the mail and the Internet, the proportion of users who were satisfied is greater then the proportions of users unsatisfied or indifferent.

8.99   a.

$$\chi^2 = \sum \frac{[n_i - E_i]^2}{E_i} = \frac{(26-23)^2}{23} + \frac{(146-136)^2}{136} + \frac{(361-341)^2}{341} + \frac{(143-136)^2}{136} + \frac{(13-23)^2}{23}$$

$$= 9.647$$

b.   From Table V, Appendix A, with df $= 5$, $\chi^2_{.05} = 11.0705$

c.   No. Since the observed value of the test statistic does not fall in the rejection region ($\chi^2 = 9.647 \not> 11.0705$), H_0 is not rejected. There is insufficient evidence to indicate the salary distribution is nonnormal for $\alpha = .05$.

d.   The p-value $= P(\chi^2 \geq 9.647)$.

Using Table V, Appendix A, with df $= 5$, $.05 < P(\chi^2 \geq 9.647) < .10$.

8.101   Using MINITAB, the results of the analyses are:

Tabulated statistics: Candidate, Time

```
Using frequencies in Fr


Rows: Candidate   Columns: Time

                  1        2        3        4        5        6      All

Coppin           55       51      109       98       88      104      505
               13.89    13.56    13.37    13.98    14.08    14.04    13.82
               54.7     52.0    112.6     96.9     86.4    102.4    505.0

Montes          133      117      255      211      186      227     1129
               33.59    31.12    31.29    30.10    29.76    30.63    30.90
              122.4    116.2    251.8    216.6    193.1    229.0   1129.0

Smith           208      208      451      392      351      410     2020
               52.53    55.32    55.34    55.92    56.16    55.33    55.28
              218.9    207.9    450.5    387.5    345.5    409.6   2020.0

All             396      376      815      701      625      741     3654
              100.00   100.00   100.00   100.00   100.00   100.00   100.00
              396.0    376.0    815.0    701.0    625.0    741.0   3654.0

Cell Contents:       Count
                     % of Column
                     Expected count

Pearson Chi-Square = 2.284, DF = 10, P-Value = 0.994
Likelihood Ratio Chi-Square = 2.272, DF = 10, P-Value = 0.994
```

To determine if time period and votes received by the candidates are independent, we test:

H_0: Votes received by candidates and time period are independent
H_a: Votes received by candidates and time period are dependent

From the printout, the test statistic is $\chi^2 = 2.284$ and the p-value is $p = .994$.

Since the p-value is so large, H_0 would not be rejected for any reasonable value of α. There is insufficient evidence to indicate that the votes received by the candidates are dependent on the time period. In other words, there is no evidence that the percentages of votes received by the candidates change over time. If one looks at the column percentages (the second number in each cell above), the values are very similar for each time period.

A graph of the column percents is:

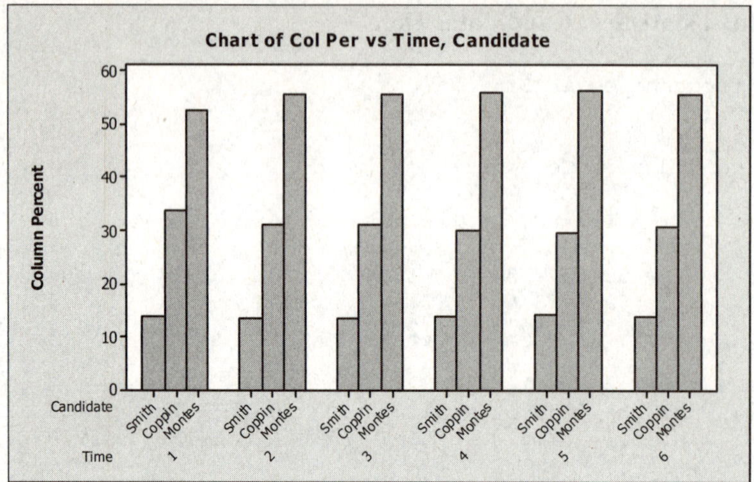

As one can see from the graph, the percent of votes received by the candidates are essentially the same for each time period.

All of the above indicates that the probability of a candidate receiving votes is independent of the time period. However, this does not necessarily imply a rigged election. If each time period is considered a random sample of voters, then the percentage of votes received by each candidate should be similar to the actual percentage of voters who favor each candidate.

Simple Linear Regression

9.1   A deterministic model does not allow for random error or variation, whereas a probabilistic model does. An example where a deterministic model would be appropriate is:

Let y = cost of a 2×4 piece of lumber and x = length (in feet)

The model would be $y = \beta_1 x$. There should be no variation in price for the same length of wood.

An example where a probabilistic model would be appropriate is:

Let y = sales per month of a commodity and
x = amount of money spent advertising

The model would be $y = \beta_0 + \beta_1 x + \varepsilon$. The sales per month will probably vary even if the amount of money spent on advertising remains the same.

9.3   The "line of means" is the deterministic component in a probabilistic model.

9.5

a.
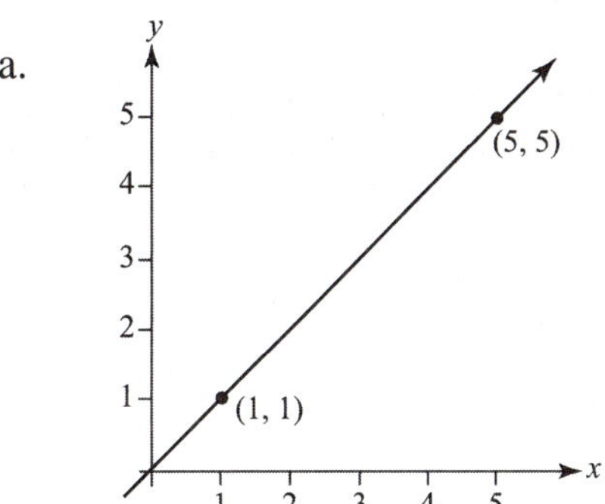

b.
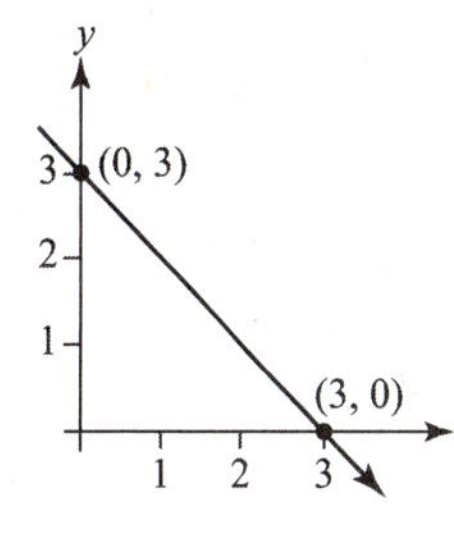

c.
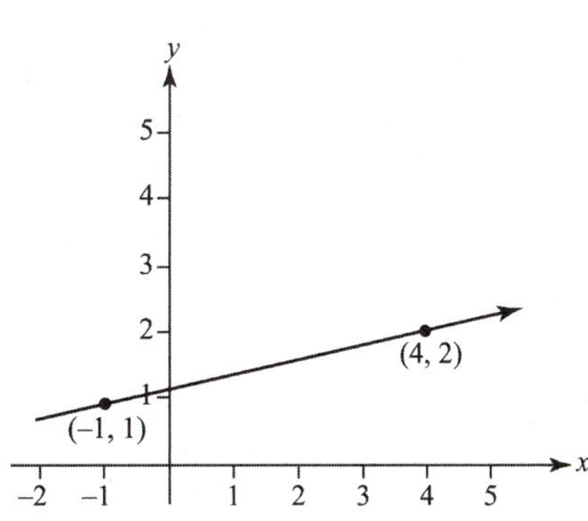

d.
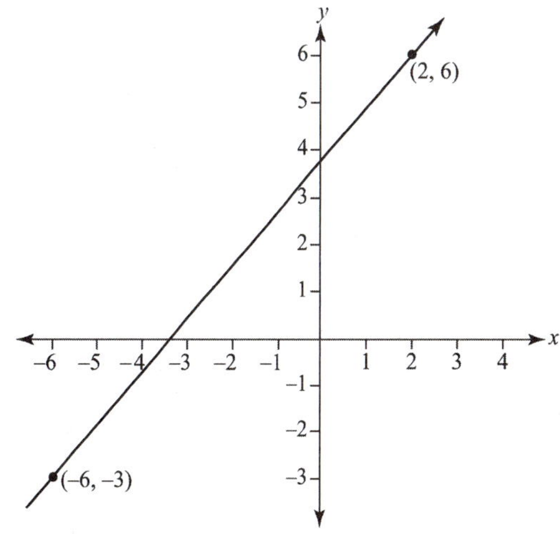

9.7   The two equations are:

$$4 = \beta_0 + \beta_1(-2) \text{ and } 6 = \beta_0 + \beta_1(4)$$

Subtracting the first equation from the second, we get

$$
\begin{aligned}
6 &= \beta_0 + 4\beta_1 \\
-(4 &= \beta_0 - 2\beta_1) \\
\hline
2 &= \ 6\beta_1 \Rightarrow \beta_1 = \frac{2}{6} = \frac{1}{3}
\end{aligned}
$$

Substituting $\beta_1 = \dfrac{1}{3}$ into the first equation, we get:

$$4 = \beta_0 + \frac{1}{3}(-2) \Rightarrow \beta_0 = 4 + \frac{2}{3} = \frac{14}{3}$$

The equation for the line is $y = \dfrac{14}{3} + \dfrac{1}{3}x$

9.9   To graph a line, we need two points. Pick two values for x, and find the corresponding y values by substituting the values of x into the equation.

a.   Let $x = 0 \Rightarrow y = 4 + (0) = 4$
and $x = 2 \Rightarrow y = 4 + (2) = 6$

b.   Let $x = 0 \Rightarrow y = 5 - 2(0) = 5$
and $x = 2 \Rightarrow y = 5 - 2(2) = 1$

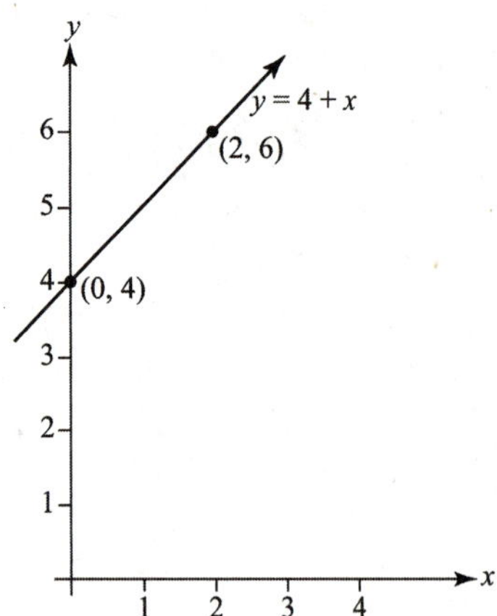

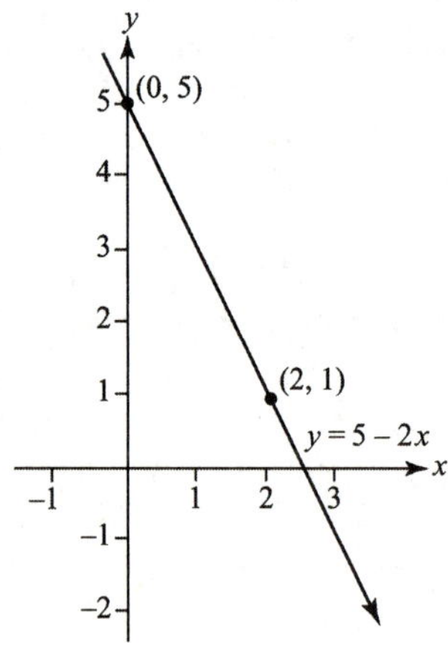

c.   Let $x = 0 \Rightarrow y = -4 + 3(0) = -4$
     and $x = 2 \Rightarrow y = -4 + 3(2) = 2$

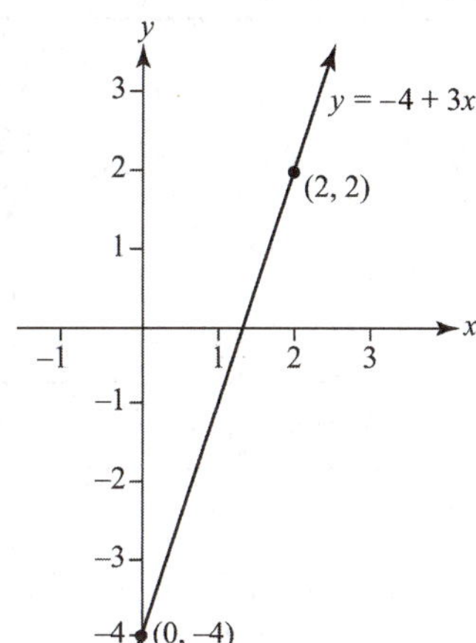

d.   Let $x = 0 \Rightarrow y = -2(0) = 0$
     and $x = 2 \Rightarrow y = -2(2) = -4$

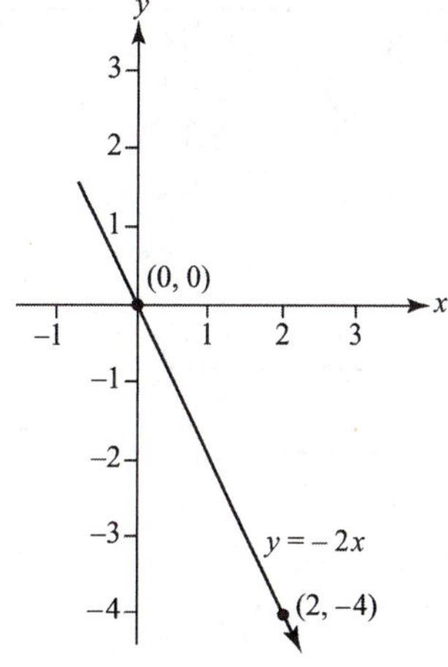

e.   Let $x = 0 \Rightarrow y = 0$
     and $x = 2 \Rightarrow y = 2$

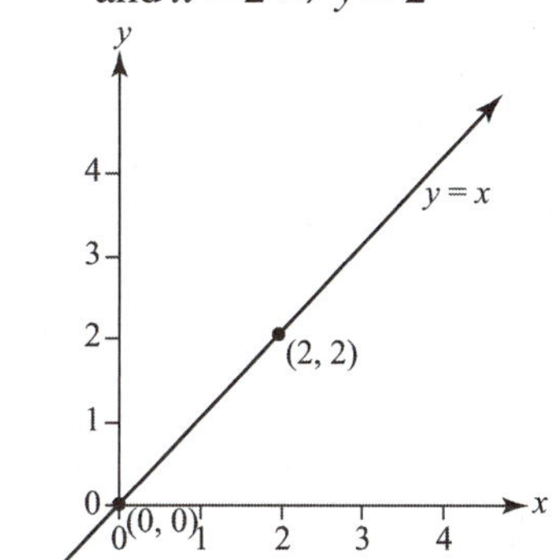

f.   Let $x = 0 \Rightarrow y = .5 + 1.5(0) = .5$
     and $x = 2 \Rightarrow y = .5 + 1.5(2) = 3.5$

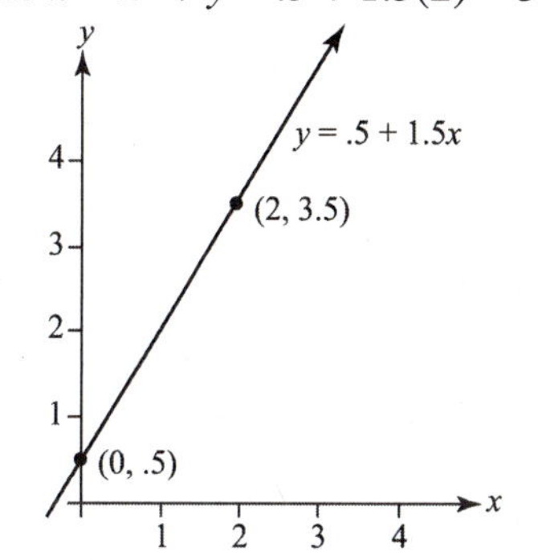

9.11   In regression, the error of prediction is the difference between the observed and predicted values of the dependent variable.

9.13   The statement "The estimates of β_0 and β_1 should be interpreted only within the sampled range of the independent variable, x." is true. We only have information about the relationship between y and x within the observed range of x.

9.15   From Exercise 9.14, $\hat{\beta}_0 = 7.10$ and $\hat{\beta}_1 = -.78$.

The fitted line is $\hat{y} = 7.10 - .78x$. To obtain values for $\hat{y}$, we substitute values of x into the equation and solve for $\hat{y}$.

a.

x	y	$\hat{y} = 7.10 - .78x$	$(y - \hat{y})$	$(y - \hat{y})^2$
7	2	1.64	.36	.1296
4	4	3.98	.02	.0004
6	2	2.42	−.42	.1764
2	5	5.54	−.54	.2916
1	7	6.32	.68	.4624
1	6	6.32	−.32	.1024
3	5	4.76	.24	.0576

$$\sum (y - \hat{y}) = 0.02 \qquad \text{SSE} = \sum (y - \hat{y})^2 = 1.2204$$

b.

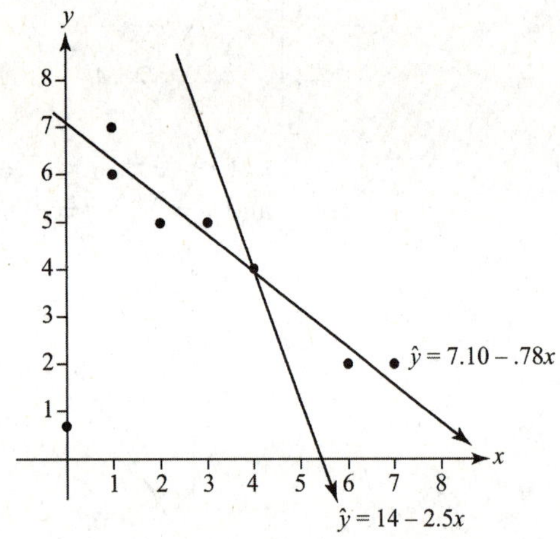

c.

x	y	$\hat{y} = 14 - 2.5x$	$(y - \hat{y})$	$(y - \hat{y})^2$
7	2	−3.5	5.5	30.25
4	4	4	0	0
6	2	−1	3	9
2	5	9	−4	16
1	7	11.5	−4.5	20.25
1	6	11.5	−5.5	30.25
3	5	6.5	−1.5	2.25

$$\sum (y - \hat{y}) = -7 \qquad \text{SSE} = 108.00$$

9.17   a.

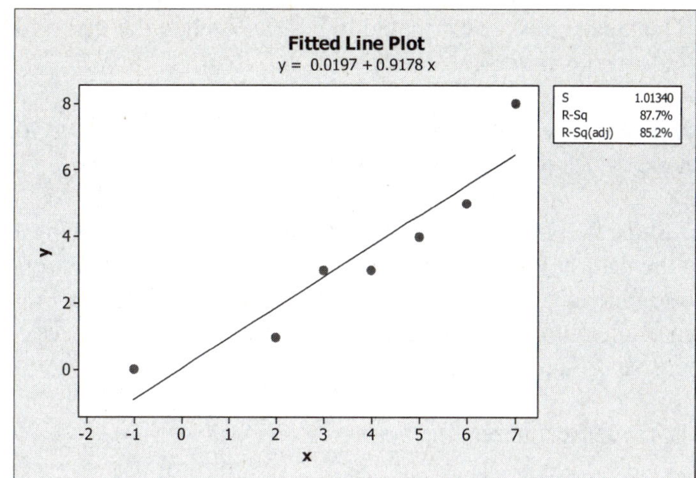

b.   As x increases, y tends to increase. Thus, there appears to be a positive, linear relationship between y and x.

c. $\hat{\beta}_1 = \dfrac{SS_{xy}}{SS_{xx}} = \dfrac{39.8571}{43.4286} = .9177616 \approx .918$

$\hat{\beta}_0 = \bar{y} - \hat{\beta}_1 \bar{x} = 3.4286 - .9177616(3.7143) = .0197581 \approx .020$

d.   The line appears to fit the data quite well.

e. $\hat{\beta}_0 = .020$ The estimated mean value of y when $x = 0$ is .020.

$\hat{\beta}_1 = .918$ The estimated change in the mean value of y for each unit change in x is .918.

These interpretations are valid only for values of x in the range from -1 to 7.

9.19   a.   The equation for the straight-line model is $y = \beta_0 + \beta_1 x + \varepsilon$.

b.   The least squares line is $\hat{y} = 19.393 - 8.036x$.

c. $\hat{\beta}_0 = 19.393$. The estimate of the mean wicking length when the concentration is 0 is 19.393 mm.

$\hat{\beta}_1 = -8.036$. For each unit increase in antibody concentration, the mean wicking length is estimated to decrease by 8.036mm.

9.21   a.   **Radiata Pine**: For each unit increase in the natural logarithm of the number of blade cycles, the mean stress is estimated to decrease by 2.50.

Hoop Pine: For each unit increase in the natural logarithm of the number of blade cycles, the mean stress is estimated to decrease by 2.36.

b.   **Radiata Pine**: The mean stress is estimated to be 97.37 when the natural logarithm of the number of blade cycles is zero.

   Hoop Pine: The mean stress is estimated to be 122.03 when the natural logarithm of the number of blade cycles is zero.

c.   Based on these results, it appears that the Hoop Pine is stronger and more fatigue resistant. When the natural logarithm of the number of blade cycles is zero, the mean stress for the Hoop Pine is greater than the mean stress for the Radiata Pine. As the natural logarithm of the number of blade cycles increases, the mean stress for the Radiata Pine decreases at a faster rate than does the Hoop Pine.

9.23   a.   Using MINITAB, the least squares line is:

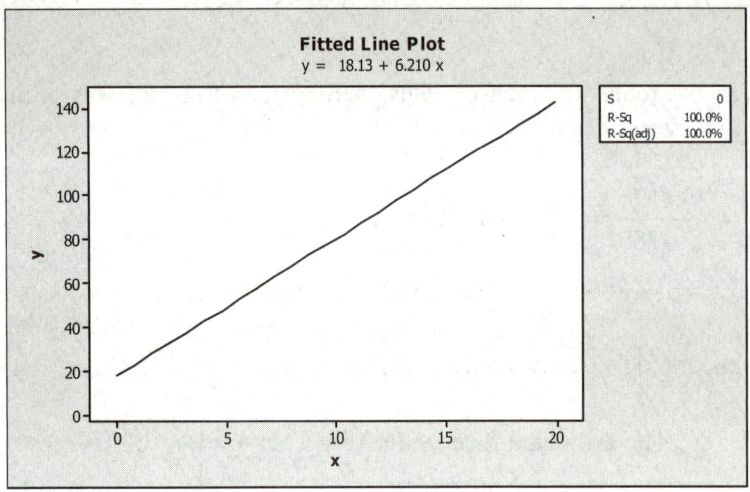

The slope of the line is positive – as x increases, y also increases.

b. $\hat{\beta}_0 = 18.13$. Since we have no information as to whether $x = 0$ is within the observed range, we will assume that it is. Then, the estimate of the mean magnitude when redshift level is 0 is 18.13.

c. $\hat{\beta}_1 = 6.21$. For each unit increase in redshift level, the mean value of magnitude is estimated to increase by 6.21.

9.25   a.   The straight-line model is $y = \beta_0 + \beta_1 x + \varepsilon$.

b.   Using MINITAB, the results are:

Regression Analysis: Accuracy versus Distance

```
The regression equation is
Accuracy = 250 - 0.629 Distance


Predictor        Coef  SE Coef        T       P
Constant       250.14    14.23    17.58   0.000
Distance      -0.62944   0.04759  -13.23   0.000


S = 2.23639    R-Sq = 82.2%    R-Sq(adj) = 81.7%


Analysis of Variance

Source           DF        SS       MS        F       P
Regression        1    874.99   874.99   174.95   0.000
Residual Error   38    190.06     5.00
Total            39   1065.04
```

The least squares prediction equation is: $\hat{y} = 250.14 - .6294x$

c.   The estimated y-intercept is $\hat{\beta}_o = 250.14$. Since 0 is not in the observed range for values of driving distance, the y-intercept has no meaning.

d.   The estimated slope of the line is $\hat{\beta}_1 = -.6294$. For each additional yard of driving distance, the mean driving accuracy value will decrease by an estimated .6294.

e.   The slope will help determine if the golfer's concern is valid. If there is no significant linear relationship between driving accuracy and driving distance, then the golfer would have nothing to worry about. However, if there is a significant linear relationship between driving accuracy and driving distance and the relationship is negative, then as the driving distance increases, the mean driving accuracy will decrease.

9.27   a.   The hypothesized model would be $y = \beta_0 + \beta_1 x + \varepsilon$.

b.   The slope will be positive because the researcher's theory indicates a linearly increasing relationship.

c.   Some preliminary calculations are:

$$\sum x_i = 300 \qquad \sum y_i = 98,494 \qquad \sum x_i y_i = 1,473,555$$
$$\sum x_i^2 = 4900 \qquad \sum y_i^2 = 456,565,950$$

$$SS_{xy} = \sum x_i y_i - \frac{\sum x_i \sum y_i}{n} = 1,473,555 - \frac{300(98,494)}{24} = 242,380$$

$$SS_{xx} = \sum x_i^2 - \frac{\left(\sum x_i\right)^2}{n} = 4900 - \frac{300^2}{24} = 1150$$

$$\hat{\beta}_1 = \frac{SS_{xy}}{SS_{xx}} = \frac{242,380}{1150} = 210.7652174$$

For each additional resonance, the mean sound wave frequency is estimated to increase by 210.765.

$$\hat{\beta}_0 = \bar{y} - \hat{\beta}_1\bar{x} = \frac{98,494}{24} - (210.7652174)\left(\frac{300}{24}\right) = 1469.351449$$

Since 0 is not in the observed range of resonances, $\hat{\beta}_0$ has no interpretation other than being the y-intercept.

9.29   a.   Using MINITAB, the results of fitting a regression model are:

Regression Analysis: IdealHt versus Height

```
The regression equation is
IdealHt = 86.0 - 0.260 Height


Predictor        Coef  SE Coef        T       P
Constant       86.018    4.786    17.97   0.000
Height        -0.26046  0.07035   -3.70   0.000


S = 3.64336    R-Sq = 8.6%    R-Sq(adj) = 8.0%


Analysis of Variance

Source             DF         SS        MS       F       P
Regression          1     181.93    181.93   13.71   0.000
Residual Error    145    1924.74     13.27
Total             146    2106.67
```

The fitted regression line is $\hat{y} = 86.0 - .260x$.

Because the parameter estimate for Height is negative (-.26046), this confirms what the researchers found.

b.   Yes, this is the correct interpretation.

c. Using MINITAB, a plot of the data is:

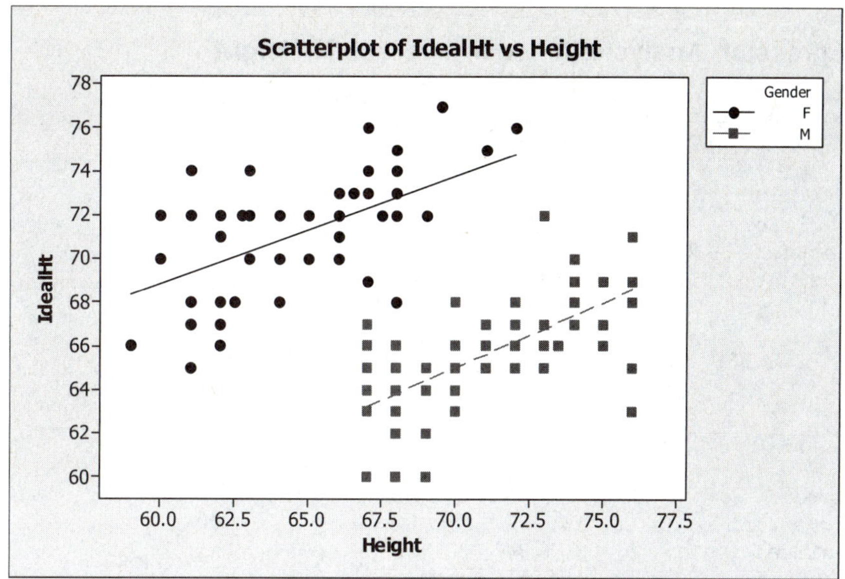

For the female students (round dots), there appears to be a positive relationship between the student's height and her ideal height. This same trend appears for the male students (squares).

d. Using MINITAB, the results of fitting the female data are:

Regression Analysis: F-IdealHt versus F-Height

```
The regression equation is
F-IdealHt = 39.3 + 0.493 F-Height


Predictor      Coef   SE Coef      T       P
Constant     39.304     6.206   6.33   0.000
F-Height    0.49261   0.09602   5.13   0.000


S = 2.32153    R-Sq = 27.3%    R-Sq(adj) = 26.3%


Analysis of Variance

Source            DF       SS       MS       F       P
Regression         1   141.84   141.84   26.32   0.000
Residual Error    70   377.27     5.39
Total             71   519.11
```

The fitted regression line is $\hat{y} = 39.3 + .493x$.

The estimated slope of the line is .493. For each additional inch of height for females, the mean ideal partner's height is estimated to increase by .493 inches.

e. Using MINITAB, the results of fitting the male data are:

Regression Analysis: M-IdealHt versus M-Height

```
The regression equation is
M-IdealHt = 23.3 + 0.596 M-Height


Predictor      Coef  SE Coef      T      P
Constant     23.274    6.333   3.67  0.000
M-Height    0.59630  0.08902   6.70  0.000


S = 2.05688    R-Sq = 38.1%    R-Sq(adj) = 37.2%


Analysis of Variance

Source           DF       SS       MS       F       P
Regression        1   189.82   189.82   44.87   0.000
Residual Error   73   308.84     4.23
Total            74   498.67
```

The fitted regression line is $\hat{y} = 23.3 + .596x$.

The estimated slope of the line is .596. For each additional inch of height for males, the mean ideal partner's height is estimated to increase by .596 inches.

f. By fitting separate lines for the males and females, the results now match the theory developed by the anthropologists.

9.31 Using MINITAB, the results are:

Regression Analysis: Mass versus Time

```
The regression equation is
Mass = 5.22 - 0.114 Time


Predictor       Coef  SE Coef        T      P
Constant      5.2207   0.2960    17.64  0.000
Time        -0.11402  0.01032   -11.05  0.000


S = 0.857257    R-Sq = 85.3%    R-Sq(adj) = 84.6%


Analysis of Variance

Source           DF        SS       MS       F       P
Regression        1    89.794   89.794  122.19   0.000
Residual Error   21    15.433    0.735
Total            22   105.227
```

The fitted regression line is $\hat{y} = 5.2207 - .11402x$. Since the estimate of the slope is negative, the data indicate that the mass of the spill tends to diminish as time increases. For each additional minute, the mean mass is estimated to decrease by .11402 pounds.

9.33   See Figure 9.7 in the text.

9.35   a. $\text{SSE} = \text{SS}_{yy} - \hat{\beta}_1 \text{SS}_{xy} = 95 - .75(50) = 57.5$ $s^2 = \dfrac{\text{SSE}}{n-2} = \dfrac{57.5}{20-2} = 3.19444$

    b. $\text{SS}_{yy} = \sum y^2 - \dfrac{\left(\sum y\right)^2}{n} = 860 - \dfrac{50^2}{40} = 797.5$

    $\text{SSE} = \text{SS}_{yy} - \hat{\beta}_1 \text{SS}_{xy} = 797.5 - .2(2700) = 257.5$

    $s^2 = \dfrac{\text{SSE}}{n-2} = \dfrac{257.5}{40-2} = 6.776315789 \approx 6.7763$

    c. $\text{SS}_{yy} = \sum (y_i - \bar{y})^2 = 58$ $\hat{\beta}_1 = \dfrac{\text{SS}_{xy}}{\text{SS}_{xx}} = \dfrac{91}{170} = .535294117$

    $\text{SSE} = \text{SS}_{yy} - \hat{\beta}_1 \text{SS}_{xy} = 58 - .535294117(91) = 9.2882353 \approx 9.288$

    $s^2 = \dfrac{\text{SSE}}{n-2} = \dfrac{9.2882353}{10-2} = 1.161029413 \approx 1.1610$

9.37 $\text{SSE} = \text{SS}_{yy} - \hat{\beta}_1 \text{SS}_{xy}$

    $\text{where } \text{SS}_{yy} = \sum y_i^2 - \dfrac{\left(\sum y_i\right)^2}{n}$

For Exercise 9.14,
    $\sum y_i^2 = 159$ $\sum y_i = 31$

    $\text{SS}_{yy} = \sum y^2 - \dfrac{\left(\sum y\right)^2}{n} = 159 - \dfrac{31^2}{7} = 159 - 137.2857143 = 21.7142857$

    $\text{SS}_{xy} = -26.2857143$ $\hat{\beta}_1 = -.779661017$

Therefore, $\text{SSE} = 21.7142857 - (-.779661017)(-26.2857143) = 1.22033896 \approx 1.2203$

    $s^2 = \dfrac{\text{SSE}}{n-2} = \dfrac{1.22033896}{7-2} = .244067792, \quad s = \sqrt{.244067792} = .4940$

We would expect most of the observations to fall within $2s = 2(.4940) = .988$ units of the least squares prediction line.

For Exercise 9.17,

    $\sum x_i = 26$ $\sum y_i = 24$ $\sum x_i y_i = 129$ $\sum x_i^2 = 140$ $\sum y_i^2 = 124$

$$SS_{xy} = \sum x_i y_i - \frac{\left(\sum x_i \sum y_i\right)}{n} = 129 - \frac{(26)(24)}{7} = 129 - 89.14285714 = 39.85714286$$

$$SS_{xx} = \sum x_i^2 - \frac{\left(\sum x_i\right)^2}{n} = 140 - \frac{(26)^2}{7} = 140 - 96.57142857 = 43.42857143$$

$$SS_{yy} = \sum y_i^2 - \frac{\left(\sum y_i\right)^2}{n} = 124 - \frac{(24)^2}{7} = 124 - 82.28571429 = 41.71428571$$

$$\hat{\beta}_1 = \frac{SS_{xy}}{SS_{xx}} = \frac{39.85714286}{43.42857143} = .917763157$$

$$SSE = SS_{yy} - \hat{\beta}_1 SS_{xy} = 41.71428571 - (.917763157)(39.85714286)$$
$$= 41.71428571 - 36.57941726 = 5.13486841 \approx 5.1349$$

$$s^2 = \frac{SSE}{n-2} = \frac{5.13486841}{7-2} = 1.026973682 \qquad s = \sqrt{1.026973682} = 1.0134$$

We would expect most of the observations to fall within $2s = 2(1.0134) = 2.0268$ units of the least squares prediction line.

9.39   a.   From the printout:

$$SSE = 22.268, \ s^2 = MSE = 5.567, \text{ and } s = \sqrt{5.567} = 2.3594 .$$

b.   We would expect most of the observations to fall within $2s = 2(2.3594) = 4.7188$ mm of the least squares prediction line.

9.41   a.   From the printout, SSE = 2760, $s^2 = 307.0$, and $s = 17.5111$.

b.   We would expect most of the observations to be within $2s = 2(17.5111) = 35.0222$ of the least squares line.

9.43   a.   From Exercise 9.26,

$$\sum y_i = 3,781.1 \qquad \sum y_i^2 = 651,612.45 \qquad SS_{xy} = -3,882.3686 \qquad \hat{\beta}_1 = -.305444503$$

$$SS_{yy} = \sum y_i^2 - \frac{\left(\sum y_i\right)^2}{n} = 651,612.45 - \frac{3,781.1^2}{22} = 1,761.6677$$

$$SSE = SS_{yy} - \hat{\beta}_1 SS_{xy} = 1,761.6677 - (-.305444503)(-3,882.3686) = 575.8195525$$

$$s^2 = \frac{SSE}{n-2} = \frac{575.8195525}{22-2} = 28.79097763$$

$$s = \sqrt{s^2} = \sqrt{28.79099763} = 5.36572247$$

We would expect most of the observations to be within $2s = 2(5.3657) = 10.7314$ of the least squares line.

b.   From Exercise 9.26,

$$\sum y_i = 3,764.2 \qquad \sum y_i^2 = 645,221.16 \qquad SS_{xy} = -3,442.16 \qquad \hat{\beta}_1 = -.270811187$$

$$SS_{yy} = \sum y_i^2 - \frac{\left(\sum y_i\right)^2}{n} = 645,221.16 - \frac{3,764.2^2}{22} = 1,166.54$$

$$SSE = SS_{yy} - \hat{\beta}_1 SS_{xy} = 1,166.54 - (-.270811187)(-3,442.16) = 234.3645646$$

$$s^2 = \frac{SSE}{n-2} = \frac{234.3645646}{22-2} = 11.71822823$$

$$s = \sqrt{s^2} = \sqrt{11.71822823} = 3.423189774$$

We would expect most of the observations to be within $2s = 2(3.4232) = 6.8464$ of   the least squares line.

c.   Since the standard deviation ($s = 3.423$) for the reading scores is smaller than the standard deviation for the math scores ($s = 5.366$), the reading score can be more accurately predicted than the math score.

9.45   a.   Using MINITAB, the results of fitting the male data are:

Regression Analysis: M-IdealHt versus M-Height

```
The regression equation is
M-IdealHt = 23.3 + 0.596 M-Height


Predictor      Coef   SE Coef      T       P
Constant     23.274     6.333   3.67   0.000
M-Height    0.59630   0.08902   6.70   0.000


S = 2.05688    R-Sq = 38.1%    R-Sq(adj) = 37.2%


Analysis of Variance

Source           DF        SS       MS       F       P
Regression        1    189.82   189.82   44.87   0.000
Residual Error   73    308.84     4.23
Total            74    498.67
```

The fitted regression line is $\hat{y} = 23.3 + .596x$.

The estimate of σ is $s = 2.05688$.

b.   Using MINITAB, the results of fitting the female data are:

Regression Analysis: F-IdealHt versus F-Height

```
The regression equation is
F-IdealHt = 39.3 + 0.493 F-Height


Predictor      Coef  SE Coef      T      P
Constant     39.304    6.206   6.33  0.000
F-Height    0.49261  0.09602   5.13  0.000


S = 2.32153   R-Sq = 27.3%   R-Sq(adj) = 26.3%


Analysis of Variance

Source           DF       SS      MS      F      P
Regression        1   141.84  141.84  26.32  0.000
Residual Error   70   377.27    5.39
Total            71   519.11
```

The fitted regression line is $\hat{y} = 39.3 + .493x$.

The estimate of σ is $s = 2.32153$.

c.   The student's height is a more accurate predictor of ideal partner's height for the males than for the females because the estimate of the standard deviation for the males is smaller than the estimate for the females.

9.47   In the equation, $E(y) = \beta_0 + \beta_1 x$, the value of β_1 is 0 if x has no linear relationship with y.

9.49   If you are running a one-tailed test of β_1 in simple linear regression, you divide the p-value from the computer printout by 2 to obtain the correct p-value of the one-tailed test.

9.51   a.   For confidence coefficient .95, $\alpha = 1 - .95 = .05$ and $\alpha / 2 = .05 / 2 = .025$. From Table IV, Appendix A, with df $= n - 2 = 12 - 2 = 10$, $t_{.025} = 2.228$.

The 95% confidence interval for β_1 is:

$$\hat{\beta}_1 \pm t_{.025} s_{\hat{\beta}_1} \Rightarrow 31 \pm 2.228(.5071) \Rightarrow 31 \pm 1.13 \Rightarrow (29.87, 32.13)$$

where $s_{\hat{\beta}_1} = \dfrac{s}{\sqrt{SS_{xx}}} = \dfrac{3}{\sqrt{35}} = .5071$

For confidence coefficient .90, $\alpha = 1 - .90 = .10$ and $\alpha / 2 = .10 / 2 = .05$. From Table IV, Appendix A, with df $= 10$, $t_{.05} = 1.812$.

The 90% confidence interval for β_1 is:

$$\hat{\beta}_1 \pm t_{.05} s_{\hat{\beta}_1} \Rightarrow 31 \pm 1.812(.5071) \Rightarrow 31 \pm .92 \Rightarrow (30.08, 31.92)$$

b.　　$s^2 = \dfrac{\text{SSE}}{n-2} = \dfrac{1960}{18-2} = 122.5 , \qquad s = \sqrt{s^2} = 11.0680$

For confidence coefficient, .95, $\alpha = 1 - .95 = .05$ and $\alpha/2 = .05/2 = .025$. From Table IV, Appendix A, with df $= n - 2 = 18 - 2 = 16$, $t_{.025} = 2.120$. The 95% confidence interval for β_1 is:

$$\hat{\beta}_1 \pm t_{.025} s_{\hat{\beta}_1} \Rightarrow 64 \pm 2.120(2.0207) \Rightarrow 64 \pm 4.28 \Rightarrow (59.72, 68.28)$$

where $s_{\hat{\beta}_1} = \dfrac{s}{\sqrt{\text{SS}_{xx}}} = \dfrac{11.0680}{\sqrt{30}} = 2.0207$

For confidence coefficient .90, $\alpha = 1 - .90 = .10$ and $\alpha/2 = .10/2 = .05$. From Table IV, Appendix A, with df $= 16$, $t_{.05} = 1.746$.

The 90% confidence interval for β_1 is:

$$\hat{\beta}_1 \pm t_{.05} s_{\hat{\beta}_1} \Rightarrow 64 \pm 1.746(2.0207) \Rightarrow 64 \pm 3.53 \Rightarrow (60.47, 67.53)$$

c.　　$s^2 = \dfrac{\text{SSE}}{n-2} = \dfrac{146}{24-2} = 6.6364 , \qquad s = \sqrt{s^2} = 2.5761$

For confidence coefficient .95, $\alpha = 1 - .95 = .05$ and $\alpha/2 = .05/2 = .025$. From Table IV, Appendix A, with df $= n - 2 = 24 - 2 = 22$, $t_{.025} = 2.074$. The 95% confidence interval for β_1 is:

$$\hat{\beta}_1 \pm t_{.025} s_{\hat{\beta}_1} \Rightarrow -8.4 \pm 2.074(.322) \Rightarrow -8.4 \pm .67 \Rightarrow (-9.07, -7.73)$$

where $s_{\hat{\beta}_1} = \dfrac{s}{\sqrt{\text{SS}_{xx}}} = \dfrac{2.5761}{\sqrt{64}} = .3220$

For confidence coefficient .90, $\alpha - 1 - .90 = .10$ and $\alpha/2 = .10/2 = .05$. From Table IV, Appendix A, with df $= 22$, $t_{.05} = 1.717$.

The 90% confidence interval for β_1 is:

$$\hat{\beta}_1 \pm t_{.05} s_{\hat{\beta}_1} \Rightarrow -8.4 \pm 1.717(.322) \Rightarrow -8.4 \pm .55 \Rightarrow (-8.95, -7.85)$$

9.53   a & c.   Using MINITAB, a scatterplot of the data is:

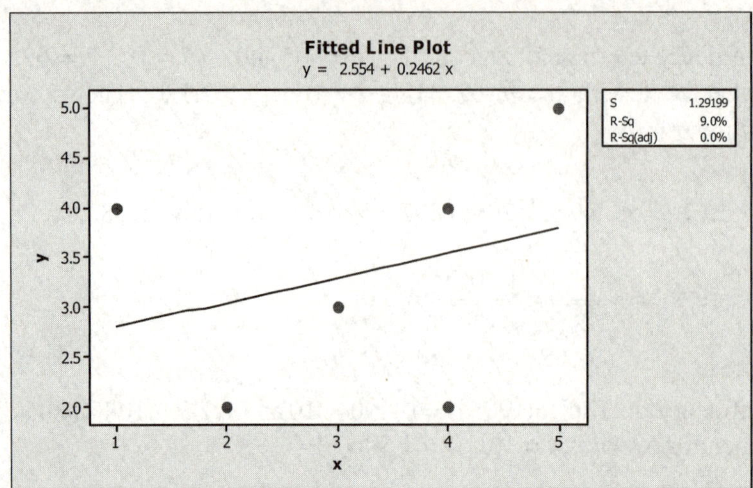

b.   Some preliminary calculations are:

$$\sum x_i = 19 \qquad \sum y_i = 20 \qquad \sum x_i y_i = 66$$

$$\sum x_i^2 = 71 \qquad \sum y_i^2 = 74$$

$$SS_{xy} = \sum x_i y_i - \frac{\sum x_i \sum y_i}{n} = 66 - \frac{19(20)}{6} = 2.666666667$$

$$SS_{xx} = \sum x_i^2 - \frac{\left(\sum x_i\right)^2}{n} = 71 - \frac{19^2}{6} = 10.833333333$$

$$\hat{\beta}_1 = \frac{SS_{xy}}{SS_{xx}} = \frac{2.66666667}{10.833333333} = .246153846$$

$$\hat{\beta}_0 = \bar{y} - \hat{\beta}_1\bar{x} = \frac{20}{6} - (.246153846)\left(\frac{19}{6}\right) = 2.553846155$$

The least squares prediction equation is $\hat{y} = 2.554 + .246x$

d.   Some preliminary calculations are:

$$SS_{yy} = \sum y_i^2 - \frac{\left(\sum y_i\right)^2}{n} = 74 - \frac{20^2}{6} = 7.3333333333$$

$$SSE = SS_{yy} - \hat{\beta}_1 SS_{xy} = 7.3333333333 - (.246153846)(2.666666667) = 6.6769$$

$$s^2 = \frac{\text{SSE}}{n-2} = \frac{6.6769}{6-2} = 1.6692 \qquad s = \sqrt{s^2} = \sqrt{1.6692} = 1.292$$

The test statistic is $t = \dfrac{\hat{\beta}_1 - 0}{s_{\hat{\beta}_1}} = \dfrac{.246}{\dfrac{1.292}{\sqrt{10.83333333}}} = .627$

e.   To determine if x and y are linearly related, we test:

H_0: $\beta_1 = 0$
H_a: $\beta_1 \neq 0$

The test statistic is $t = .627$.

The rejection region requires $\alpha / 2 = .01 / 2 = .005$ in each tail of the t distribution with df $= n - 2 = 6 - 2 = 4$. From Table IV, Appendix A, $t_{.005} = 4.604$. The rejection region is $t < -4.604$ or $t > 4.604$.

Since the observed value of the test statistic does not fall in the rejection region ($t = .627 \not> 4.604$), H_0 is not rejected. There is insufficient evidence to indicate that x and y are linearly related at $\alpha = .01$.

f.   For confidence coefficient .99, $\alpha = 1 - .99 = .01$ and $\alpha / 2 = .01 / 2 = .005$. From Table IV, Appendix A, with df $= n - 2 = 6 - 2 = 4$, $t_{.005} = 4.604$. The 99% confidence interval is:

$$\hat{\beta}_1 \pm t_{.005} s_{\hat{\beta}_1} \Rightarrow \hat{\beta}_1 \pm t_{.005} \frac{s}{\sqrt{\text{SS}_{xx}}} \Rightarrow .246 \pm 4.604 \frac{1.292}{\sqrt{10.8333333}}$$

$$\Rightarrow .246 \pm 1.807 \Rightarrow (-1.561,\ 2.053)$$

We are 99% confident that the true value of β_1 is between -1.561 and 2.053.

9.55   a.   Using MINITAB, a scatterplot of the data is:

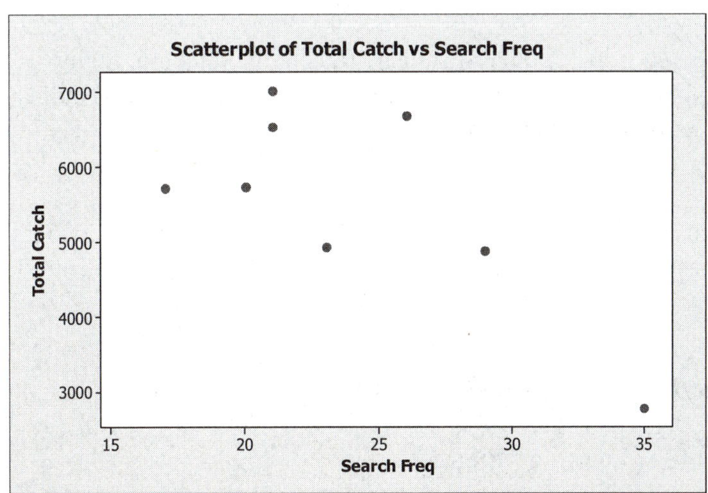

There appears to be a negative linear trend – as search frequency increases, the total catch tends to decrease.

b.   From the printout, the least squares prediction equation is: $\hat{y} = 9{,}658.24 - 171.573x$. For each additional unit increase in search frequency, the mean total catch is estimated to decrease by 171.573 kilograms.

c.   To determine if the total catch is negatively linearly related to search frequency, we test:

H_0: $\beta_1 = 0$
H_a: $\beta_1 < 0$

d.   From the printout, the p-value is $p = .0402/2 = .0201$.

e.   Since the p-value is less than α ($p = .0201 < .05$), H_0 is rejected. There is sufficient evidence to indicate that the total catch is negatively linearly related to search frequency at $\alpha = .05$.

9.57   a.   From Exercises 9.26 and 9.43, $SS_{xx} = 12{,}710.55318$, $\hat{\beta}_1 = -.305$, and $s = 5.3657$,

To determine if math score and percentage of students below the poverty level are negatively linearly related, we test:

H_0: $\beta_1 = 0$
H_a: $\beta_1 < 0$

The test statistic is $t = \dfrac{\hat{\beta}_1 - 0}{s_{\hat{\beta}_1}} = \dfrac{-.305}{\dfrac{5.3657}{\sqrt{12{,}710.55318}}} = -6.41$

The rejection region requires $\alpha = .01$ in the lower tail of the t distribution with df $= n - 2 = 22 - 2 = 20$. From Table IV, Appendix A, $t_{.01} = 2.582$. The rejection region is $t < -2.582$.

Since the observed value of the test statistic falls in the rejection region ($t = -6.41 < -2.582$), H_0 is rejected. There is sufficient evidence to indicate the math score and percentage of students below the poverty level are negatively linearly related at $\alpha = .01$.

b.   For confidence coefficient .99, $\alpha = 1 - .99 = .01$ and $\alpha/2 = .01/2 = .005$. From Table IV, Appendix A, with df $= n - 2 = 22 - 2 = 20$, $t_{.005} = 2.845$. The 99% confidence interval is:

$$\hat{\beta}_1 \pm t_{.005} s_{\hat{\beta}_1} \Rightarrow \hat{\beta}_1 \pm t_{.005} \dfrac{s}{\sqrt{SS_{xx}}} \Rightarrow -.305 \pm 2.845 \dfrac{5.3657}{\sqrt{12{,}710.55318}}$$

$$\Rightarrow -.305 \pm .135 \Rightarrow (-.440,\ -.170)$$

We are 99% confident that the change in the mean value of math scores for each unit change in percentage of students below the poverty level is between $-.440$ and $-.170$.

9.59 Some preliminary calculations are:

$$SS_{yy} = \sum y_i^2 - \frac{\left(\sum y_i\right)^2}{n} = 769.72 - \frac{(135.8)^2}{24} = 1.3183333$$

$$SSE = SS_{yy} - \hat{\beta}_1 SS_{xy} = 1.3183333 - (-.002310625)(-130.441667) = 1.016931523$$

$$s^2 = \frac{SSE}{n-2} = \frac{1.016931523}{22} = .046224 \qquad s = \sqrt{.046224} = .215$$

$$s_{\hat{\beta}_1} = \sqrt{\frac{s^2}{SS_{xx}}} = \sqrt{\frac{.046224}{56,452.958}} = .000905$$

For confidence level .95, $\alpha = .05$ and $\alpha/2 = .05/2 = .025$. From Table IV, Appendix A with $df = n - 2 = 24 - 2 = 22$, $t_{.025} = 2.074$.

The confidence interval is:

$$\hat{\beta}_1 \pm t_{.025} s_{\hat{\beta}_1} \Rightarrow -.0023 \pm 2.074(.000905) \Rightarrow -.0023 \pm .0019 \Rightarrow (-.0042, -.0004)$$

We are 95% confident that the change in the mean sweetness index for each one unit change in the pectin is between $-.0042$ and $-.0004$.

9.61 a. The model relating Anthropogenic Index, y, to Natural Origin Index, x, is:

$$y = \beta_0 + \beta_1 x + \varepsilon$$

b. Some preliminary calculations are:

$$\sum x_i = 1,047.22 \qquad \sum y_i = 1,283.46 \qquad \sum x_i y_i = 32,048.2$$

$$\sum x_i^2 = 24,607.5 \qquad \sum y_i^2 = 54,184.8$$

$$SS_{xy} = \sum x_i y_i - \frac{\sum x_i \sum y_i}{n} = 32,048.2 - \frac{1,047.22(1,283.46)}{54} = 7,158.10776$$

$$SS_{xx} = \sum x_i^2 - \frac{\left(\sum x_i\right)^2}{n} = 24,607.5 - \frac{1,047.22^2}{54} = 4,298.80133$$

$$\hat{\beta}_1 = \frac{SS_{xy}}{SS_{xx}} = \frac{7,158.10776}{4,298.80133} = 1.665140399$$

$$\hat{\beta}_0 = \bar{y} - \hat{\beta}_1\bar{x} = \frac{1,283.46}{54} - (1.665140399)\left(\frac{1,047.22}{54}\right) = -8.524228315$$

The least squares prediction equation is $\hat{y} = -8.524 + 1.665x$

c. $\hat{\beta}_1 = 1.665$. For each unit increase in the Natural Origin Index, the mean Anthropogenic Index is estimated to increase by 1.665.

$\hat{\beta}_0 = 2.522$. Since $x = 0$ is not in the observed range of the Natural Origin Index, $\hat{\beta}_0$ has no interpretation other than the y-intercept.

d.   Some preliminary calculations are:

$$SS_{yy} = \sum y_i^2 - \frac{\left(\sum y_i\right)^2}{n} = 54,184.8 - \frac{1,283.46^2}{54} = 23,679.80793$$

$$SSE = SS_{yy} - \hat{\beta}_1 SS_{xy} = 23,679.80793 - (1.665140399)(7,158.10776) = 11,760.55352$$

$$s^2 = \frac{SSE}{n-2} = \frac{11,760.55352}{54-2} = 226.1644908$$

$$s = \sqrt{s^2} = \sqrt{226.1644908} = 15.03876627$$

To determine if the natural origin index and anthropogenic index are positively linearly related, we test:

H_0: $\beta_1 = 0$
H_a: $\beta_1 > 0$

The test statistic is $t = \dfrac{\hat{\beta}_1 - 0}{s_{\hat{\beta}_1}} = \dfrac{1.665}{\dfrac{15.0388}{\sqrt{4,298.80133}}} = 7.259$

The rejection region requires $\alpha = .05$ in the upper tail of the t distribution with df $= n - 2$ $= 54 - 2 = 52$. From Table IV, Appendix A, $t_{.05} \approx 1.678$. The rejection region is $t > 1.678$.

Since the observed value of the test statistic falls in the rejection region $(t = 7.259 > 1.678)$, H_0 is rejected. There is sufficient evidence to indicate the natural origin index and anthropogenic index are positively linearly related at $\alpha = .05$.

e.   For confidence coefficient .95, $\alpha = 1 - .95 = .05$ and $\alpha / 2 = .05 / 2 = .025$. From Table IV, Appendix A, with df $= n - 2 = 54 - 2 = 52$, $t_{.025} \approx 2.011$. The 95% confidence interval is:

$$\hat{\beta}_1 \pm t_{.025} s_{\hat{\beta}_1} \Rightarrow \hat{\beta}_1 \pm t_{.025} \frac{s}{\sqrt{SS_{xx}}} \Rightarrow 1.665 \pm 2.011 \frac{15.0688}{\sqrt{4,298.80133}}$$

$$\Rightarrow 1.665 \pm .461 \Rightarrow (1.204, \quad 2.126)$$

We are 95% confident that the change in the anthropogenic index for each unit increase in the natural origin index is between 1.204 and 2.126.

9.63   a.   To determine if body-plus-head rotation and active head movement are positively linearly related, we test:

$$H_0: \ \beta_1 = 0$$
$$H_a: \ \beta_1 > 0$$

The test statistic is $t = \dfrac{\hat{\beta}_1 - 0}{s_{\hat{\beta}_1}} = \dfrac{.88 - 0}{.14} = 6.286$

The rejection region requires $\alpha = .05$ in the upper tail of the t distribution with df $= n - 2 = 39 - 2 = 37$. From Table IV, Appendix A, $t_{.05} \approx 1.687$. The rejection region is $t > 1.687$.

Since the observed value of the test statistic falls in the rejection region ($t = 6.286 > 1.687$), H_0 is rejected. There is sufficient evidence to indicate that body-plus-head rotation and active head movement are positively linearly related at $\alpha = .05$.

    b.   For confidence level .90, $\alpha = .10$ and $\alpha / 2 = .10 / 2 = .05$. From Table IV, Appendix A, with df $= n - 2 = 39 - 2 = 37$, $t_{.05} \approx 1.687$. The confidence interval is:

$$\hat{\beta}_1 \pm s_{\hat{\beta}_1} \Rightarrow .88 \pm 1.687(.14) \Rightarrow .88 \pm .2362 \Rightarrow (.6438, 1.1162)$$

We are 90% confident that the true value of β_1 is between .6438 and 1.1162.

    c.   Because the interval in part b contains the value 1, there is no evidence that the true slope of the line differs from 1.

9.65   a.   From the printout, $\hat{\beta}_o = .515$, $\hat{\beta}_1 = .000021$, and $s = .0370$.

    b.   To determine if there is a positive linear relationship between elevation and slugging percentage, we test:

$$H_0: \ \beta_1 = 0$$
$$H_a: \ \beta_1 > 0$$

From the printout, the test statistic is $t = 2.89$ and the p-value is $p = .008/2 = .004$.

Since the *p*-value is less than $\alpha = .01$ $(p = .004 < .01)$, H_o is rejected. There is sufficient evidence to indicate a positive linear relationship between elevation and slugging percentage at $\alpha = .01$.

c.   Using MINITAB, the scatterplot is:

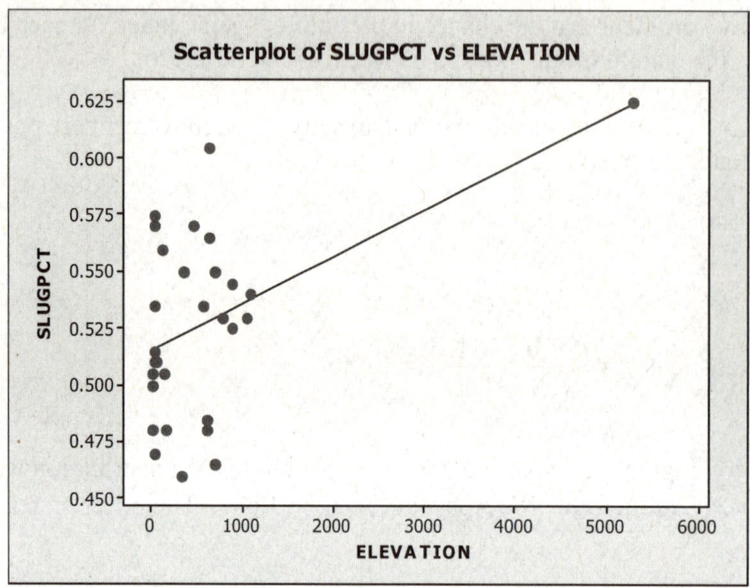

The data point corresponding to Denver is the data point furthest to the right. This point is very influential. If this point were deleted, there probably would not be a relationship between slugging percentage and elevation.

d.   Using MINITAB with Denver removed, the results are:

Regression Analysis: SLUGPCT2 versus ELEVATION2

```
The regression equation is
SLUGPCT2 = 0.515 + 0.000020 ELEVATION2


Predictor          Coef       SE Coef       T       P
Constant        0.51537       0.01066   48.33   0.000
ELEVATION2   0.00002012   0.00002034    0.99   0.332


S = 0.0376839    R-Sq = 3.6%    R-Sq(adj) = 0.0%


Analysis of Variance

Source             DF          SS         MS      F       P
Regression          1    0.001389   0.001389   0.98   0.332
Residual Error     26    0.036922   0.001420
Total              27    0.038311
```

From the printout, $\hat{\beta}_o = .515$, $\hat{\beta}_1 = .000020$, and $s = .0377$.

To determine if there is a positive linear relationship between elevation and slugging percentage, we test:

H_0: $\beta_1 = 0$
H_a: $\beta_1 > 0$

From the printout, the test statistic is $t = 0.99$ and the p-value is $p = .332/2 = .166$.

Since the p-value is not less than $\alpha = .01$ ($p = .166 > .01$), H_0 is not rejected. There is insufficient evidence to indicate a positive linear relationship between elevation and slugging percentage when Denver is deleted using $\alpha = .01$.

The "thin air" theory appears to be valid. No other city has an elevation that is close to that of Denver's. If Denver is not included in the analysis, there is no relationship between the slugging percentage and the elevation. Without Denver, the elevations of the cities range from 10 feet to 1082 feet. The elevation of Denver is 5277. The slugging percentage at Denver stadium is also higher than any other city. Even if we changed the value of the elevation of Denver to 1500 feet, the linear relationship between slugging percentage and elevation is still statistically significant.

9.67   The statement "The correlation coefficient is a measure of the strength of the linear relationship between x and y." is a true statement.

9.69   a. $r = 1$ implies x and y have a perfect positive linear relationship.

b. $r = -1$ implies x and y have a perfect negative linear relationship.

c. $r = 0$ implies x and y are not linearly related.

d. $r = .90$ implies x and y have a positive linear relationship. Since r is close to 1, the strength of the relationship is very high.

e. $r = .10$ implies x and y have a positive linear relationship. Since r is close to 0, the relationship is very weak.

f. $r = -.88$ implies x and y have a negative linear relationship. Since r is close to -1, the relationship is fairly strong.

9.71   a.   Using MINITAB, the scatterplot of the data is:

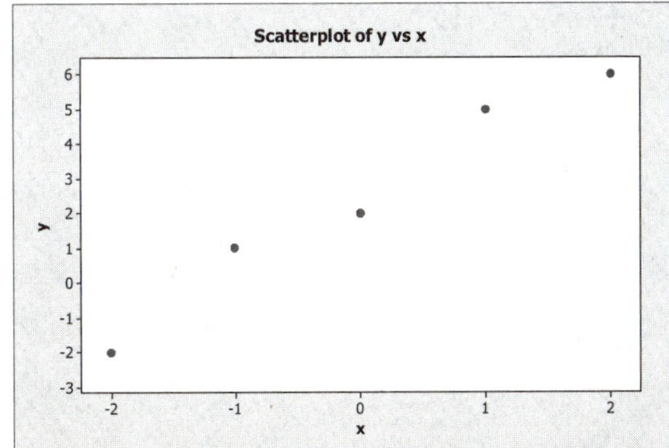

Some preliminary calculations are:

$$\sum x = 0 \qquad \sum x^2 = 10 \qquad \sum xy = 20 \qquad \sum y = 12 \qquad \sum y^2 = 70$$

$$SS_{xy} = \sum xy - \frac{\sum x \sum y}{n} = 20 - \frac{0(12)}{5} = 20$$

$$SS_{xx} = \sum x^2 - \frac{\left(\sum x\right)^2}{n} = 10 - \frac{0^2}{5} = 10$$

$$SS_{yy} = \sum y^2 - \frac{\left(\sum y\right)^2}{n} = 70 - \frac{12^2}{5} = 41.2$$

$$r = \frac{SS_{xy}}{\sqrt{SS_{xx}SS_{yy}}} = \frac{20}{\sqrt{10(41.2)}} = .9853 \qquad r^2 = .9853^2 = .9709$$

b.   Using MINITAB, the scatterplot of the data is:

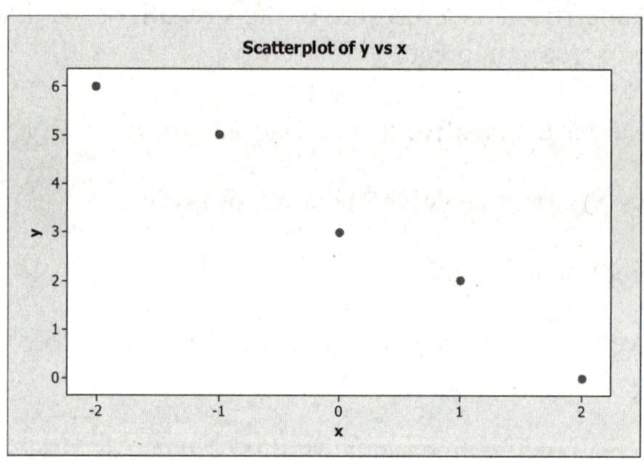

Some preliminary calculations are:

$$\sum x = 0 \qquad \sum x^2 = 10 \qquad \sum xy = -15 \qquad \sum y = 16 \qquad \sum y^2 = 74$$

$$SS_{xy} = \sum xy - \frac{\sum x \sum y}{n} = -15 - \frac{0(16)}{5} = -15$$

$$SS_{xx} = \sum x^2 - \frac{\left(\sum x\right)^2}{n} = 10 - \frac{0^2}{5} = 10$$

$$SS_{yy} = \sum y^2 - \frac{\left(\sum y\right)^2}{n} = 74 - \frac{16^2}{5} = 22.8$$

$$r = \frac{SS_{xy}}{\sqrt{SS_{xx}SS_{yy}}} = \frac{-15}{\sqrt{10(22.8)}} = -.9934 \qquad r^2 = (-.9934)^2 = .9868$$

c. Using MINITAB, the scatterplot is:

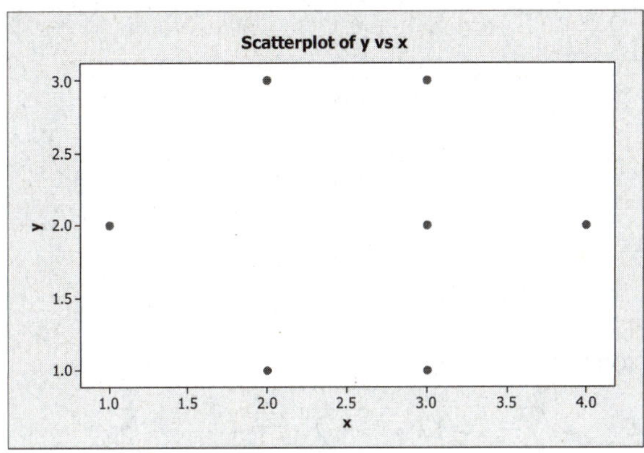

Some preliminary calculations are:

$$\sum x = 18 \qquad \sum x^2 = 52 \qquad \sum xy = 36 \qquad \sum y = 14 \qquad \sum y^2 = 32$$

$$SS_{xy} = \sum xy - \frac{\sum x \sum y}{n} = 36 - \frac{18(14)}{7} = 0$$

$$SS_{xx} = \sum x^2 - \frac{\left(\sum x\right)^2}{n} = 52 - \frac{18^2}{7} = 5.71428571$$

$$SS_{yy} = \sum y^2 - \frac{\left(\sum y\right)^2}{n} = 32 - \frac{14^2}{7} = 4$$

$$r = \frac{SS_{xy}}{\sqrt{SS_{xx}SS_{yy}}} = \frac{0}{\sqrt{5.71428571(4)}} = 0 \qquad r^2 = 0^2 = 0$$

d.   Using MINITAB, the scatterplot of the data is:

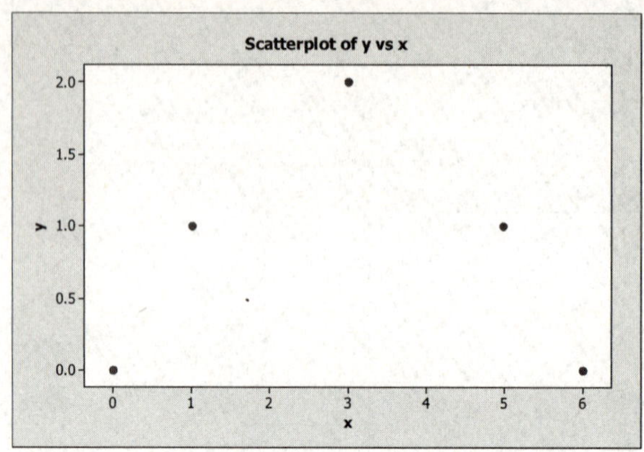

Some preliminary calculations are:

$$\sum x = 15 \qquad \sum x^2 = 71 \qquad \sum xy = 12 \qquad \sum y = 4 \qquad \sum y^2 = 6$$

$$SS_{xy} = \sum xy - \frac{\sum x \sum y}{n} = 12 - \frac{15(4)}{5} = 0$$

$$SS_{xx} = \sum x^2 - \frac{\left(\sum x\right)^2}{n} = 71 - \frac{15^2}{5} = 26$$

$$SS_{yy} = \sum y^2 - \frac{\left(\sum y\right)^2}{n} = 6 - \frac{4^2}{5} = 2.8$$

$$r = \frac{SS_{xy}}{\sqrt{SS_{xx}SS_{yy}}} = \frac{0}{\sqrt{26(2.8)}} = 0 \qquad r^2 = 0^2 = 0$$

9.73   From Exercises 9.17 and 9.37,

$$r^2 = 1 - \frac{SSE}{SS_{yy}} = 1 - \frac{5.13486841}{41.7142857} = .8769$$

87.69% of the total sample variability around $\bar{y}$ is explained by the linear relationship between y and x.

9.75   a. $r^2 = .18$. 18% of the sample variability around the sample mean number of points scored is explained by the linear relationship between the number of points scored and number of yards from the opposing goal line.

b. $r = -\sqrt{.18} = -.424$. The value of r is negative because the estimated slope of the regression line is negative.

9.77   a.   The correlation coefficient of $r = .50$ indicates that the strength of the positive linear relationship between baseline and follow-up physical activity among the obese adults is moderate. Since the p-value is not less than α $(p = .07 \not< .05)$, there is no evidence to indicate correlation coefficient differs from 0 at $\alpha = .05$.

   b.   Using MINITAB, a possible scatterplot with 13 data points that would yield a value of $r = .50$ is:

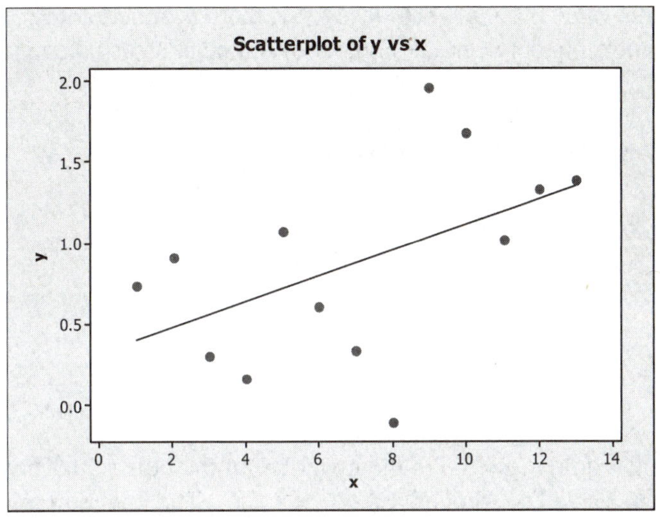

   c.   Since the p-value is not small $(p = .66)$ there is no evidence to indicate a linear relationship between baseline and follow-up physical activity among the normal adults. The correlation coefficient of $r = -.12$ indicates that the strength of the negative linear relationship is extremely weak.

   d.   Using MINITAB, a possible scatterplot with 15 data points that would yield a value of $r = -.12$ is:

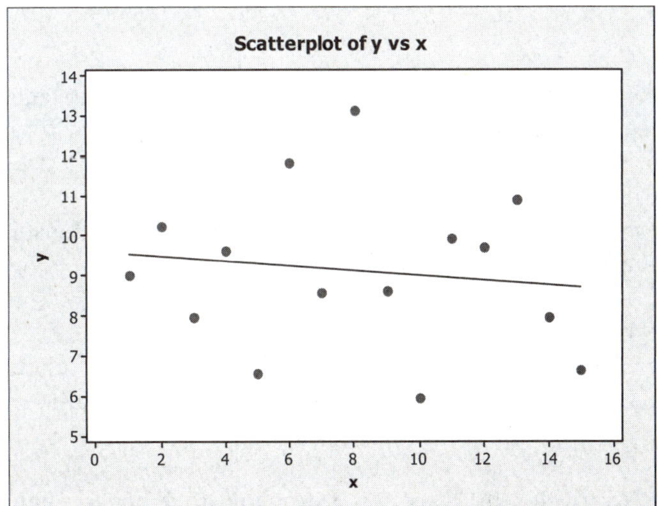

9.79   a. $r = .43$. Since the value is not particularly close to 1, there is a moderately weak positive linear relationship between time allotted to sports and audience ratings.

b. $r^2 = .43^2 = .1849$. 18.49% of the total sample variability around the sample mean audience rating is explained by the linear relationship between audience rating and total time allotted to sports.

9.81   a. $r = -.726$. Since this value is somewhat close to -1, there is a moderately strong negative linear relationship between the number of online courses taken and the weekly quiz grade.

b.   To determine if a negative correlation exists, we test:

$$H_0: \ \rho = 0$$
$$H_a: \ \rho < 0$$

The test statistic is $t = \dfrac{r\sqrt{n-2}}{\sqrt{1-r^2}} = \dfrac{-.726\sqrt{24-2}}{\sqrt{1-(-.726)^2}} = -4.95$.

The rejection region requires $\alpha = .05$ in the lower tail of the t distribution with df $= n - 2 = 24 - 2 = 22$. From Table IV, Appendix A, $t_{.05} = 1.717$. The rejection region is $t < -1.717$.

Since the observed value of the test statistic falls in the rejection region ($t = -4.95 < -1.717$), H_o is rejected. There is sufficient evidence to indicate a negative linear relationship between the number of online courses taken and the weekly quiz grade.

9.83   a.   **Piano**: $r = .447$
Because this value is near .5, there is s slight positive linear relationship between recognition exposure time and goodness of view for piano.

Bench: $r = -.057$
Because this value is extremely close to 0, there is an extremely weak negative linear relationship between recognition exposure time and goodness of view for bench.

Motorbike: $r = .619$
Because this value is near .5, there is a moderate positive linear relationship between recognition exposure time and goodness of view for motorbike.

Armchair: $r = .294$
Because this value is fairly close to 0, there is a weak positive linear relationship between recognition exposure time and goodness of view for armchair.

Teapot: $r = .949$
Because this value is very close to 1, there is a strong positive linear relationship between recognition exposure time and goodness of view for teapot.

b.   **Piano**: $r^2 = (.447)^2 = .1998$
19.98% of the total sample variability around the sample mean recognition exposure time is explained by the linear relationship between the recognition exposure time and the goodness of view for piano.

Bench: $r^2 = (-.057)^2 = .0032$
.32% of the total sample variability around the sample mean recognition exposure time is explained by the linear relationship between the recognition exposure time and the goodness of view for bench.

Motorbike: $r^2 = (.619)^2 = .3832$
38.32% of the total sample variability around the sample mean recognition exposure time is explained by the linear relationship between the recognition exposure time and the goodness of view for motorbike.

Armchair: $r^2 = (.294)^2 = .0864$
8.64% of the total sample variability around the sample mean recognition exposure time is explained by the linear relationship between the recognition exposure time and the goodness of view for armchair.

Teapot: $r^2 = (.949)^2 = .9006$
90.06% of the total sample variability around the sample mean recognition exposure time is explained by the linear relationship between the recognition exposure time and the goodness of view for teapot.

c.   The test is:

$$H_0:\ \beta_1 = 0$$
$$H_a:\ \beta_1 \neq 0$$

The following are the values of α and $t_{\alpha/2}$ that correspond to df $= n - 2 = 25 - 2 = 23$ from Table IV, Appendix A.

α	.20	.10	.05	.02	.01	.002	.001
$t_{\alpha/2}$	1.319	1.714	2.069	2.500	2.807	3.485	3.767

Piano: $t = 2.40$
$2.069 < 2.40 < 2.500$. Thus, $p \approx .025$
For levels of significance greater than $\alpha = .025$, H_0 can be rejected. There is sufficient evidence to indicate that there is a linear relationship between goodness of view and recognition exposure time for piano for $\alpha > .025$.

Bench: $t = .27$
$.27 < 1.319$. Thus, $p > .2$
H_0 is not rejected. There is insufficient evidence to indicate that there is a linear relationship between goodness of view and recognition exposure time for bench for $\alpha \leq .2$.

Motorbike: $t = 3.78$
$3.78 > 3.767$. Thus, $p < .001$

H_0 can be rejected for $\alpha \geq .001$. There is sufficient evidence to indicate that there is a linear relationship between goodness of view and recognition exposure time for motorbike for $\alpha \geq .001$.

Armchair: $t = 1.47$
$1.319 < 1.47 < 1.714$. Thus, $p \approx .15$
H_0 cannot be rejected for levels of significance $\alpha < .15$. There is insufficient evidence to indicate that there is a linear relationship between goodness of view and recognition exposure time for armchair for $\alpha < .15$.

Teapot: $t = 14.50$
$14.50 > 3.767$. Thus, $p < .001$
H_0 can be rejected for $\alpha \geq .001$. There is sufficient evidence to indicate that there is a linear relationship between goodness of view and recognition exposure time for teapot for $\alpha \geq .001$.

9.85   a.   To determine whether average payoff and punishment use are negatively correlated, we test:

$$H_0: \ \beta_1 = 0$$
$$H_a: \ \beta_1 < 0$$

b.   Since the p-value is very small ($p = .001$), we would reject H_0 for any value of α greater than .001. There is sufficient evidence to indicate average payoff and punishment use are negatively correlated for $\alpha > .001$.

c.   No. Just because two variables are significantly correlated does not mean that one variable causes the other. Something else could be causing both of the variables to react in certain ways.

9.87   Using the values computed in Exercise 9.60:

$$r = \frac{SS_{xy}}{\sqrt{SS_{xx}SS_{yy}}} = \frac{48.125}{\sqrt{379.9375(18.75)}} = .5702$$

Because r is moderately small, there is a rather weak positive linear relationship between blood lactate concentration and perceived recovery.

$$r^2 = .5702^2 = .3251.$$

32.51% of the sample variance of blood lactate concentration around the sample mean is explained by the linear relationship between blood lactate concentration and perceived recovery.

9.89   For a given x, y is the actual value of the dependent variable and $E(y)$ is the mean value of y at the given value of x.

9.91   The statement "The greater the deviation between x and $\bar{x}$, the wider the prediction interval for y will be." is true. The further x gets from $\bar{x}$, the less precise the interval will be.

9.93   a. $\hat{\beta}_1 = \dfrac{SS_{xy}}{SS_{xx}} = \dfrac{28}{32} = .875$

$\hat{\beta}_0 = \bar{y} - \hat{\beta}_1 \bar{x} = 4 - .875(3) = 1.375$

The least squares line is $\hat{y} = 1.375 + .875x$.

b.   The least squares line is:

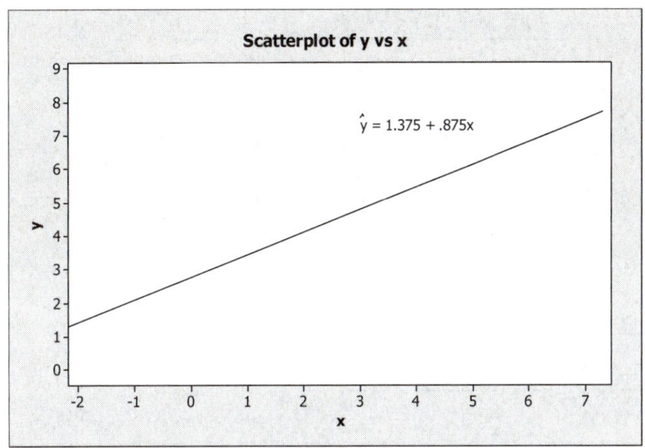

c. $SSE = SS_{yy} - \hat{\beta}_1 SS_{xy} = 26 - .875(28) = 1.5$

d. $s^2 = \dfrac{SSE}{n-2} = \dfrac{1.5}{10-2} = .1875$

e. $s = \sqrt{.1875} = .4330$

The form of the confidence interval is $\hat{y} \pm t_{\alpha/2}\, s \sqrt{\dfrac{1}{n} + \dfrac{(x_p - \bar{x})^2}{SS_{xx}}}$

For $x_p = 2.5$, $\hat{y} = 1.375 + .875(2.5) = 3.5625$

For confidence coefficient .95, $\alpha = .05$ and $\alpha/2 = .05/2 = .025$. From Table IV, Appendix A, with df $= n - 2 = 10 - 2 = 8$, $t_{.025} = 2.306$. The confidence interval is:

$$3.5625 \pm 2.306(.4330)\sqrt{\dfrac{1}{10} + \dfrac{(2.5-3)^2}{32}} \Rightarrow 3.5625 \pm .3279 \Rightarrow (3.2346,\ 3.8904)$$

f.   The form of the prediction interval is $\hat{y} \pm t_{\alpha/2}\, s \sqrt{1 + \dfrac{1}{n} + \dfrac{(x_p - \bar{x})^2}{SS_{xx}}}$

For $x_p = 4$, $\hat{y} = 1.375 + .875(4) = 4.875$

For confidence coefficient .95, $\alpha = .05$ and $\alpha / 2 = .05 / 2 = .025$. From Table IV, Appendix A, with df $= n - 2 = 10 - 2 = 8$, $t_{.025} = 2.306$. The prediction interval is:

$$4.875 \pm 2.306(.4330)\sqrt{1 + \frac{1}{10} + \frac{(4-3)^2}{32}} \Rightarrow 4.875 \pm 1.062 \Rightarrow (3.813, \, 5.937)$$

9.95   a, b. The scattergram is:

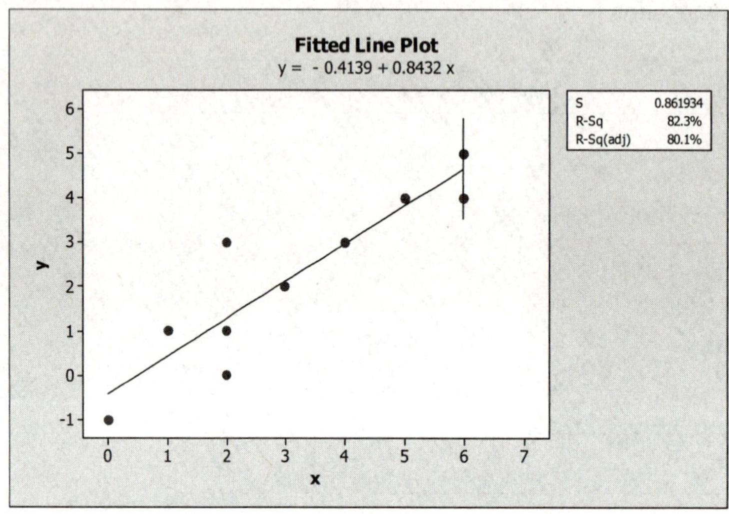

c. $\text{SSE} = \text{SS}_{yy} - \hat{\beta}_1 \text{SS}_{xy} = 33.6 - .843(32.8) = 5.9496$

$$s^2 = \frac{\text{SSE}}{n-2} = \frac{5.9496}{10-2} = .7437 \qquad s = \sqrt{.7437} = .8623$$

$$\bar{x} = \frac{31}{10} = 3.1$$

The form of the confidence interval is $\hat{y} \pm t_{\alpha/2} \, s \sqrt{\dfrac{1}{n} + \dfrac{(x_p - \bar{x})^2}{\text{SS}_{xx}}}$

For $x_p = 6$, $\hat{y} = -.414 + .843(6) = 4.644$

For confidence coefficient .95, $\alpha = .05$ and $\alpha / 2 = .05 / 2 = .025$. From Table IV, Appendix A, with df $= n - 2 = 10 - 2 = 8$, $t_{.025} = 2.306$. The confidence interval is:

$$4.644 \pm 2.306(.8623)\sqrt{\frac{1}{10} + \frac{(6-3.1)^2}{38.9}} \Rightarrow 4.644 \pm 1.118 \Rightarrow (3.526, \, 5.762)$$

d.   For $x_p = 3.2$, $\hat{y} = -.414 + .843(3.2) = 2.284$

The confidence interval is:

$$2.284 \pm 2.306(.8623)\sqrt{\frac{1}{10} + \frac{(3.2-3.1)^2}{38.9}} \Rightarrow 2.284 \pm .630 \Rightarrow (1.654, \, 2.914)$$

For $x_p = 0$, $\hat{y} = -.414 + .843(0) = -.414$

The confidence interval is:

$$-.414 \pm 2.306(.8623)\sqrt{\frac{1}{10} + \frac{(0-3.1)^2}{38.9}} \Rightarrow -.414 \pm 1.171 \Rightarrow (-1.585, .757)$$

e. The width of the confidence interval for the mean value of y depends on the distance x_p is from $\bar{x}$. The width of the interval for $x_p = 3.2$ is the smallest because 3.2 is the closest to $\bar{x} = 3.1$. The width of the interval for $x_p = 0$ is the widest because 0 is the farthest from $\bar{x} = 3.1$.

9.97 a. To predict the average payoff for a single player who used punishment 10 times, you would form a prediction interval for the individual payoff when punishment is 10.

b. To predict the mean of the average payoff for all players who used punishment 10 times, you would form a confidence interval for the mean payoff when punishment is 10.

9.99 For observation 7, the 95% prediction interval for the actual annual rainfall when the maximum daily temperature is 11.4 is (92.298, 125.104). We are 95% confident that the actual rainfall in a site with a maximum daily temperature of 11.4 is between 92.298 and 125.104.

9.101 Answers may vary. One possible answer is:

The 90% confidence interval for $x = 220.00$ is (5.64898, 5.83848). We are 90% confident that the mean sweetness index of all orange juice samples will be between 5.64898 and 5.83848 parts per million when the pectin value is 220.00.

9.103 a. From Exercise 9.29 d, $\hat{y} = 39.304 + .493x$ and $s = 2.32153$. From Exercise 9.58 b, $SS_{xx} = 584.6528$.

For $x = 66$, $\hat{y} = 39.304 + .493(66) = 71.842$.

For confidence coefficient .95, $\alpha = .05$ and $\alpha / 2 = .05 / 2 = .025$. From Table IV, Appendix A, with $df = n - 2 = 72 - 2 = 70$, $t_{.025} \approx 2.00$. The 95% prediction interval is:

$$\hat{y} \pm t_{\alpha/2} s \sqrt{1 + \frac{1}{n} + \frac{(x_p - \bar{x})^2}{SS_{xx}}} \Rightarrow 71.842 \pm 2.00(2.32153)\sqrt{1 + \frac{1}{72} + \frac{(66 - 64.5694)^2}{584.6528}}$$

$$\Rightarrow 71.842 \pm 4.683 \Rightarrow (67.159, 76.525)$$

We are 95% confident that the actual ideal partner's height is between 67.159 and 76.525 inches when a female's height is 66 inches.

b. From Exercise 9.29 e, $\hat{y} = 23.274 + .596x$ and $s = 2.05688$. From Exercise 9.58 a, $SS_{xx} = 534.3467$.

For $x = 66$, $\hat{y} = 23.274 + .596(66) = 62.61$.

For confidence coefficient .95, $\alpha = .05$ and $\alpha / 2 = .05 / 2 = .025$. From Table IV, Appendix A, with $df = n - 2 = 75 - 2 = 73$, $t_{.025} \approx 2.00$. The 95% prediction interval is:

$$\hat{y} \pm t_{\alpha/2} s \sqrt{1 + \frac{1}{n} + \frac{\left(x_p - \overline{x}\right)^2}{SS_{xx}}} \Rightarrow 62.61 \pm 2.00(2.05688)\sqrt{1 + \frac{1}{75} + \frac{\left(66 - 71.0933\right)^2}{534.3467}}$$

$$\Rightarrow 62.61 \pm 4.239 \Rightarrow (58.371, 66.849)$$

We are 95% confident that the actual ideal partner's height is between 58.371 and 66.849 inches when a male's height is 66 inches.

c. The prediction for the ideal partner's height for males might be invalid because 66 is outside the range of observed values for male heights. The smallest observed value of male heights is 67 inches. We are not sure if the relationship between the two variables remains the same outside the observed range.

9.105   a. Using MINITAB, the output is:

```
Predicted Values for New Observations

New
Obs    Fit   SE Fit      99% CI          99% PI
 1   3.510   0.196   (2.955, 4.066)   (1.020, 6.000)


Values of Predictors for New Observations

New
Obs  Time
  1  15.0
```

From the printout, the 99% confidence interval for the mean mass of all spills with an elapsed time of 15 minutes is (2.955, 4.066). We are 99% confident that the mean mass of all spills with an elapsed time of 15 minutes is between 2.955 and 4.066.

b. From the printout, the 99% prediction interval for the actual mass of a spill with an elapsed time of 15 minutes is (1.020, 6.000). We are 99% confident that the actual mass of a spill with an elapsed time of 15 minutes is between 1.020 and 6.000.

c. The prediction interval in part b is larger than the confidence interval in part a. This will always be the case. The confidence interval gives a range of values for the mean of a distribution. The prediction interval gives a range of values for the actual value of a distribution. Once a mean is located the actual values can still vary around it. This is why the prediction interval is always wider than the confidence interval for the mean.

9.107   a. From Exercise 9.46, $SS_{xx} = 3000$ and $\overline{x} = 50$.

Also, for Brand A, $s = 1.211$; for Brand B, $s = .610$.

For Brand A, $\hat{y} = 6.62 - .0727(45) = 3.349$, while for Brand B, $\hat{y} = 9.31 - .1077(45) = 4.464$.

The degrees of freedom for both brands is $n - 2 = 15 - 2 = 13$. For confidence coefficient .90, (i.e., for all parts of this question), $\alpha = .10$ and $\alpha / 2 = .10 / 2 = .05$. From Table IV, Appendix A, with df = 13, $t_{.05} = 1.771$.

The form of both confidence intervals is $\hat{y} \pm t_{\alpha/2}\, s \sqrt{\dfrac{1}{n} + \dfrac{(x_p - \bar{x})^2}{SS_{xx}}}$

For Brand A, we obtain:

$$3.349 \pm 1.771(1.211)\sqrt{\frac{1}{15} + \frac{(45 - 50)^2}{3000}} \Rightarrow 3.349 \pm .587 \Rightarrow (2.762, 3.936)$$

For Brand B, we obtain:

$$4.464 \pm 1.771(.610)\sqrt{\frac{1}{15} + \frac{(45 - 50)^2}{3000}} \Rightarrow 4.464 \pm .296 \Rightarrow (4.168, 4.760)$$

The interval for Brand A is wider, caused by the larger value of s.

b.   The form of both prediction intervals is $\hat{y} \pm t_{\alpha/2}\, s \sqrt{1 + \dfrac{1}{n} + \dfrac{(x_p - \bar{x})^2}{SS_{xx}}}$

For Brand A, we obtain:

$$3.349 \pm 1.771(1.211)\sqrt{1 + \frac{1}{15} + \frac{(45 - 50)^2}{3000}} \Rightarrow 3.349 \pm 2.224 \Rightarrow (1.125, 5.573)$$

For Brand B, we obtain:

$$4.464 \pm 1.771(.610)\sqrt{1 + \frac{1}{15} + \frac{(45 + 50)^2}{3000}} \Rightarrow 4.464 \pm 1.120 \Rightarrow (3.344, 5.584)$$

Again, the interval is wider for Brand A, caused by the larger value of s. Each of these intervals is wider than its counterpart from part **a**, since, for the same x, a prediction interval for an individual y is always wider than a confidence interval for the mean of y. This is due to an individual observation having a greater variance than the variance of the mean of a set of observations.

c.   To obtain a confidence interval for the life of a brand A cutting tool that is operated at 100 meters per minute, we use:

$$\hat{y} \pm t_{\alpha/1} s \sqrt{1 + \frac{1}{n} + \frac{(x_p - \bar{x})}{SS_{xx}}}$$

For $x = 100$, $\hat{y} = 6.62 - .0727(100) = -.65$.

The degrees of freedom are $n - 2 = 15 - 2 = 13$. For confidence coefficient .95, $\alpha = .05$ and $\alpha / 2 = .05 / 2 = .025$. From Table IV, Appendix A, with df = 13, $t_{.025} = 2.160$.

Here, we obtain:

$$-.65 \pm 2.160(1.211)\sqrt{1+\frac{1}{15}+\frac{(100-50)^2}{3000}} \Rightarrow -.65 \pm 3.606 \Rightarrow (-4.256,\ 2.956)$$

The additional assumption would be that the straight line model fits the data well for the x's actually observed all the way up to the value under consideration, 100. Clearly from the estimated value of $-.65$, this is not true (usually, negative "useful lives" are not found).

9.109   Using MINITAB, a scatterplot of the data is:

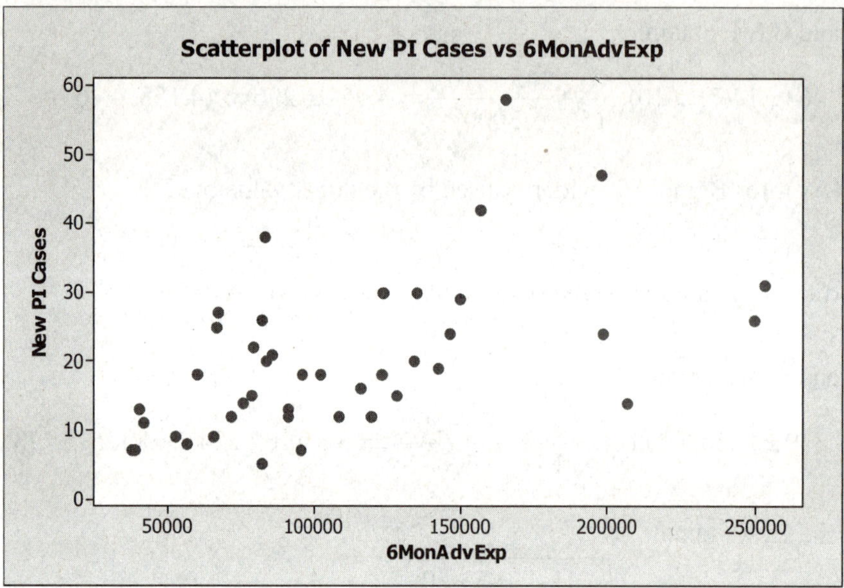

It appears that there is a positive linear relationship between the 6-month cumulative advertising expenditure and the number of new PI cases at the law firm.

The descriptive statistics for the two variables are:

Descriptive Statistics: New PI Cases, 6MonAdvExp

```
Variable         N    Mean   StDev  Minimum      Q1  Median       Q3  Maximum
New PI Cases    42   20.05   11.34     5.00   12.00   18.00    26.00    58.00
6MonAdvExp      42  108779   54104    37490   70293   93059   136611   253011
```

Using MINITAB, the results of fitting the regression line are:

Regression Analysis: New PI Cases versus 6MonAdvExp

```
The regression equation is
New PI Cases = 7.77 + 0.000113 6MonAdvExp


Predictor            Coef       SE Coef      T      P
Constant            7.767         3.385   2.29  0.027
6MonAdvExp      0.00011289    0.00002793   4.04  0.000


S = 9.67521    R-Sq = 29.0%    R-Sq(adj) = 27.2%


Analysis of Variance

Source            DF        SS        MS      F      P
Regression         1    1529.5    1529.5  16.34  0.000
Residual Error    40    3744.4      93.6
Total             41    5273.9
```

The fitted regression line is $\hat{y} = 7.77 + .000113x$.

To determine if there is a positive linear relationship between the 6-month cumulative advertising expenditures and the number of new PI cases, we test:

$$H_0:\ \beta_1 = 0$$
$$H_a:\ \beta_1 > 0$$

The test statistic is $t = 4.04$ and the p-value is $p = 0.000/2 = 0.000$. Since the p-value is so small, H_0 is rejected for any reasonable value of α. There is sufficient evidence to indicate a positive linear relationship between the 6-month cumulative advertising expenditure and the number of new PI cases. The more money spent on advertising, the greater the number of new PI cases.

From the printout, the value of r^2 is 29%. Thus, 29% of the total sample variability around the sample mean number of new PI cases is explained by the linear relationship between the number of new PI cases and the 6-month cumulative advertising expenditure. With only 29% of the variability explained, we know that there might be other factors not considered that can help explain the variability in the number of new PI cases.

9.111   The conditions required for a valid Spearman's test are:

1.   The sample of experimental units on which the two variables are measured is randomly selected.
2.   The probability distributions of the two variables are continuous.

9.113   a.   From Table XII with $n = 10$, $r_{s,\alpha} = r_{s,.025} = .648$. The rejection region is $r_s > .648$
or $r_s < -.648$.

b. From Table XII with $n = 20$, $r_{s,\alpha} = r_{s,.025} = .450$. The rejection region is $r_s > .450$.

c. From Table XII with $n = 30$, $r_{s,\alpha} = r_{s,.01} = .432$. The rejection region is $r_s < -.432$.

9.115 a. H_0: $\rho_s = 0$

H_a: $\rho_s \neq 0$

b. The test statistic is $r_s = \dfrac{SS_{uv}}{\sqrt{SS_{uu}SS_{vv}}}$

x	Rank, u	y	Rank, v	u^2	v^2	uv
0	3	0	1.5	9	2.25	45
3	5.5	2	5	30.25	25	27.5
0	3	2	5	9	25	15
−4	1	0	1.5	1	2.25	1.5
3	5.5	3	7	30.25	49	38.5
0	3	1	3	9	9	9
4	7	2	5	49	25	35
$\sum u = 28$		$\sum v = 28$		$\sum u^2 = 137.5$	$\sum v^2 = 137.5$	$\sum uv = 131$

$$SS_{uv} = \sum uv - \frac{\left(\sum u\right)\left(\sum v\right)}{n} = 131 - \frac{28(28)}{7} = 19$$

$$SS_{uu} = \sum u^2 - \frac{\left(\sum u\right)^2}{n} = 137.5 - \frac{(28)^2}{7} = 25.5$$

$$SS_{vv} = \sum v^2 - \frac{\left(\sum v\right)^2}{n} = 137.5 - \frac{(28)^2}{7} = 25.5$$

$$r_s = \frac{19}{\sqrt{25.5(25.5)}} = \frac{SS_{uv}}{\sqrt{SS_{uu}SS_{vv}}} = \frac{19}{\sqrt{25.5(25.5)}} = .745$$

Reject H_0 if $r_s < -r_{s,\alpha/2}$ or $r_s > r_{s,\alpha/2}$ where $\alpha / 2 = .025$ and $n = 7$:

Reject H_0 if $r_s < -.786$ or $r_s > .786$ (from Table XII, Appendix A).

Since the observed value of the test statistic does not fall in the rejection region, ($r_s = .745 \not> .786$), H_0 is not rejected. There is insufficient evidence to indicate x and y are correlated at $\alpha = .05$.

c. The p-value is $P(r_s \geq .745) + P(r_s \leq -.745)$. For $n = 7$, $r_s = .745$ is above $r_{s,.025}$ where $\alpha / 2 = .025$ and below $r_{s,.05}$ where $\alpha / 2 = .05$. Therefore, $2(.025) = .05 < p\text{-value} < 2(.05) = .10$.

d. The assumptions of the test are that the samples are randomly selected and the probability distributions of the two variables are continuous.

9.117   a, b.   The ranks and Spearman correlation coefficient are:

Porosity	Rank, u	Diameter	Rank, v	u^2	v^2	uv
18.8	5	12.0	5	25	25	25
18.3	4	9.7	3	16	9	12
16.3	2	7.3	2	4	4	4
6.9	1	5.3	1	1	1	1
17.1	3	10.9	4	9	16	12
20.4	6	16.8	6	36	36	36
	$\sum u = 21$		$\sum v = 21$	$\sum u^2 = 91$	$\sum v^2 = 91$	$\sum uv = 90$

$$SS_{uv} = \sum uv - \frac{\left(\sum u\right)\left(\sum v\right)}{n} = 90 - \frac{21(21)}{6} = 16.5$$

$$SS_{uu} = \sum u^2 - \frac{\left(\sum u\right)^2}{n} = 91 - \frac{21^2}{6} = 17.5$$

$$SS_{vv} = \sum v^2 - \frac{\left(\sum v\right)^2}{n} = 91 - \frac{21^2}{6} = 17.5$$

$$r_s = \frac{SS_{uv}}{\sqrt{SS_{uu} SS_{vv}}} = \frac{16.5}{\sqrt{17.5(17.5)}} = .9429$$

Since r_s is very close to 1, apparent porosity and mean pore diameter are strongly, positively related.

c.   To determine if apparent porosity and mean pore diameter have a positive rank correlation, we test:

H_o: $\rho_s = 0$

H_a: $\rho_s > 0$

The test statistic is $r_s = .9429$.

Reject H_0 if $r_s > r_{s,\alpha}$ where $\alpha = .01$ and $n = 6$.

Reject H_0 if $r_s > .943$ (From Table XII, Appendix A)

Since the observed value of r_s does not fall in the rejection region ($r_s = .9429 \not> .943$), H_0 is not rejected. There is insufficient evidence of a positive rank correlation between apparent porosity and mean pore diameter at $\alpha = .01$.

9.119 a.   Some preliminary calculations are:

Blood Lactate Level	Rank u	Perceived Recovery	Rank v	u^2	v^2	uv
3.8	3	7	1.5	9	2.25	4.5
4.2	5.5	7	1.5	30.25	2.25	8.25
4.8	7	11	3	49	9	21
4.1	4	12	5	16	25	20
5.0	8	12	5	64	25	40
5.3	9.5	12	5	90.25	25	47.5
4.2	5.5	13	7	30.25	49	38.5
2.4	1	17	9	1	81	9
3.7	2	17	9	4	81	18
5.3	9.5	17	9	90.25	81	85.5
5.8	12	18	11.5	144	132.25	138
6.0	14	18	11.5	196	132.25	161
5.9	13	21	14.5	169	210.25	188.5
6.3	15	21	14.5	225	210.25	217.5
5.5	12	20	13	121	169	143
6.5	16	24	16	256	256	256

$$\sum u = 136 \qquad \sum v = 136 \qquad \sum u^2 = 1495 \qquad \sum v^2 = 1490.5 \qquad \sum uv = 1396.25$$

$$SS_{uv} = \sum uv - \frac{(\sum u)(\sum v)}{n} = 1396.25 - \frac{136(136)}{16} = 240.25$$

$$SS_{uu} = \sum u^2 - \frac{(\sum u)^2}{n} = 1495 - \frac{136^2}{16} = 339$$

$$SS_{vv} = \sum v^2 - \frac{(\sum v)^2}{n} = 1490.5 - \frac{136^2}{16} = 334.5$$

a.   The ranks of the blood lactate levels are stored in the "u" column above.

b.   The ranks of the perceived recovery values are stored in the "v" column above.

c. $r_s = \dfrac{SS_{uv}}{\sqrt{SS_{uu}SS_{vv}}} = \dfrac{240.25}{\sqrt{339(334.5)}} = .713$

Since r_s is relatively close to 1, blood lactate level and perceived recovery have a fairly strong, positive relationship.

d.   Reject H_0 if $r_s < -r_{s,\alpha/2}$ or if $r_s > r_{s,\alpha/2}$ where $\alpha/2 = .05$ and $n = 16$;

Reject H_0 if $r_s < -.425$ or $r_s > .425$ (from Table XII, Appendix A.)

e.   To determine if blood lactate level and perceived recovery are rank correlated, we test:

H_0: Blood lactate level and Perceived recovery are not rank correlated
H_a: Blood lactate level and Perceived recovery are rank correlated

The test statistic is $r_s = .713$.

Since the observed value of the test statistic falls in the rejection region, ($r_s = .713 >$ .425), H_0 is rejected. There is sufficient evidence to indicate that blood lactate level and perceived recovery are rank correlated at $\alpha = .10$.

9.121   a.   Some preliminary calculations are:

Pair	Rank-Child	Rank-Parent	d	d^2
1	1	1	0	0
2	2	2	0	0
3	3	3	0	0
4	4	4	0	0
5	5	5	0	0
6	6	6	0	0
7	7	7	0	0
8	8	8	0	0
9	9	9	0	0
10	10	10	0	0
				$\Sigma d^2 = 0$

$$r_s = 1 - \frac{6\sum d^2}{n(n^2 - 1)} = 1 - \frac{6(0)}{10(100 - 1)} = 1$$

b.   There is a perfect positive rank correlation between the BMI of children and the BMI of their parents.

c.   The correlation found in part a is the rank correlation. This indicates that the ranks have a perfect linear relationship. However, the relationship between the actual child BMI values and parent BMI values could be curvilinear.

9.123   a.   Some preliminary calculations are:

Change KSADS-MRS	Rank, u	Improve CGI-BP	Rank, v	u^2	v^2	uv
80	18	6	18	324	324	324
65	17	5	17	289	289	289
20	15.5	4	15	240.25	225	232.5
−15	13	4	15	169	225	195
−50	9	4	15	81	225	135
20	15.5	3	12	240.25	144	186
−30	11	3	12	121	144	132
−70	5.5	3	12	30.25	144	66
−10	14	2	6	196	36	84
−25	12	2	6	144	36	72
−35	10	2	6	100	36	60
−65	7.5	2	6	56.25	36	45
−65	7.5	2	6	26.25	36	45
−70	5.5	2	6	30.25	36	33
−80	4	2	6	16	36	24
−90	2.5	2	6	6.25	36	15
−95	1	2	6	1	36	6
−90	2.5	1	1	6.25	1	2.5
$\sum u = 171$		$\sum v = 171$		$\sum u^2 = 2107$	$\sum v^2 = 2045$	$\sum uv = 1946$

$$SS_{uv} = \sum uv - \frac{\sum u \sum v}{n} = 1946 - \frac{171(171)}{18} = 321.5$$

$$SS_{uu} = \sum u^2 - \frac{\left(\sum u\right)^2}{n} = 2107 - \frac{171^2}{18} = 482.5$$

$$SS_{vv} = \sum v^2 - \frac{\left(\sum v\right)^2}{n} = 2045 - \frac{171^2}{18} = 420.5$$

$$r_s = \frac{SS_{uv}}{\sqrt{SS_{uu}SS_{vv}}} = \frac{321.5}{\sqrt{482.5(420.5)}} = .714$$

b.   To determine if there is a positive rank correlation between the two test score changes in the population of all pediatric patients with manic symptoms, we test:

H_0: $\rho_s = 0$

H_a: $\rho_s > 0$

The test statistic is $r_s = .714$.

Reject H_0 if $r_s > r_{s,\,.05}$ with $n = 18$

Reject H_0 if $r_s > .399$ (From Table XII, Appendix A)

Since the observed value of the test statistic falls in the rejection region
($r_s = .714 > .399$), H_0 is rejected. There is sufficient evidence to indicate there is a
positive rank correlation between the two test score changes in the population of all
pediatric patients with manic symptoms at $\alpha = .05$.

9.125 a.   Some preliminary calculations are:

FCAT-Math	Rank, u	% Below Poverty	Rank, v	u^2	v^2	uv
166.4	7	91.7	22	49	484	154
159.6	4	90.2	21	16	441	84
159.1	3	86.0	20	9	400	60
155.5	1	83.9	19	1	361	19
164.3	5	80.4	18	25	324	90
169.8	9	76.5	17	81	289	153
155.7	2	76.0	16	4	256	32
165.2	6	75.8	15	36	225	90
175.4	13	75.6	14	169	196	182
178.1	16	75.0	13	256	169	208
167.1	8	74.7	12	64	144	96
177.0	15	63.2	11	225	121	165
174.2	11	52.9	10	121	100	110
175.6	14	48.5	9	196	81	126
170.8	10	39.1	8	100	64	80
175.1	12	38.4	7	144	49	84
182.8	20.5	34.3	6	420.25	36	123
180.3	18	30.3	4.5	324	20.25	81
178.8	17	30.3	4.5	289	20.25	76.5
181.4	19	29.6	3	361	9	57
182.8	20.5	26.5	2	420.25	4	41
186.1	22	13.8	1	484	1	22
$\sum u = 253$			$\sum v = 253$	$\sum u^2 = 3794.5$	$\sum v^2 = 3794.5$	$\sum uv = 2133.5$

$$SS_{uv} = \sum uv - \frac{\sum u \sum v}{n} = 2133.5 - \frac{253(253)}{22} = -776$$

$$SS_{uu} = \sum u^2 - \frac{\left(\sum u\right)^2}{n} = 3794.5 - \frac{253^2}{22} = 885$$

$$SS_{vv} = \sum v^2 - \frac{\left(\sum v\right)^2}{n} = 3794.5 - \frac{253^2}{22} = 885$$

$$r_s = \frac{SS_{uv}}{\sqrt{SS_{uu} SS_{vv}}} = \frac{-776}{\sqrt{885(885)}} = -.877$$

Since r_s is very close to -1, there is a fairly strong, negative relationship between FCAT math scores and the percentage of students below the poverty level.

b.   Some preliminary calculations are:

FCAT-Read	Rank, u	% Below Poverty	Rank, v	u^2	v^2	uv
165.0	7.5	91.7	22	56.25	484	165
157.2	1	90.2	21	1	441	21
164.4	5	86.0	20	25	400	100
162.4	3	83.9	19	9	361	57
162.5	4	80.4	18	16	324	72
164.9	6	76.5	17	36	289	102
162.0	2	76.0	16	4	256	32
165.0	7.5	75.8	15	56.25	225	112.5
173.7	14	75.6	14	196	196	196
171.0	11	75.0	13	121	169	143
169.4	9	74.7	12	81	144	108
172.9	13	63.2	11	169	121	143
172.7	12	52.9	10	144	100	120
174.9	16	48.5	9	256	81	144
174.8	15	39.1	8	225	64	120
170.1	10	38.4	7	100	49	70
181.4	20	34.3	6	400	36	120
180.6	19	30.3	4.5	361	20.25	85.5
178.0	18	30.3	4.5	324	20.25	81
175.9	17	29.6	3	289	9	51
181.6	21	26.5	2	441	4	42
183.8	22	13.8	1	484	1	22
$\sum u = 253$		$\sum v = 253$		$\sum u^2 = 3794.5$	$\sum v^2 = 3794.5$	$\sum uv = 2107$

$$SS_{uv} = \sum uv - \frac{\sum u \sum v}{n} = 2107 - \frac{253(253)}{22} = -802.5$$

$$SS_{uu} = \sum u^2 - \frac{\left(\sum u\right)^2}{n} = 3794.5 - \frac{253^2}{22} = 885$$

$$SS_{vv} = \sum v^2 - \frac{\left(\sum v\right)^2}{n} = 3794.5 - \frac{253^2}{22} = 885$$

$$r_s = \frac{SS_{uv}}{\sqrt{SS_{uu}SS_{vv}}} = \frac{-802.5}{\sqrt{885(885)}} = -.907$$

Since r_s is very close to -1, there is a fairly strong, negative relationship between FCAT reading scores and the percentage of students below the poverty level.

c.   To determine whether the FCAT math scores and the percentage of students below the poverty level are negatively rank correlated, we test:

H_0: $\rho_s = 0$
H_a: $\rho_s < 0$

The test statistic is $r_s = -.877$.

Reject H_0 if $r_s < -r_{s,.01}$ with $n = 22$

Reject H_0 if $r_s < -.508$ (From Table XII, Appendix A)

Since the observed value of the test statistic falls in the rejection region ($r_s = -.877 < -.508$), H_0 is rejected. There is sufficient evidence to indicate FCAT math scores and the percentage of students below the poverty level are negatively rank correlated at $\alpha = .01$.

d.   To determine whether the FCAT reading scores and the percentage of students below the poverty level are negatively rank correlated, we test:

H_0: $\rho_s = 0$
H_a: $\rho_s < 0$

The test statistic is $r_s = -.907$.

Reject H_0 if $r_s < -r_{s,.01}$ with $n = 22$

Reject H_0 if $r_s < -.508$ (From Table XII, Appendix A)

Since the observed value of the test statistic falls in the rejection region ($r_s = -.907 < -.508$), H_0 is rejected. There is sufficient evidence to indicate FCAT reading scores and the percentage of students below the poverty level are negatively rank correlated at $\alpha = .01$.

9.127   a.   Using MINITAB, histograms of the two data sets are:

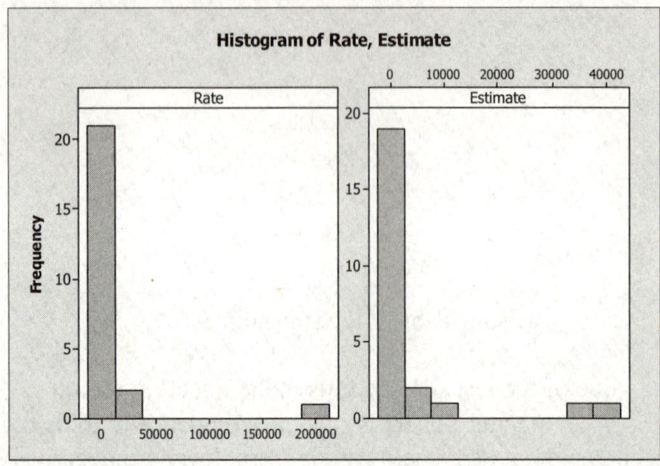

From the graphs, both data sets are skewed to the right and are not normal.

b.   The analysis will not be valid because the data are not normal.

c.   Some preliminary calculations are:

Incidence rate	Rank, u	Estimate	Rank, v	u^2	v^2	uv
.25	1	300	9	1	81	9
1	2	1000	16.5	4	272.25	33
1.75	3	691	14	9	196	42
2	4	20	6	16	36	24
3	5	17.5	2	25	4	10
5	6	.8	1	36	1	6
9	7	1000	16.5	49	272.25	115.5
10	8	150	4	64	16	32
22	10	326.5	10	100	100	100
23	11	146.5	3	121	9	33
39	12	370	11	144	121	132
98	13	400	12.5	169	156.25	162.5
119	14	225	8	196	64	112
152	15	200	6	225	36	90
179	16	200	6	256	36	96
936	17	400	12.5	289	156.25	212.5
1514	18	1500	19.5	324	380.25	351
1627	19	1000	16.5	361	272.25	313.5
2926	20	6000	22	400	484	440
4019	21	5000	21	441	441	441
12619	22	1500	19.5	484	380.25	429
14889	23	1000	16.5	529	272.25	379.5
203864	24	37000	23	576	529	552
15	9	37500	24	81	576	216
	$\sum u = 300$		$\sum v = 300$	$\sum u^2 = 4900$	$\sum v^2 = 4892$	$\sum uv = 4331.5$

$$SS_{uv} = \sum uv - \frac{\left(\sum u\right)\left(\sum v\right)}{n} = 4331.5 - \frac{300(300)}{24} = 581.5$$

$$SS_{uu} = \sum u^2 - \frac{\left(\sum u\right)^2}{n} = 4900 - \frac{300^2}{24} = 1150$$

$$SS_{vv} = \sum v^2 - \frac{\left(\sum v\right)^2}{n} = 4892 - \frac{300^2}{24} = 1142$$

$$r_s = \frac{SS_{uv}}{\sqrt{SS_{uu}SS_{vv}}} = \frac{581.5}{\sqrt{1150(1142)}} = .5074$$

Since the value of r_s is about half way between 0 and 1, there is a moderate positive relationship between incidence rate and estimate.

d.   To determine if there is a positive association between actual incidence and estimated incidence, we test:

H_0: $\rho_s = 0$

H_a: $\rho_s > 0$

The test statistic is $r_s = .5074$.

From Table XII, with $n = 24$ and $\alpha = .01$, the rejection region is $r_s > .485$.

Since the observed value of r_s falls in the rejection region ($r_s = .5074 > .485$), H_0 is rejected. There is sufficient evidence to indicate a positive association between incidence rate and estimated rate at $\alpha = .01$.

9.129   The general form of the straight-line model for $E(y)$ is $E(y) = \beta_0 + \beta_1 x$.

9.131   The statement "In simple linear regression, about 95% of the y-values in the sample will fall within 2s of their respective predicted values." is a true statement.

9.133   a.

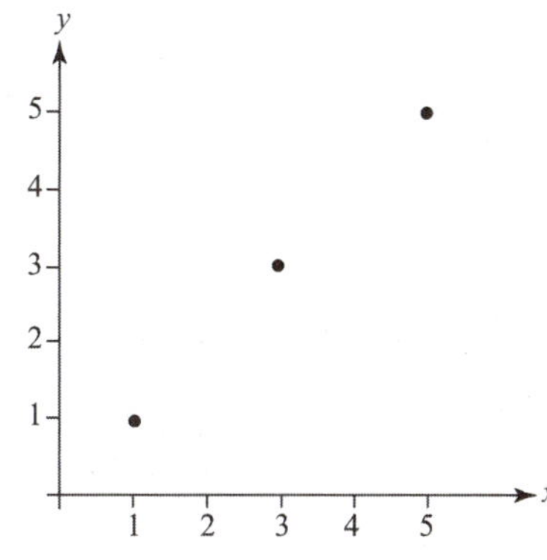

b. One possible line is $\hat{y} = x$.

x	y	$\hat{y}$	$y - \hat{y}$
1	1	1	0
3	3	3	0
5	5	5	0
			0

For this example $\sum(y - \hat{y}) = 0$

A second possible line is $\hat{y} = 3$.

x	y	$\hat{y}$	$y - \hat{y}$
1	1	3	-2
3	3	3	0
5	5	3	2
			0

For this example $\sum(y - \hat{y}) = 0$

c. Some preliminary calculations are:

$$\sum x_i = 9 \qquad \sum x_i^2 = 35 \qquad \sum x_i y_i = 35 \qquad \sum y_i = 9 \qquad \sum y_i^2 = 35$$

$$SS_{xy} = \sum x_i \sum y_i - \frac{\sum x_i \sum y_i}{n} = 35 - \frac{9(9)}{3} = 8$$

$$SS_{xx} = \sum x_i^2 - \frac{\left(\sum x_i\right)^2}{n} = 35 - \frac{9^2}{3} = 8$$

$$SS_{yy} = \sum y_i^2 - \frac{\left(\sum y_i\right)^2}{n} = 35 - \frac{9^2}{3} = 8$$

$$\hat{\beta}_1 = \frac{SS_{xy}}{SS_{xx}} = \frac{8}{8} = 1 \qquad\qquad \hat{\beta}_0 = \bar{y} - \hat{\beta}_1\bar{x} = \frac{9}{3} - 1\left(\frac{9}{3}\right) = 0$$

The least squares line is $\hat{y} = 0 + 1x = x$.

d. For $\hat{y} = x$, $SSE = SS_{yy} - \hat{\beta}_1 SS_{xy} = 8 - 1(8) = 0$

For $\hat{y} = 3$, $SSE = \sum(y_i - \hat{y}_i)^2 = (1-3)^2 + (3-3)^2 + (5-3)^2 = 8$

The least squares line has the smallest SSE of all possible lines.

9.135 a. The straight-line model is $y = \beta_0 + \beta_1 x + \varepsilon$. Based on the theory, we would expect the metal level to decrease as the distance increases. Thus, the slope should be negative.

b. Yes. As the distance from the plant increases, the concentration of calcium tends to decrease.

c. No. As the distance from the plant increases, the concentration of arsenic tends to increase.

9.137 a. We would expect the slope of the line from modeling the relationship between the mean number of games won and the team's batting average to be positive. As the team's batting average increases, we would expect that the number of games won should also increase.

  b. Using MINITAB, a scatterplot of the data is:

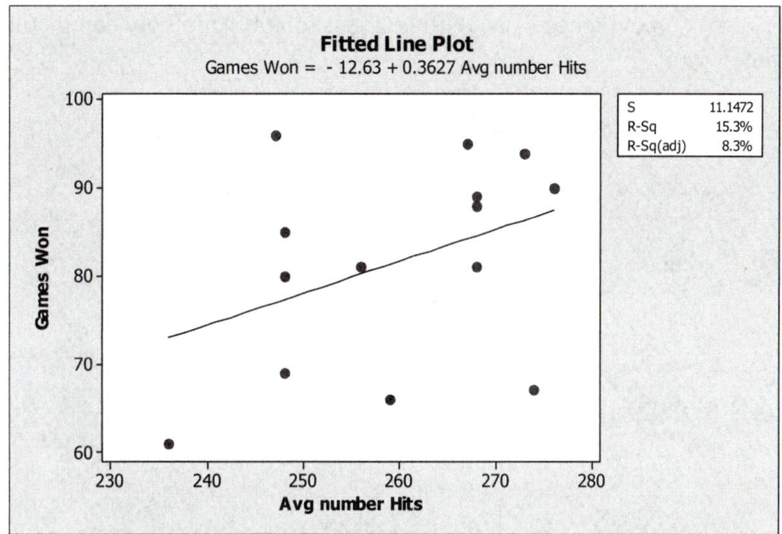

The graph indicates that batting average and games won have a positive linear relationship. As batting average increases, the number of games won tends to increase.

  c. From the printout, $\hat{\beta}_0 = -12.62556$ and $\hat{\beta}_1 = 0.36269$. The least squares line is:
$\hat{y} = -12.62556 + .36269x.$

  d. The least squares line is plotted on the graph in part **a**. The data appear to fit the data somewhat well.

  e. $\hat{\beta}_0 = -12.62556$. Since there is no team with a batting average of 0, this has no meaning other than the y-intercept.
$\hat{\beta}_1 = .36269$. For each additional hit per 1000 at bats, the mean number of games won is estimated to increase by .36269.

  f. To determine if the mean number of games won by a major league baseball team is positively linearly related to the team's batting average, we test:

H_0: $\beta_1 = 0$
H_a: $\beta_1 > 0$

The test statistic is $t = 1.47$ and the p-value is $p = 0.1660$. Since we are running only a one-tailed test, we must divide the p-value by 2. Thus, the p-value of this test is

$p = 0.1660/2 = 0.0830$. Since the p-value is not less than $\alpha = .05$ ($p = 0.0830 > .05$), H_0 is not rejected. There is insufficient evidence to indicate the mean number of games won by a major league baseball team is positively linearly related to the team's batting average at $\alpha = .05$

g.   From the printout, $r^2 = $ R-Square $= .1535$. 15.35% of the sample variability around the sample mean number of games won is explained by the linear relationship between the number of games won and the batting average of the team.

h.   No. There is no evidence of a positive linear relationship between the number of hits and games won.

9.139   a.   The equation for the straight-line model is $y = \beta_0 + \beta_1 x + \varepsilon$.

b.   Some preliminary calculations are:

$$\bar{y} = \frac{\sum y}{n} = \frac{398}{9} = 44.2222$$

$$\bar{x} = \frac{\sum x}{n} = \frac{1,444}{9} = 160.4444$$

$$SS_{xy} = \sum xy - \frac{\sum x \sum y}{n} = 60,428 - \frac{1,444(398)}{9} = -3,428.88889$$

$$SS_{xx} = \sum x^2 - \frac{(\sum x)^2}{n} = 235,866 - \frac{1,444^2}{9} = 4,184.2222$$

$$\hat{\beta}_1 = \frac{SS_{xy}}{SS_{xx}} = \frac{-3,428.88889}{4,184.2222} = -.819480593$$

$$\hat{\beta}_0 = \bar{y} - \hat{\beta}_1 \bar{x} = \frac{398}{9} - (-.819480593)\left(\frac{1,444}{9}\right) = 175.7033307$$

The fitted model is $\hat{y} = 175.7033 - .8195x$.

c. $\hat{\beta}_0 = 175.7033$. Since $x = 0$ (age of fish) is not in the observed range, $\hat{\beta}_0$ has no practical meaning.

d. $\hat{\beta}_1 = -.8195$. For each additional day of age, the mean number of strikes is estimated to decrease by .8195 strikes.

e.   Some additional calculations are:

$$\sum y_i = 398 \qquad\qquad \sum y_i^2 = 22,078$$

$$SS_{yy} = \sum y_i^2 - \frac{\left(\sum y_i\right)^2}{n} = 22,078 - \frac{398^2}{9} = 4,477.55556$$

$$SSE = SS_{yy} - \hat{\beta}_1 SS_{xy} = 4,477.55556 - (-.819480593)(-3,428.88889) = 1,667.647659$$

$$s^2 = \frac{SSE}{n-2} = \frac{1,667.647659}{9-2} = 238.2353799$$

$$s = \sqrt{s^2} = \sqrt{238.2353799} = 15.43487544$$

$$H_o: \ \beta_1 = 0$$
$$H_a: \ \beta_1 < 0$$

The test statistic is $t = \dfrac{\hat{\beta}_1 - 0}{s_{\hat{\beta}_1}} = \dfrac{-.8195 - 0}{15.4349 \Big/ \sqrt{4,184.2222}} = -3.434$

The rejection region requires $\alpha = .10$ in the lower tail of the t distribution with df $= n - 2$ $= 9 - 2 = 7$. From, Table IV, Appendix A, $t_{.10} = 1.415$. The rejection region is $t < -1.415$.

Since the observed value of the test statistic falls in the rejection region $(t = -3.434 < -1.415)$, H_0 is rejected. There is sufficient evidence to indicate a negative linear relationship exists between the number of strikes and the age of the fish at $\alpha = .10$.

f. Some preliminary calculations:

Number of Strikes	Rank, u	Age of Fish	Rank, v	u^2	v^2	uv
85	9	120	1	81	1	
63	8	136	2	64	4	
34	3	150	3	9	9	
39	5	155	4	25	16	
58	7	162	5	49	25	
35	4	169	6	16	36	
57	6	178	7	36	49	
12	1	184	8	1	64	
15	2	190	9	4	81	
	$\sum u = 45$		$\sum v = 45$	$\sum u^2 = 285$	$\sum v^2 = 285$	$\sum uv = 181$

$$SS_{uv} = \sum uv - \frac{\sum u \sum v}{n} = 181 - \frac{45(45)}{9} = -44$$

$$SS_{uu} = \sum u^2 - \frac{\left(\sum u\right)^2}{n} = 285 - \frac{45^2}{9} = 60$$

$$SS_{vv} = \sum v^2 - \frac{\left(\sum v\right)^2}{n} = 285 - \frac{45^2}{9} = 60$$

$$r_s = \frac{SS_{uv}}{\sqrt{SS_{uu}SS_{vv}}} = \frac{-44}{\sqrt{60(60)}} = -.733$$

g.   To determine if the number of strikes and age are negatively rank correlated, we test:

$$H_o: \ \rho_s = 0$$
$$H_a: \ \rho_s \neq 0$$

The test statistic is $r_s = -.733$.

Reject H_0 if $r_s < -r_{s,\,.01}$ with $n = 9$

Reject H_0 if $r_s < -.783$ (From Table XII, Appendix A)

Since the observed value of the test statistic does not fall in the rejection region ($r_s = -.733 \not< -.783$), H_0 is not rejected. There is insufficient evidence to indicate that the number of strikes and age are negatively rank correlated at $\alpha = .01$.

9.141   a.   The simple linear regression model relating y to x is $y = \beta_0 + \beta_1 x + \varepsilon$.

b. $r^2 = .92$. 92% of the total sample variability around the sample mean metal uptake is explained by the linear relationship between metal uptake and the final concentration of metal in the solution.

9.143   a.   Using MINITAB, the results are:

Regression Analysis: Cost versus JIF

```
The regression equation is
Cost = 560 + 63 JIF

Predictor    Coef   SE Coef      T       P
Constant    560.1     168.4   3.33   0.003
JIF          63.3     113.2   0.56   0.581

S = 454.248    R-Sq = 1.2%    R-Sq(adj) = 0.0%

Analysis of Variance

Source            DF        SS       MS      F       P
Regression         1     64381    64381   0.31   0.581
Residual Error    26   5364863   206341
Total             27   5429244
```

The fitted regression equation is $\hat{y} = 560.1 + 63.3x$.

For each additional unit increase in the Journal Impact Factor (JIF), the mean cost of the journal is estimated to increase by \$63.30.

b.   From the printout, $s = 454.248$. We would expect to predict cost within $2s = 2(454.248) = \$908.50$.

c.   For confidence level .95, $\alpha = .05$ and $\alpha / 2 = .05 / 2 = .025$. From Table IV, Appendix A with df $= n - 2 = 28 - 2 = 26$, $t_{.025} = 2.056$.

The confidence interval is:

$$\hat{\beta}_1 \pm t_{.025} s_{\hat{\beta}_1} \Rightarrow 63.3 \pm 2.056(113.2) \Rightarrow 63.3 \pm 232.7 \Rightarrow (-169.4,\ 296.0)$$

We are 95% confident that for each unit increase in JIF, the mean cost of the journal is estimated to change from -\$169.40 to \$296.00.

d.   Using MINITAB, the results are:

Regression Analysis: Cost versus Cites

```
The regression equation is
Cost = 326 + 1.48 Cites


Predictor     Coef   SE Coef      T       P
Constant     326.5     109.1   2.99   0.006
Cites       1.4801    0.3953   3.74   0.001


S = 368.331    R-Sq = 35.0%    R-Sq(adj) = 32.5%

Analysis of Variance

Source           DF        SS        MS       F       P
Regression        1   1901880   1901880   14.02   0.001
Residual Error   26   3527364    135668
Total            27   5429244
```

The fitted regression equation is $\hat{y} = 326.5 + 1.48x$.

For each additional unit increase in the number of citations for the journal over the past 5 years, the mean cost of the journal is estimated to increase by \$1.48.

From the printout, $s = 368.331$. We would expect to predict cost within $2s = 2(368.331) = \$736.66$.

For confidence level .95, $\alpha = .05$ and $\alpha / 2 = .05 / 2 = .025$. From Table IV, Appendix A with df $= n - 2 = 28 - 2 = 26$, $t_{.025} = 2.056$.

The confidence interval is:

$$\hat{\beta}_1 \pm t_{.025} s_{\hat{\beta}_1} \Rightarrow 1.48 \pm 2.056(.3953) \Rightarrow 1.48 \pm .81 \Rightarrow (.67,\, 2.29)$$

We are 95% confident that for each additional citation, the mean cost of the journal is estimated to increase from $.67 to $2.29.

e.   Using MINITAB, the results are:

Regression Analysis: Cost versus RPI

```
The regression equation is
Cost = 339 + 197 RPI


27 cases used, 1 cases contain missing values


Predictor    Coef  SE Coef     T       P
Constant    338.9    156.6  2.16   0.040
RPI        197.21    86.24  2.29   0.031


S = 423.363   R-Sq = 17.3%   R-Sq(adj) = 14.0%


Analysis of Variance

Source          DF        SS       MS      F       P
Regression       1    937326   937326   5.23   0.031
Residual Error  25   4480916   179237
Total           26   5418242
```

The fitted regression equation is $\hat{y} = 338.9 + 197.21x$.

For each additional unit increase in the Relative Price Index (RPI), the mean cost of the journal is estimated to increase by $197.21.

From the printout, $s = 423.363$. We would expect to predict cost within $2s = 2(423.363) = \$846.73$.

For confidence level .95, $\alpha = .05$ and $\alpha / 2 = .05 / 2 = .025$. From Table IV, Appendix A with df $= n - 2 = 28 - 2 = 26$, $t_{.025} = 2.056$.

The confidence interval is:

$$\hat{\beta}_1 \pm t_{.025} s_{\hat{\beta}_1} \Rightarrow 197.21 \pm 2.056(86.24) \Rightarrow 197.21 \pm 177.31 \Rightarrow (19.90,\, 374.52)$$

We are 95% confident that for each unit increase in RPI, the mean cost of the journal is estimated to increase from $19.90 to $374.52.

9.145   a.   Using MINITAB, the scattergram of the data is:

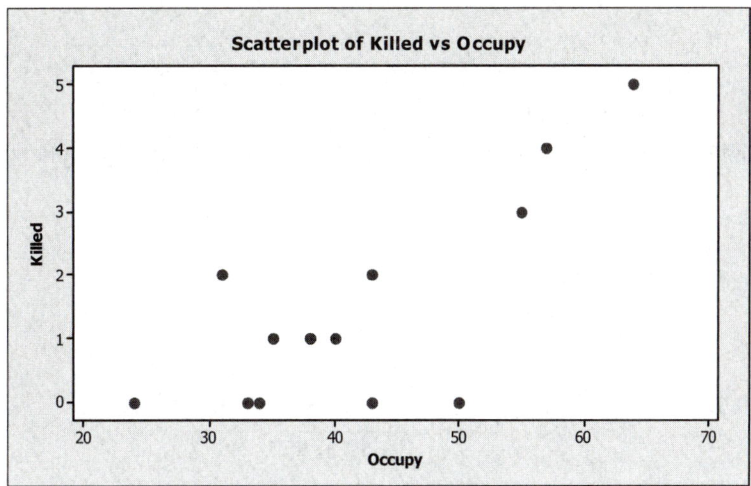

From the graph, it appears that as nest box tit occupancy increases, the number of flycatchers killed also increases.

b.   Some preliminary calculations are:

$$\sum x_i = 582 \qquad \sum y_i = 20 \qquad \sum x_i y_i = 1{,}009$$

$$\sum x_i^2 = 25{,}844 \qquad \sum y_i^2 = 62$$

$$SS_{xy} = \sum x_i y_i - \frac{\sum x_i \sum y_i}{n} = 1{,}009 - \frac{582(20)}{14} = 177.5714286$$

$$SS_{xx} = \sum x_i^2 - \frac{\left(\sum x_i\right)^2}{n} = 25{,}844 - \frac{582^2}{14} = 1{,}649.42857$$

$$\hat{\beta}_1 = \frac{SS_{xy}}{SS_{xx}} = \frac{177.5714286}{1{,}649.42857} - .107656331$$

For each unit increase in the nest box tit occupancy percent, the mean number of flycatchers killed is estimated to increase by .1077.

$$\hat{\beta}_0 = \bar{y} - \hat{\beta}_1 \bar{x} = \frac{20}{14} - (.107656331)\left(\frac{582}{14}\right) = -3.046856046$$

Since $x = 0$ is not in the observed range, $\hat{\beta}_0$ has no meaning other than the y-intercept.

c.   Some additional calculations are:

$$SS_{yy} = \sum y_i^2 - \frac{\left(\sum y_i\right)^2}{n} = 62 - \frac{20^2}{14} = 33.42857143$$

$$\mathrm{SSE} = \mathrm{SS}_{yy} - \hat{\beta}_1 \mathrm{SS}_{xy} = 33.42857143 - (.107656331)(177.5714286) = 14.31188294$$

$$s^2 = \frac{\mathrm{SSE}}{n-2} = \frac{14.31188294}{14-2} = 1.192656912 \qquad s = \sqrt{s^2} = \sqrt{1.192656912} = 1.0921$$

To determine if the model is useful in predicting the number of flycatchers killed, we test:

$$H_0: \ \beta_1 = 0$$
$$H_a: \ \beta_1 \neq 0$$

The test statistic is $t = \dfrac{\hat{\beta}_1 - 0}{s_{\hat{\beta}_1}} = \dfrac{.1077 - 0}{1.0921 \big/ \sqrt{1,649.42857}} = 4.005$

The rejection region requires $\alpha / 2 = .05 / 2 = .025$ in each tail of the t distribution. From Table IV, Appendix A, with df $= n - 2 = 14 - 2 = 12$, $t_{.025} = 2.179$. The rejection region is $t < -2.179$ or $t > 2.179$.

Since the observed value of the test statistic falls in the rejection region $(t = 4.005 > 2.179)$, H_0 is rejected. There is sufficient evidence to indicate that the model is useful in predicting the number of flycatchers killed at $\alpha = .05$.

d. $r = \dfrac{\mathrm{SS}_{xy}}{\sqrt{\mathrm{SS}_{xx}\mathrm{SS}_{yy}}} = \dfrac{177.5714286}{\sqrt{1,649.42857(33.42857143)}} = .7562$

The value of r is moderately close to 1. Thus, there is a moderately strong positive linear relationship between the number of flycatchers killed and the nest box tit occupancy.

$r^2 = .7562^2 = .572$. 57.2% of the sample variability around the sample mean number of flycatchers killed is explained by the linear relationship between the number of flycatchers killed and the nest box tit occupancy.

e.   From part c, $s = 1.0921$. We would expect most of the observed number of flycatchers killed to fall within $2s = 2(1.0921) = 2.1842$ units of their predicted values.

f.   Since there is a significant linear relationship between the number of flycatchers killed and the nest box tit occupancy, we would recommend using this model. In addition, the r^2 value is .572. Although this is not great, over half of the sample variation is explained by the model.

9.147   Using MINITAB, the output for fitting the least squares line is:

Regression Analysis: Impur versus Temp

```
The regression equation is
Impur = - 13.5 - 0.0528 Temp


Predictor          Coef    SE Coef        T       P
Constant        -13.490      2.074    -6.51   0.000
Temp          -0.052829    0.007728    -6.84   0.000


S = 0.133347    R-Sq = 85.4%    R-Sq(adj) = 83.6%


Analysis of Variance

Source            DF         SS        MS       F       P
Regression         1    0.83089   0.83089   46.73   0.000
Residual Error     8    0.14225   0.01778
Total              9    0.97315


Unusual Observations

Obs   Temp    Impur      Fit   SE Fit   Residual   St Resid
  2   -265   0.2020   0.5093   0.0492    -0.3073     -2.48R
  3   -256   0.2040   0.0339   0.1038     0.1701      2.03RX

R denotes an observation with a large standardized residual.
X denotes an observation whose X value gives it large influence.


Predicted Values for New Observations

New
Obs     Fit   SE Fit          95% CI              95% PI
  1  0.9320   0.0558   (0.8034, 1.0605)   (0.5987, 1.2652)


Values of Predictors for New Observations

New
Obs   Temp
  1   -273
```

a. $\hat{\beta}_0 = -13.490$. Since $x = 0$ is not in the observed range, $\hat{\beta}_0$ has no meaning since. It is the y-intercept.

$\hat{\beta}_1 = -.052829$. For each additional degree, the mean proportion of impurity passing through helium is estimated to decrease by .052829 units.

b.   For confidence coefficient .95, $\alpha = .05$ and $\alpha / 2 = .05 / 2 = .025$. From Table IV, Appendix A, with df $= n - 2 = 10 - 2 = 8$, $t_{.025} = 2.306$. The confidence interval is:

$$\hat{\beta}_1 \pm t_{\alpha/2} s_{\hat{\beta}_1} \Rightarrow -.0528 \pm 2.306(.007728) \Rightarrow -.0528 \pm .0178 \Rightarrow (-.0706, -.0350)$$

We are 95% confident that the change in mean proportion of impurity passing through helium for each additional degree is between $-.0706$ and $-.0350$. Since 0 is not in the interval, there is evidence to indicate that temperature contributes information about the proportion of impurity passing through helium.

c. $r^2 = R\text{-sq} = .854$. 85.4% of the total sample variation around the mean proportion of impurity is explained by the linear relationship between proportion of impurity and temperature.

d.   From the printout, the 95% prediction interval is (.5987, 1.2652). We are 95% confident that the actual percentage of impurity passing through solid helium at -273°C is between 59.87% and 126.52%. Since we cannot have more than 100% of the impurity passing through solid helium, the interval is from 59.87% to 100%.

e.   We cannot be sure that the relationship between the proportion of impurity passing through helium and temperature is the same outside the observed range.

9.149   a.   A scattergram of the data is:

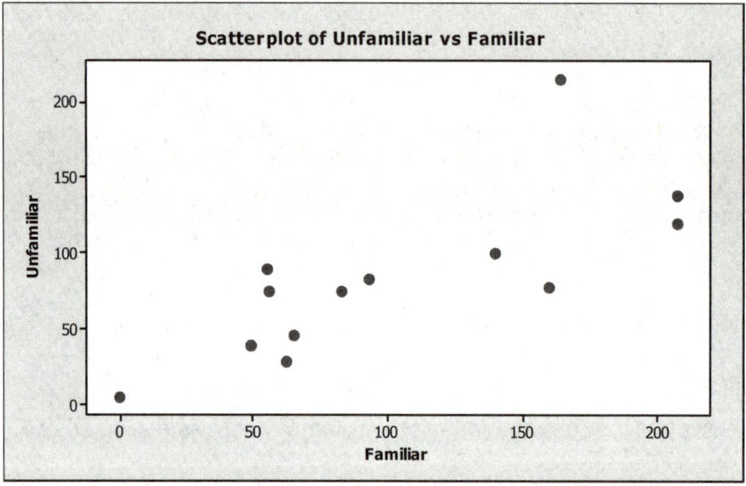

It appears that there is a positive linear relationship between total number of words with unfamiliar partner and with familiar partner.

b.   A straight line model would be $y = \beta_0 + \beta_1 x + \varepsilon$.

c.   Some preliminary calculations are:

$$\sum x_i = 1{,}341 \qquad \sum y_i = 1{,}099 \qquad \sum x_i y_i = 145{,}607$$

$$\sum x_i^2 = 189{,}963 \qquad \sum y_i^2 = 127{,}397$$

$$SS_{xy} = \sum x_i y_i - \frac{\sum x_i \sum y_i}{n} = 145{,}607 - \frac{1{,}341(1{,}099)}{13} = 32{,}240.9231$$

$$SS_{xx} = \sum x_i^2 - \frac{\left(\sum x_i\right)^2}{n} = 189,963 - \frac{1,341^2}{13} = 51,633.6923$$

$$\hat{\beta}_1 = \frac{SS_{xy}}{SS_{xx}} = \frac{32,240.9231}{51,633.6923} = .624416377$$

$$\hat{\beta}_0 = \bar{y} - \hat{\beta}_1\bar{x} = \frac{1,099}{13} - (.624416377)\left(\frac{1,341}{13}\right) = 20.12751061$$

The least squares prediction equation is $\hat{y} = 20.1275 + .6244x$

d. $\hat{\beta}_0 = 20.1275$. Since $x = 0$ is in the observed range, the mean number of words in conversation with an unfamiliar partner is estimated to be 20.1275 when the number of words in conversation with a familiar partner is 0.

     $\hat{\beta}_1 = .6244$. For each additional word in conversation with a familiar partner, the mean number of words in conversation with an unfamiliar partner is estimated to increase by .6244.

9.151   Some preliminary calculations are:

$$\sum x = 4305 \qquad \sum x^2 = 1,652,025 \qquad \sum xy = 76,652,695$$

$$\sum y = 201,558 \qquad \sum y^2 = 3,571,211,200$$

a. $\hat{\beta}_1 = \dfrac{\sum xy}{\sum x^2} = \dfrac{76,652,695}{1,652,025} = 46.39923427 \approx 46.3992$

The least squares line is $\hat{y} = 46.3992x$.

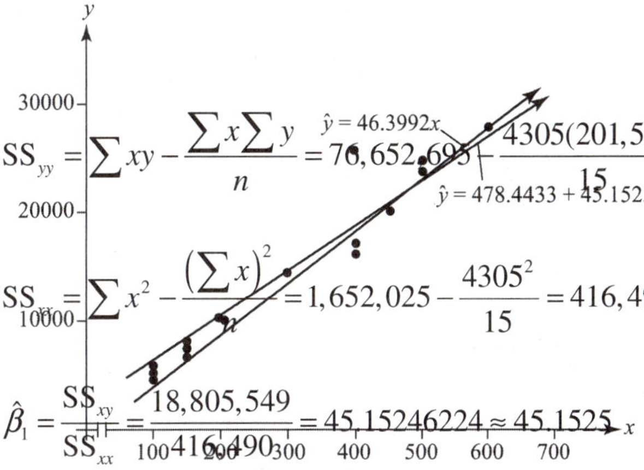

b. $SS_{yy} = \sum xy - \dfrac{\sum x \sum y}{n} = 76,652,695 - \dfrac{4305(201,558)}{15} = 18,805,549$

$$SS_{xx} = \sum x^2 - \frac{\left(\sum x\right)^2}{n} = 1,652,025 - \frac{4305^2}{15} = 416,490$$

$$\hat{\beta}_1 = \frac{SS_{xy}}{SS_{xx}} = \frac{18,805,549}{416,490} = 45.15246224 \approx 45.1525$$

$$\hat{\beta}_0 = \bar{y} - \hat{\beta}_1 \bar{x} = \frac{201,558}{15} - 45.15246224\left(\frac{4305}{15}\right) = 478.4433$$

The least squares line is $\hat{y} = 478.4433 + 45.1525x$.

c.   Because $x = 0$ is not in the observed range, we are trying to represent the data on the observed interval with the best fitting line. We are not concerned with whether the line goes through $(0, 0)$ or not.

d.   Some preliminary calculations are:

$$SS_{yy} = \sum y^2 - \frac{\left(\sum y\right)^2}{n} = 3,571,211,200 - \frac{201,558^2}{15} = 862,836,042$$

$$SSE = SS_{yy} - \hat{\beta}_1 SS_{xy} = 862,836,042 - 45.15246224(18,805,549) = 13,719,200.88$$

$$s^2 = \frac{SSE}{n-2} = \frac{13,719,200.88}{15-2} = 1,055,323.145 \qquad s = \sqrt{s^2} = 1027.2892$$

$$H_0: \ \beta_0 = 0$$
$$H_a: \ \beta_0 \neq 0$$

The test statistic is $t = \dfrac{\hat{\beta}_0 - 0}{s\sqrt{\dfrac{1}{n} + \dfrac{\bar{x}^2}{SS_{xx}}}} = \dfrac{478.443}{1027.2892\sqrt{\dfrac{1}{15} + \dfrac{287^2}{416,490}}} = .906$

The rejection region requires $\alpha / 2 = .10 / 2 = .05$ in each tail of the t distribution with $df = n - 2 = 15 - 2 = 13$. From Table IV, Appendix A, $t_{.05} = 1.771$. The rejection region is $t < -1.771$ or $t > 1.771$.

Since the observed value of the test statistic does not fall in the rejection region ($t = .906$ $\not> 1.771$), H_0 is not rejected. There is insufficient evidence to indicate β_0 is different from 0 at $\alpha = .10$. Thus, β_0 should not be included in the model.

9.153   a.   Some preliminary calculations are:

$$\sum x_i = 1,098 \qquad\qquad \sum y_i = 121.8 \qquad\qquad \sum x_i y_i = 4,294.6$$

$$\sum x_i^2 = 56,712 \qquad\qquad \sum y_i^2 = 597.52$$

$$SS_{xy} = \sum x_i y_i - \frac{\sum x_i \sum y_i}{n} = 4,294.6 - \frac{1,098(121.8)}{27} = -658.6$$

$$SS_{xx} = \sum x_i^2 - \frac{\left(\sum x_i\right)^2}{n} = 56,712 - \frac{1,098^2}{27} = 12,060$$

$$\hat{\beta}_1 = \frac{SS_{xy}}{SS_{xx}} = \frac{-658.6}{12,060} = -.054610281$$

$$\hat{\beta}_o = \bar{y} - \hat{\beta}_1\bar{x} = \frac{121.8}{27} - (-.054610281)\left(\frac{1,098}{27}\right) = 6.731929244$$

$$SS_{yy} = \sum y_i^2 - \frac{\left(\sum y_i\right)^2}{n} = 597.52 - \frac{121.8^2}{27} = 48.06666667$$

$$SSE = SS_{yy} - \hat{\beta}_1 SS_{xy} = 48.06666667 - (-.054610281)(-658.6) = 12.10033563$$

$$r^2 = 1 - \frac{SSE}{SS_{yy}} = 1 - \frac{12.10033563}{48.066666667} = .7483$$ 74.83% of the total sample variability around

the sample mean fertility rate is explained by the linear relationship between the fertility rate and the contraceptive prevalence. This is not 90%.

b.   From above, the fitted regression line is $\hat{y} = 6.732 - .0546x$

If the contraceptive use increases by 18 percent, then the mean fertility rate will decrease by an estimated $(.0546)(18) = .9828$. This is approximately 1. The researchers' statement is supported by the data.